Contents

Introduction   *Leone Burton and Barbara Jaworski*   1

Section One: Setting the Scene   11

1   Fractals: from Ferns to Fireworks   *Philip J. Rippon*   15

2   Designing Tasks for Learning Geometry in a   35
     Computer-based Environment   *Colette Laborde*

3   Collaborating via Boxer   *Andrea A. diSessa*   69

Section Two: Why Technology?   95

4   Personal Technology and New   97
     Horizons   *Adrian J Oldknow*

5   Software for Mathematics   109
     Education   *Eric Love*

6   Less may be more on a Screen   *John Mason*   119

7   Working for Change in Teacher   135
     Education   *Hilary Povey*

8   Looking At, Through, and Back At: Useful Ways of   153
     Viewing Mathematical Software   *Liddy Nevile*

Section Three: Using Calculators   171

9   Calculators and Young Children: 173
     A bridge to number?   *Pat Perks*

10   Visualisation, Confidence and   195
     Magic   *Teresa Smart*

Contents

11 Research Results on the Effect of the Graphic   213
Calculator on Female Students' Cognitive Levels and
Visual Thinking   *Mary Margaret
Shoaf-Grubbs*

12 Pressing On   *Kenneth Ruthven*   231

Section Four: How may technology support school algebra?   257

13 An Analysis of the Relationship between Spreadsheet   261
and Algebra   *Guiliana Dettori,
Rossella Garuti,
Enrica Lemut and
Ljuba Netchitailova*

14 Algebraic Thinking: the role of   275
the computer   *Rosamund Sutherland*

15 Bridging a Gap from Computer Science   289
to Algebra   *Jean-Baptiste Lagrange*

16 Using a Computer Algebra System with 14–15 year old   307
Students   *Mark Hunter,
Paul Marshall,
John Monaghan and
Tom Roper*

Section Five: Advanced Mathematics – Some Perspectives   325

17 Computer Algebra and the Structure of the Mathematics   329
Curriculum   *Andrew Rothery*

18 Prime (Iterating) Number Generators   345
with Derive   *Patrick Wild*

19 Using Supercalculators to Affect Change in Linear   353
Algebra   *Donald R. LaTorre*

Leone Burton
Barbara Jaworski (editors)

Technology in Mathematics Teaching

a bridge between teaching and learning

7

WITHDRAWN
UNIV... FROM
PLYMOUTH
...ICES

 Chartwell-Bratt   Studentlitteratur

90 0253710 4

British Library Cataloguing in Publication Data
A catalogue record for this book is available from the British Library

All rights reserved. No part of this publication may be reproduced or transmitted in any form or by any means, electronic or mechanical, including photocopying, recording, or any information storage and retrieval system, without permission in writing from the publisher.

© Leone Burton, Barbara Jaworski, the authors and Chartwell-Bratt Ltd, 1995

Cover picture courtesy of Macsyma symbolic mathematics software
Chartwell-Bratt (Publishing and Training) Ltd
ISBN 0-86238-401-X

Printed in Sweden
Studentlitteratur, Lund
ISBN 91-44-60711-3

Printing:	1 2 3 4 5 6 7 8 9 10	1999 98 97 96 95

20 Models of Technology   *Chris Bissell*   365

21 Using Technology In Examinations *Jim Tabor*   387

Section Six: Innovative Uses of Technology   405

22 Mathematics and Technology   *Michele Emmer*   407

23 Children Making Maths Videos *Geoff Sheath*   423

24 Planning for Portability   *Janet Ainley and*   435
                                  Dave Praft

25 Arithmetic Microworlds in a Hypermedia System for 449
    Problem Solving   *Rosa Maria Bottino,*
                                  Giampaolo Chiappini
                                  and Pier Luigi Ferrari

List of Contributors   469

List of Reviewers   477

Subject Index   478

Name Index   487

UNIVERSITY OF PLYMOUTH
LIBRARY SERVICES

Item No.	253710 4
Class No.	374. 51 TEC
Contl No.	0 - 86238 - 401 - x

Introduction

Leone Burton and Barbara Jaworski

In September, 1993, a conference called *Technology and Mathematics Teaching* was held at the University of Birmingham, U.K. Although a regular event in the United States of America, initiated and supported by Professor Bert Waits and his colleagues at the University of Ohio, it was the first occasion that this conference had been held outside the United States. It provided an opportunity for all those with interests in the use of technology, for the teaching of mathematics to gather, share their enthusiasms, alert one another to their experiences, and reflect on the state-of-the-art. Participants came from sixteen different countries, mostly in Europe (East and West) but a few from Australia, and individuals from Canada, Malaysia, Nigeria, the Republic of China, Singapore, South Africa, and Venezuela. A large contingent came from the United States of America. (Copies of Proceedings of the Conference are available from Room 416, School of Education, University of Birmingham, Edgbaston, Birmingham B15 2TT, U.K.)

The conference aroused so much interest, as well as many questions, that it was decided to put together a book reflecting the current situation in the use of technology for the teaching of mathematics. Participants were invited to signify their interest in submitting a paper which might become a chapter in the book, after a peer-review process had been conducted. This volume, *Technology - a Bridge between Teaching and Learning Mathematics* is the result. It contains original chapters some of which were the basis for presented papers at the conference, some of which were commissioned to fill a perceived gap or because the Editors were aware of some work which they felt would interest the wider community. Authors from the United Kingdom predominate,

1

probably because both the conference and the book originated in the U.K. but also because the Editors were aware of innovative work that deserves a wider audience. However, it is interesting to note the degree of consistency between authors across the globe, across languages and cultures. Many express concerns about the impact of technology on mathematics curricula as currently conceived. All speak of the learner's need to control the technology in order to enter the mathematical·world that it makes available and of the profound effect that this has on their consequent understanding.

The call for papers resulted in over 50 responses. The main criterion for selection of papers was that they should speak overtly to the theme of the book – that of technology acting as a *bridge* between the *teaching* and *learning* of mathematics. Many papers were rejected because they spoke of mathematics and technology, but said little about the associated issues in teaching and learning. We made an initial scan of the papers ourselves, identifying those which clearly did not satisfy the above criterion. This was communicated directly to the authors concerned. Each of the remaining papers was sent to two reviewers, seeking their critical judgment on its suitability for inclusion in the book with reference to the following questions.

1. Does the paper address mathematics specifically, and if so what aspects?

2. Is the mathematical content assumed unquestioningly or does the use of the technology provoke the author(s) to problematise the content?

3. Do the author(s) address contexts of learning – eg. individual, small group, whole class?

4. Do the author(s) make explicit the expectations held about learning and are they linked to the teaching approaches discussed?

Some papers presented innovative features of a curriculum development which relied upon the use of technology. The following questions were associated particularly with such papers:

5.   To what degree does the paper, in an informative or challenging way, present potential readers with possibilities for uses to be made of technology by those teaching mathematics?

6.   Does the paper justify these uses by consideration of advantages/disadvantages, similarities/differences etc?

7.   Does the paper address issues for the teacher with regard to learning and the learning environment?

We were very dependent on the good will and critical judgment of our reviewers and take this opportunity to record our thanks to them. (See names listed at the end of the book). Their recommendations were followed carefully. If both had recommended acceptance (or rejection) of the paper, we accepted (or rejected) it. If there was disagreement between the two reviewers, one or both of us then reviewed the paper and we made the final judgment. If reviewers recommended modifications to the paper, these were communicated to the author(s) and a revised paper requested. Revised papers were sent to the reviewers for their approval. This process was necessarily lengthy and we thank all concerned for their cooperation and patience.

Once we had an agreed set of papers for inclusion, we read them all ourselves, making editorial suggestions to the authors, and seeking appropriate groupings. We settled, finally, on the six sections which you find here. This inevitably involved some compromise. We could have grouped according to types of software being discussed. For example, a section on *spreadsheets* would have been a possibility. We could also have grouped according to the ages of students taught. However, there were themes which seemed to transcend these divisions, and the resulting groupings are designed to highlight these.

Section One consists of three chapters invited from the plenary presenters at the conference. The content of these chapters differs in varying degrees from the papers presented at the conference. The chapter by Phil. Rippon is not overtly about teaching and learning but presents a new area in mathematics which is technologically dependent, visually very exciting and which, we feel, despite its great learning potential has, as yet, largely been unexploited in schools. Also, Phil's presentational style has covert messages for a reader concerned about teaching and learning. We found his images so exciting that we decided to use one for the cover of the book, and we are greatly indebted to him for this. Chapters in Section Two address general issues relating teaching, learning and technology. Calculator use, clearly an important theme, is pursued in Section Three, where chapters span all phases of education from lower primary to undergraduate level. Section Four contains research reports focussing on the use of software packages to support the learning of algebra at school level. Section Five includes chapters which address the learning of higher level mathematics from several perspectives. In Section Six are chapters which look at innovative uses of technology and their implications for learning and teaching. Each section is introduced with a brief account of the characteristics of the grouping with reference to the papers it includes.

For us, the process of editing this volume raised a number of issues which are not necessarily dealt with in the chapters of the book but remain pertinent to a consideration of the impact of technology on the learning and teaching of mathematics. First of all is the issue of **enthusiasts**. The strength of enthusiasts is that they innovate, they try new ideas, they explore where more cautious colleagues might wait and see. Their weakness is that the outcomes of their enthusiastic endeavours are often poorly contextualised, theorised or evaluated. In reading submitted chapters, the excited energy of the author was often palpable; however, frequently, little supportive evidence was provided from the work of students nor any critical stance taken about the results of the innovation. Can, and will, every teacher of mathematics similar to this enthusiast obtain the same impressive results? This question is rarely addressed. We hope that readers find that the chapters in this volume are **considered**. Far from articulating the next big fashion, the authors represented here often raise caveats or point to issues which are potential causes of concern for the learner or teacher of mathematics. They are not slow to articulate questions which demand careful research. Many, implicitly, draw attention to the failure of curriculum developers to work closely with those who are generating what we might consider to be a 21st century use of technology for learning mathematics, rather than the drill and practice scene of the most replicative computer software. Also, the enthusiasm for computers and calculators could be said to have begun in the United States of America, a country whose **resources** are unmatchable elsewhere. Even within the U.S.A. itself, technology is not as widely spread and used as enthusiasts would suggest. If we are moving into an era where the possession of technology is a necessary component of the learning of mathematics, we are setting up further barriers to learning for citizens in many countries and even for less well-resourced pupils in rich countries. Our enthusiasm must, of necessity therefore, be tempered by social justice considerations. It was, consequently, of interest to us that a number of authors raised **gender** as a descriptive category of those about whose learning they were reporting, or as an organising principle through which to analyse. Although many myths about the issue of female/male learning of mathematics are being challenged, myths are also in the making as we observe technology in our classrooms being gendered. It is noticeable that for most of our authors this was not a central concern even though it might have been affecting what they were observing in the classrooms on which they were reporting.

The major issues

The metaphor of a *bridge,* as used in the title of the book, is offered as a way of viewing the relationship between technology and the learning and teaching of mathematics. The major issues which the book addresses are therefore firmly in the domains of learning and teaching. As most experienced teachers are aware, it often takes more than a well-formed exposition or explanation from the teacher for a student to develop a good conceptual understanding of the mathematical content of a lesson. John Mason points out that 'Getting students to make connections is not achieved merely by the teacher telling them the connections'. A seminal research study by Denvir and Brown (1986) showed that children's learning in the primary phase followed quite a different route to that planned by their teachers, or expected according to suggested hierarchies of mathematics. The Cockcroft Report (DES, 1982) made clear that problem-solving, appropriate practical work, discussion and investigational work were necessary features of mathematical learning environments at all levels. The Report addressed *school* mathematics, but it is our personal belief that this advice applies also to the teaching of mathematics beyond school level, and that many of the problems faced by learners at higher levels are due to limited approaches in the teaching experienced. This view is supported by authors in this collection who address mathematical learning in higher education (e.g. LaTorre, Shoaf-Grubbs)

Technology in mathematics learning and teaching can be seen to support an ethos of students' and teacher interactivity supported by appropriate materials. The technological hardware and software offers motivating, dynamic and exciting possibilities to stimulate students' involvement with and understanding of mathematics. This rhetoric derives from various authors in the collection. We are aware of our need to problematise such assertions and to ask what evidence we see for such claims. Adrian Oldknow draws attention to common experience that 'materials alone cannot be relied upon to bring about change in practice'. Together with Kenneth Ruthven he emphasises the importance of research in highlighting the roles and curricular implications of differents forms of technology, and the recognition that teachers need support in becoming aware of possibilities and problems.

The chapters highlight many roles for technology. For example, Don LaTorre speaks of calculators carrying 'the computational burden'. Andy DiSessa makes broad claims for the *Boxer* software environment, including its 'success as a collaborative medium'.

6

Many authors (e.g. Bottino, Emmer) use the metaphor of a 'tool' to describe a computer's role. Eric Love problematises this metaphorical use. He asks who are the users of such tools and what is the context of their use, arguing that software designed by certain users, such as professional mathematicians in a research context, may not provide appropriate tools for inexperienced learners of mathematics in a school context. For example, what are the implications for students, using packages such as *Maple* or *Mathematica,* who are not able to appreciate the sophisticated and often subtle assumptions of the designers? Some of the submitted papers were rejected because they offered little more than an unquestioning exposition of the wonders of such packages for mathematics teaching at various levels. In encouraging teachers to contemplate using such items of hardware and software, we need to address, honestly and profoundly, their implications for all learners and learning.

Liddy Nevile seeks to elucidate the nature of the tool, by addressing what she, and John Mason, call its *dimension*. Accordingly, videos and films (as discussed by Michele Emmer) are of *zero* dimension. The user has no scope for varying what is seen other than start, stop, rewind and replay. The more scope a tool provides, the more power it affords the learner, the greater its dimension. However, the corresponding sacrifice must be in complexity and the demands this places on the learner in terms of learning to operate the tool, rather than seeing through to the mathematics. A claim made for the perceived success of the *Boxer* environment (DiSessa), for example, is its transparency for the learner alongside its power as a tool.

The relationship between the technology and the mathematics to which it provides access must be seen as problematic. The research reported by Janet Ainley and Dave Pratt draws attention to the use of software, in this case a graph-drawing package, without understanding of the mathematics involved. A child draws beautiful but meaningless graphs. This emphasises the crucial role of the teacher in being aware of such potential, and in being able to bring to bear appropriate teaching strategies. The dangers of seeing software

7

packages as teacher-free zones are striking! Important too is the philosophy a teacher brings to teaching mathematics using technology. If mathematics teaching is seen as a *transmissive* process, then technology is likely to be used to expound, demonstrate and clarify the mathematical ideas the teacher seeks to convey to the learner; if seen as a *constructive* process, then technology is likely to be used to involve the learner, to encourage autonomy and a spirit of questioning and reflection. Teacher awareness of these positions thus changes the pedagogical environment allowing for more knowledgeable, and therefore appropriate, use of technology.

Diverse areas of mathematics, to which technology is claimed to provide access, for example, elementary statistics, basic algebra, functions, calculus, linear algebra (vectors and matrices), differential equations, are addressed by chapters in this collection. In some cases (as for example in the learning of basic algebra), research shows salient features of software (notably spreadsheets, programming languages and CAS [Computer Algebra Systems]) in engaging learners with fundamental conceptual structures (for example the understanding and use of variables). There is recognition (see, for example, Dettori et al) that disparity between mathematical conceptual structures, and their software manifestations might lead to problems for learners. Such research, and the examples it reports, seems essential in alerting teachers to the possibilities *and* the problems their students might encounter. There are also important consequences for the mathematics curriculum. Eric Love points out the propensity of school mathematics to teach students to carry out techniques. If software is designed to fulfil this role, what curricular influences might we expect? For example, if supercalculators carry out matrix-manipulation techniques (Don LaTorre), what alternative foci might the curriculum offer students who no longer need to spend time practising the techniques? Andrew Rothery and Chris. Bissell both recognise the powerful curricular implications of software packages for their teaching (of, respectively, calculus and mathematical models for engineering).

It is clear from the above paragraphs that issues for the teacher are abundant. Many authors recognise the implications of these issues for pedagogy and for teacher development. It is not *just* a case of resourcing – although the case is strong for persuading governments and other funding agencies to invest in teacher development as well as in the provision of hardware and software. It is also a case of what *forms* of teacher-development are likely to be effective, and, moreover, what teacher awarenesses to promote. Hilary Povey talking of her own role as a teacher of teachers, makes a strong case for the development of a 'critical pedagogy'. Central to such a concept is the fostering of learner-autonomy and collective responsibility, whether the learners are students or teachers. Chapters by Geoff. Sheath and by Patrick Wild provide examples of associated practices. The implications of such a philosophy for teacher-educators are salutary, consistency of approach being a central feature, and John Mason's *didactic tension* a potential obstacle.

It is John Masons' chapter which makes it clear that such a *critical pedagogy* is essential, rather than a luxury, if effective learning is to be achieved. Starting from the premise that 'screen images do not guarantee abstraction', he reminds us of Buber's (1970) elaboration of the difference between the I-Thou and I-It relationship, and the transformation by teachers of the first into the second. From teaching *students* we, all too easily and readily, move into teaching *mathematics*. As teachers, we carry our generalised mathematical ideas, offer students examples (specific and particular), and expect, often without further work, that they will reach the generalities we have in mind. Computer screens are prime examples of particularity and we cannot assume that any abstractions or generalities our students create will bear resemblance to those we (unreasonably?) expect. From this recognition, the importance of learners articulating their understandings (often through interactive reconstruction of their mental images related to screen images) and questioning their own learning seem central to the effective pedagogy highlighted by various authors. Having learners reflect critically on their own learning, just as teachers or teacher-educators we need to reflect critically on our own teaching, is the basis of such a pedagogy, which might aim, ambitiously, for a common discourse between these levels of reflection.

References

Buber, M.: 1923, *Ich und Du*, translated Kaufman, 1970, *I and Thou*. Edinburgh: Clark.

Denvir, B. & Brown, M.: 1986 'Understanding of Number Concepts in Low Attaining 7-9 Year Olds: Part I. Development of Descriptive Framework and Diagnostic Instrument', *Educational Studies in Mathematics*, 17, 15-36.

Denvir, B. & Brown, M.: 1986 'Understanding of Number Concepts in Low Attaining 7-9 Year Olds: Part II. The Teaching Studies', *Educational Studies in Mathematics*, 17, 143-164.

Department of Education and Science: 1982, *Mathematics Counts*, Her Majesty's Stationery Office, London.

Section One: Setting the Scene

As already explained in the Introduction, this book arose as the result of a conference, Technology in Mathematics Teaching. Section One consists of chapters commissioned from the three plenary speakers to the conference. It is not intended to be an overview but, rather, a taster for ways in which technology might be seen as a bridge between teaching and learning mathematics. The first chapter concentrates on the mathematics which can be provoked through using the computer to generate a particular class of objects. The second and third chapters look at the pedagogic implications of utilising particular computer software.

Chapter 1, *Fractals: from Ferns to Fireworks*, is by Phil. Rippon, a mathematician who has entered the world of mathematics education through his lectures and demonstrations about fractals. He was invited to address the conference because this new mathematical area is a product of the development of computer technology, it has high aesthetic appeal and yet has not become widely used by teachers nor ensured a place in the learning of mathematics at school or undergraduate level. His chapter requires interactive reading as he frequently suggests to the reader, and by extension to the teacher in the classroom, ways of developing the argument he is making. The mathematics which lies behind the intriguing pictures is introduced in a manner which is a model for such writing. The reader's fingers itch to pursue some of the suggestions in order to see the results. The chapter is not overtly about learning mathematics. However, we feel that, for all the reasons given, it is an excellent way of setting the scene and, implicitly, has many messages for the classroom which the active reader can elicit.

Colette Laborde and Andy diSessa, in Chapters 2, *Designing Tasks for Learning Geometry in a Computer-based Environment* and 3,

Collaborating via Boxer write of the application of specific computer software to the learning and teaching of mathematics. In the case of Colette Laborde, this is dynamic software known as Cabri-Géomètre, developed at the University of Grenoble in France. Colette Laborde makes clear that such software poses truly new learning problems in the mathematics classroom as it enables the learner to change features of a drawing in order to come to an appreciation of the properties embedded within it, or, as she puts it, to move from drawing to figure. She comes to important conclusions for mathematical pedagogy both with respect to teacher resistance to using computers in their classrooms and the potential for pupils to misconstrue the purposes of a paper and pencil construction task. Her argument, supported by pupil material, is compelling for a re-consideration of the learning of geometry making effective use of technology to provide experiences which were formerly unavailable.

Andy diSessa helped to create a piece of software called Boxer. In Chapter 3 he explores how he used Boxer with pupils some examples of whose work enriches the text. It was found to be a medium which led to surprising results when used collaboratively. He says: "Boxer was designed with an eye towards (1) learnability and comprehensibility, and (2) general expressiveness" although its designers wanted "to create a genuinely new and powerful medium-like written text only extended by essentially new capabilities computers can add." For us, a crucial lesson learnt by the designers from the observation of pupil collaboration was not simply about more effective learning of mathematics but that learners surprised them both in the strategies they acquired and used, and in the mathematical purposes to which these strategies were put. This lesson is to be found repeatedly throughout the book and underlies the degree to which bringing to the classroom a closed mind about pupils' mathematical potential acts as a brake on the development of that potential. Andy di Sessa's argument is, then, also about new learning which takes place from collaboration with this particular medium, that is it links the structure of Boxer to the social learning context.

12

As is always the case with written material, the excitement of the oral presentations and the sense of involvement and enthusiasm which these three plenary lecturers provoked in their audience are difficult to capture. Nonetheless, in their different ways, all three authors demonstrate technology acting as a bridge between the teaching and learning of mathematics. They set the scene by querying the mathematics that the technology provokes as well as the pedagogy which is appropriate to its learning.

1 Fractals: from Ferns to Fireworks

Philip J. Rippon

An outline is given of one route through the subject usually called 'fractals', visiting on the way iteration of real quadratic sequences, complex quadratic sequences, Julia sets, the Mandelbrot set, iterated function systems and images obtained from piecewise affine mappings.

1. Introduction

It is remarkable that the simple real recurrence relation

$$x_{n+1} = x_n^2 + c, \qquad n = 0, 1, 2, \ldots, \qquad (1)$$

leads eventually to the construction of 'fractal' images such as the fern and the firework display in Figure 1.

Figure 1(a) Fern

Figure 1(b) Fireworks

The subject of 'fractals', and in particular the wonderful images, has been well-publicised over the last few years, and much has been written about it; see, for example, *Fractals for the classroom, Part Two* (Peitgen, Jürgens and Saupe, 1992) and the references therein. This article will attempt to highlight some of the milestones along one route through this material, using computer-generated pictures to illustrate the main results, and outlining a few elementary proofs.

2. The Basic Real Quadratic Family

An iteration sequence x_n is a sequence defined by a recurrence relation of the form

$$x_{n+1} = f(x_n), \qquad n = 0, 1, 2, \dots,$$

where f is a function; x_0 is called the **initial term** (or **seed**) of the iteration sequence, and x_n, $n = 0, 1, 2, \dots$, is often referred to as the **orbit** of x_0 under f. In (1) the function being iterated is the **basic quadratic function** $P_c(x) = x^2 + c$.

If $f(x) = ax + b$, then it can easily be proved (by induction, for example) that

$$x_n = a^n\left(x_0 - \frac{b}{1-a}\right) + \frac{b}{1-a}, \qquad \text{for } n = 0, 1, 2, \dots,$$

unless $a = 1$, in which case $x_n = x_0 + nb$. Thus the behaviour of such sequences, as $n \to \infty$, depends in a rather simple way on the parameters a and b.

The story is very different if $f(x) = ax^2 + bx + c$, where $a \neq 0$. By making the change of variables $x'_n = ax_n + \frac{1}{2}b$, all such quadratic iteration sequences are equivalent (in the sense of conjugacy of functions) to (1). For example, the sequences

$$x_{n+1} = kx_n(1-x_n), \qquad n = 0, 1, 2, \dots, \tag{2}$$

which are frequently studied because of their modelling applications, are equivalent to ones of the form (1) with

$c = \frac{1}{2}k(1-\frac{1}{2}k)$. Some advantages of using the sequences (1) in the classroom, rather than (2), are as follows:

(a) squaring and then adding c is 'easy';
(b) for each c (except $c = 1/4$) there are two equivalent values of k;
(c) plotting an accurate graph $y = x^2 + c$ by hand requires only one template;
(d) all the functions P_c have the same derivative $P_c'(x) = 2x$;
(e) generalisation to complex sequences is straightforward.

The best way to 'see' how an iteration sequence (1) behaves for a given x_0 and c is to use **graphical iteration**, illustrated in Figure 2.

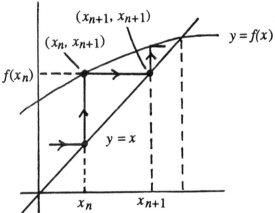

Figure 2 Graphical iteration

This is a geometric procedure for finding the position on the x-axis of the term $x_{n+1} = f(x_n)$, given the position of the term x_n:

(a) draw a vertical line to meet $y = f(x)$ at (x_n, x_{n+1});
(b) draw a horizontal line to meet $y = x$ at (x_{n+1}, x_{n+1}).

Figure 3 shows the result of performing graphical iteration with a range of values of c. In each case $x_0 = 0$ has been used, but you should experiment with other initial terms.

Various different types of long-term behaviour of x_n are usually observed:

(a) x_n tends to infinity ($c = 0.3, c = -2.1$);

(b) x_n tends to a cycle of P_c ($c = -0.6, c = -1, c = -1.75$).

To begin to explain these observations, it is essential to make a few definitions.

A point a is a **fixed point** of a function f if $f(a) = a$. The fixed point a is

(a) **attracting**, if $|f'(a)| < 1$;

(b) **repelling**, if $|f'(a)| > 1$;

(c) **indifferent**, if $|f'(a)| = 1$;

(d) **super-attracting**, if $f'(a) = 0$.

For example, the function $P_{1/4}(x) = x^2 + 1/4$ has the indifferent fixed point 1/2, because

$$P_{1/4}(\tfrac{1}{2}) = (\tfrac{1}{2})^2 + \tfrac{1}{4} = \tfrac{1}{2} \text{ and } P'_{1/4}(\tfrac{1}{2}) = 2 \times \tfrac{1}{2} = 1.$$

In order to classify cycles in the same sort of way, it is helpful to define the **nth iterate** of a function f to be the function

$$f^n = f \circ f \circ \ldots \circ f, \qquad \text{obtained by applying } f \text{ exactly } n \text{ times,}$$

including f^0 as the identity! Thus if $x_{n+1} = f(x_n), n = 0, 1, 2, \ldots,$ then

$$x_n = f^n(x_0), \qquad \text{for } n = 0, 1, 2, \ldots .$$

A point a is a **periodic point**, with **period** p, of a function f if

$$f^p(a) = a, \text{ but } f(a), f^2(a), \ldots, f^{p-1}(a) \neq a;$$

that is, a is a fixed point of f^p but of no lower iterate of f. The points

$$a, f(a), f^2(a), \ldots, f^{p-1}(a),$$

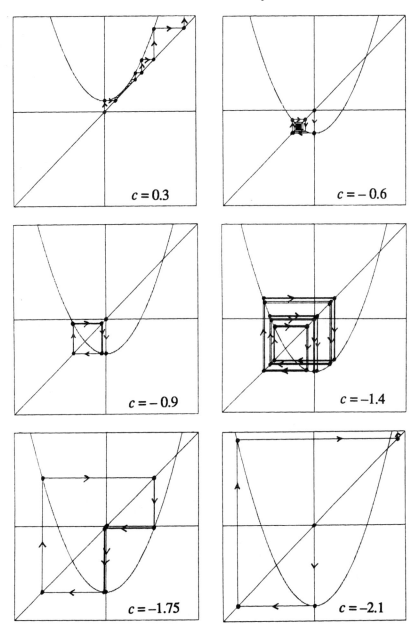

Figure 3 Graphical iteration with $P_c(x) = x^2 + c$

19

then form a **cycle of period** p, or a p-**cycle**, of f. (A fixed point may be called a 1-cycle.) The type of a cycle is determined by the type of the corresponding fixed point of f^P.

For example, the function $P_{-1}(x) = x^2 - 1$ has the super-attracting 2-cycle 0, -1, because $P_{-1}^2(0) = P_{-1}(-1) = 0$ and

$$(P_{-1}^2)'(0) = \frac{d}{dx} P_{-1} (P_{-1} (x))\Big|_{x=0}$$

$$= (2P_{-1}(x)) (2x)\Big|_{x=0} \qquad \text{(by the chain rule)}$$

$$= 0.$$

The names 'attracting' and 'repelling' indicate the effects that these fixed points and cycles have on nearby orbits. For example, if a is an attracting fixed point of the function f, then $f(a) = a$ and $|f'(a)| < 1$, so that, by the definition of $f'(a)$,

$$\left|\frac{x_{n+1} - a}{x_n - a}\right| = \left|\frac{f(x_n) - f(a)}{x_n - a}\right| \leq k < 1,$$

whenever x_n is close (but not equal) to a. It follows that $x_n \to a$ whenever x_0 is close to a. More generally, if a belongs to an attracting p-cycle of f, then a is an attracting fixed point of f^P, and so the subsequence

$$x_{np} = f^{np}(x_0) = (f^P)^n(x_0) \to a \quad \text{as } n \to \infty,$$

whenever x_0 is close to a. Therefore, the sequence $x_n = f^n(x_0)$, $n = 0, 1, 2, \ldots$, forms itself into p convergent subsequences, with one subsequence tending to each point of the attracting p-cycle.

The values of c for which the function P_c has either an attracting fixed point or an attracting 2-cycle are:

(a) $-3/4 < c < 1/4$: attracting fixed point $1/2 - \sqrt{1/4 - c}$;

(b) $-5/4 < c < -3/4$: attracting 2-cycle $-1/2 \pm \sqrt{-3/4 - c}$.

These values of c are justified later (in the complex case).

Such ranges of values of c cannot be determined so explicitly for other attracting p-cycles. However, the following result can be used to investigate them experimentally:

if P_c has an attracting p-cycle, then $P_c^n(0)$ converges to it.

This result is tricky because the initial term 0 need not be close to the attracting cycle (though sometimes 0 is part of the cycle, as when $c = -1$); see Bruce, Giblin and Rippon, 1990, p393, for a proof with the family (2). Even for $-3/4 < c < 0$, the proof presents a challenge, though it looks obvious from Figure 3 ($c = -0.6$). The proof depends on the fact that 0 is the only **critical point** (that is, stationary point) of P_c for all c. One consequence of the above result is that, for each c, the function P_c can have at most one attracting cycle, because the sequence $P_c^n(0)$ cannot converge to two different cycles!

Using this result, the attracting cycle structure of the functions P_c can be investigated experimentally, by considering a large number of values of c between -2 and $1/4$ (outside this interval the sequence $P_c^n(0)$ is easily seen to tend to infinity by graphical iteration). For each such c, the corresponding sequence $P_c^n(0)$ is plotted above and below a horizontal c-axis, leaving out the first 200 terms, say, in order to detect any attracting cycle which is present (see Figure 4(a)).

As predicted above, the sequence $P_c^n(0)$ tends to a fixed point for $-3/4 < c < 1/4$ and tends to a 2-cycle for $-5/4 < c < -3/4$. To the left of $-5/4$, the sequence tends to a 4-cycle, then to an 8-cycle, and so on. The values of c at which these changes take place are called **period-doubling bifurcation points** and they form a sequence which tends to approximately -1.401. From -1.401 to -2, the situation is less clear. Attracting 3-cycles appear just to the left of -1.75 (see Figure 3 again), together with an accompanying sequence of period-doublings.

21

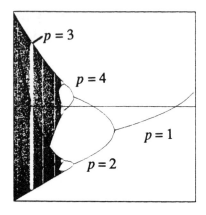
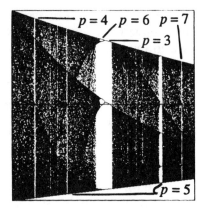

Figure 4(a) $-2 \le c \le 0.25$ *Figure 4(b)* $-2 \le c \le -1.54$

Enlarging the scale on the c-axis (as in Figure 4(b)), shows the presence of many more so-called **periodic windows**, each with an abrupt cut-off on the right and period-doubling on the left. However, these periodic windows become extremely thin and rather sparse, and so finding them is quite a challenge (for example, try finding a period 9 window to the right of −1.75). In spite of their sparseness, it is widely believed that the periodic windows are dense in [−2, 1/4]; that is, they appear in every subinterval of [−2, 1/4]. This seems to be an extremely difficult problem, not least because it *is* known that the complement of the periodic windows in [−2, 1/4] has non-zero length, but a proof of their density (over 100 pages long!) has recently been claimed by Swiatek.

Notice that each of the visible periodic windows in Figure 4 contains a point c such that 0 belongs to the corresponding cycle. In fact, a p-cycle a, $P_c(a)$, . . . , $P_c^{p-1}(a)$ of P_c is super-attracting if and only if it includes the point 0, because

$$(P_c^p)'(a) = P_c'(P_c^{p-1}(a)) \ldots P_c'(P_c(a)) \, P_c'(a)$$

$$= (2P_c^{p-1}(a)) \ldots (2P_c(a))(2a),$$

which is 2^p times the product of the points in the p-cycle. Thus, such values of c can be found by determining the solutions of the equation $P_c^p(0) = 0$, which has degree 2^{p-1}. For example,

$$P_c(0) = c = 0$$

has solution $c = 0$ (super-attracting fixed point);

$$P_c^2(0) = c^2 + c = 0$$

has solutions $c = 0$ and $c = -1$ (super-attracting 2-cycle);

$$P_c^3(0) = (c^2 + c)^2 + c = c^4 + 2c^3 + c^2 + c = 0$$

has real solutions $c = 0$ and $c = -1.755$ (super-attracting 3-cycle).

As a result of detailed calculations of many of these points (for the family (2), and others), M. Feigenbaum was led to discover the universal nature of the *rate* at which period-doubling occurs. Such calculations can readily be repeated (see Bruce, Giblin and Rippon, p374 or Peitgen, Jürgens and Saupe, p227) and the Feigenbaum constant 4.6692 . . . obtained.

3. The Complex Quadratic Family

Now consider the quadratic iteration sequences

$$z_{n+1} = z_n^2 + c, \qquad n = 0, 1, 2, \ldots,$$

where c is complex. Once again, take $z_0 = 0$ and plot the corresponding orbits for various values of c (see Figure 5).

As in the real case, one of the following usually occurs:
(a) z_n tends to infinity;
(b) z_n tends to a cycle.

Cycles of the complex function $P_c(z) = z^2 + c$ are defined as in the real case and also classified into various types in the same way.

For example, the function $P_i(z) = z^2 + i$ has the repelling 2-cycle $-i, -1+i$ because

$$P_i(-i) = -1+i, \quad P_i^2(-i) = P_i(-1+i) = -i$$

and

$$|(P_i^2)'(-i)| = |(2P_i(-i))(-2i)| = |4+4i| = 4\sqrt{2} > 1.$$

Once again, it can be checked that an attracting p-cycle attracts nearby orbits. Moreover, it was proved by Fatou in 1905 that:

if P_c has an attracting p-cycle, then $P_c^n(0)$ converges to it.

The proof is a nice application of results in a standard complex analysis course (see Baker and Rippon, 1985, for an elementary treatment of the argument applied to complex exponential functions).

The analogue of the real bifurcation diagram in Figure 4, would be a plot of the set

$$\{c : P_c \text{ has an attracting cycle}\},$$

found by monitoring the sequence $P_c^n(0)$ for many complex values of c (using, perhaps, a different colour for each different period). Such a plot was made by Brooks and Matelski in 1979, using asterisks to represent points in the set! They had arrived at quadratic iteration sequences while studying groups of Möbius transformations. Also in 1979, Mandelbrot plotted the larger set

$$M = \{c : P_c^n(0) \text{ does not tend to } \infty\},$$

which became known as the Mandelbrot set (see Figure 6). Mandelbrot had arrived at the set M by studying and plotting what are known as Julia sets.

For a function of the form $P_c(z) = z^2 + c$, the **Julia set** J_c is most simply defined to be the boundary of the **keep set**

$$K_c = \{z : P_c^n(z) \text{ does not tend to } \infty\}.$$

24

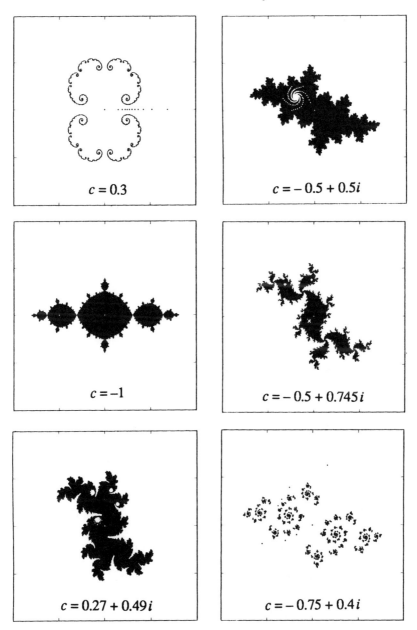

Figure 5 Keep sets for $P_c(z) = z^2 + c$, and orbits of 0.

For example, $K_0 = \{z : |z| \leq 1\}$ and so $J_0 = \{z : |z| = 1\}$. Figure 5 shows plots of various keep sets K_c with the corresponding orbits $P_c^n(0)$ included for comparison. These plots illustrate the truth of the following fundamental result of Fatou and Julia:

$$J_c \text{ is connected} \Leftrightarrow K_c \text{ is connected} \Leftrightarrow 0 \in K_c,$$

so that $M = \{c : J_c \text{ is connected}\}$.

Now, it is straightforward to show that, for each c,

$$K_c \subseteq \{z : |z| \leq r_c\},$$

where $r_c = 1/2 + \sqrt{1/4 + |c|}$ is the positive solution of $r^2 = r + |c|$. Indeed, if $z_n = P_c^n(z_0)$, for $n = 1, 2, \ldots$, then, by the Triangle Inequality,

$$|z_{n+1}| \geq |z_n|^2 - |c| = \left(\frac{r_c + |c|}{r_c^2}\right)|z_n|^2 - |c|.$$

Thus, if $|z_n| > r_c$ then

$$|z_{n+1}| \geq \frac{|z_n|^2}{r_c} > r_c.$$

Hence, whenever $|z_0| > r_c$ it follows that $|z_n| \geq (|z_0|/r_c)^{2^n} r_c$, for $n = 0, 1, 2, \ldots$, so that z_0 does not belong to K_c.

Therefore, the following deductions can be made:

$$c \in M \Rightarrow |P_c^n(0)| \leq r_c, \quad \text{for } n = 0, 1, 2, \ldots,$$
$$\Rightarrow |c| \leq r_c, \quad (\text{since } P_c^1(0) = c)$$
$$\Rightarrow r_c \leq 2 \quad (\text{by inequality manipulation})$$
$$\Rightarrow |P_c^n(0)| \leq 2, \quad \text{for } n = 0, 1, 2, \ldots.$$

It follows that

$$M = \{c : |P_c^n(0)| \leq 2, \text{ for } n = 0, 1, 2, \ldots\},$$

and this is the basis for the standard 'naive' algorithm for plotting M (see Figure 6(a)):

$$\text{plot } c \text{ if} \left| P_c^n(0) \right| \le 2, \text{ for } n = 0, 1, 2, \dots, N, \tag{3}$$

where N is large. More cunning algorithms show that the naive algorithm misses many thin parts of M. See Figure 6(b), which was produced by plotting the point c if either (3) holds or: the number of iterations needed for $P_c^n(0)$ to escape from $\{z : |z| \le 2\}$ is 'significantly greater' than for a nearby c.

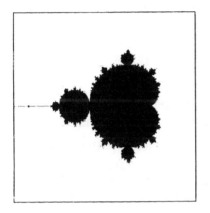

Figure 6(a) M-the naive algorithm

Figure 6(b) M-showing connectedness

Figure 6(b) suggests that, in spite of its immense complexity, the set M is connected. This was proved by Douady and Hubbard in 1982. It is not yet known whether M is pathwise connected, though this would follow if it could be shown that M is locally connected, as is widely believed.

The interior of M contains the Brooks-Matelski set, mentioned earlier, which consists of infinitely many **periodic regions** that generalise the periodic windows on the real axis. Two of these periodic regions are easily determined (see Figure 7):

(a) P_c has an attracting fixed point $\Leftrightarrow c \in \{z - z^2 : |z| < \frac{1}{2}\}$, because $P_c(z) = z \Leftrightarrow c = z - z^2$ and a fixed point z is attracting if and only if $|P_c'(z)| = |2z| < 1$ (this periodic region is the interior of a cardioid, with equation $(8|c|^2 - 3/2)^2 + 8\text{Re } c = 3$);

(b) P_c has an attracting 2-cycle $\Leftrightarrow |c+1| < 1/4$, because the factorisation

$$P_c^2(z) - z = (P_c(z) - z)(z^2 + z + c + 1),$$

shows that the solutions z_1, z_2 of $z^2 + z + c + 1 = 0$ form a 2-cycle of P_c, and this 2-cycle is attracting if and only if

$$|(P_c^2)'(z_1)| = |(2P_c(z_1))(2z_1)| = |4 z_1 z_2| = 4|c+1| < 1.$$

Information about the location of other periodic regions (whose boundaries are either roughly circular or cardioid-shaped) may be obtained by solving the polynomial equations $P_c^p(0) = 0$ to find super-attracting p-cycles. For example, the largish 3-periodic region in M above the main cardioid (see Figure 6) contains the point $c = -0.123 + 0.745i$, for which P_c has a super-attracting 3-cycle, because this c satisfies $P_c^3(0) = 0$.

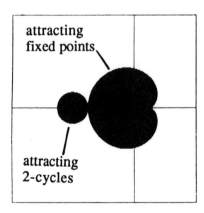

Figure 7 Two periodic regions

4. Plotting Julia sets

For a given complex number c, the Julia set J_c acts as the 'watershed' between those points which escape to ∞ under iteration of P_c and those which are kept. Figure 5 indicates that these sets are usually extremely complicated and display self-similarity properties. It *is* possible to plot J_c by using its definition as the boundary of K_c, but there are other methods. One approach is based on the property (proved by Julia) that :

> J_c is the smallest closed set which contains all the backward iterates of any given point of J_c.

A point z is a **backward iterate** of z_1 under the function P_c if $P_c^n(z) = z_1$, for some positive integer n. This property explains the self-similar nature of Julia sets.

Now, since $w = z^2 + c \Leftrightarrow z = \pm\sqrt{w - c}$, the backward iterates of a point form themselves into a binary tree, as indicated in Figure 8 .

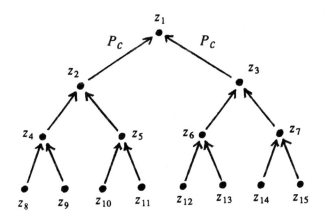

Figure 8 Backward iterates

Plotting all of these backward iterates down to a given level requires some care, and so it is tempting to cheat by making a random choice (of square root) at each stage, and hope for the best. In fact this method often gives a good rendering of J_c, especially if you experiment with the random weightings of the two choices; see Figure 9, where z_1 is the non-attracting fixed point $1/2 + \sqrt{1/4 - c}$.

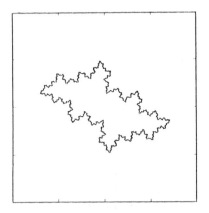

Figure 9(a) $c = -0.5 + 0.5i$ *Figure 9(b)* $c = i$

This process of backward iteration with random choice amounts to iterating the two rather special complex functions $f_1(z) = \sqrt{z-c}$ and $f_2(z) = -\sqrt{z-c}$, making at each stage a random choice between them. In the early 1980s, various authors (see references to Barnsley and Hutchinson in Peitgen, Jürgens and Saupe, 1992) began to experiment in this way with other pairs of functions from $\mathbb{R}^2$ to $\mathbb{R}^2$, and, more generally, with finite sets of such functions and associated random weightings, which they called **iterated function systems**. They used **affine mappings** of the plane, which take the form

$$t\begin{pmatrix} x \\ y \end{pmatrix} = \begin{pmatrix} a & b \\ c & d \end{pmatrix}\begin{pmatrix} x \\ y \end{pmatrix} + \begin{pmatrix} e \\ f \end{pmatrix},$$

where a, b, c, d, e, f are real. These combine simplicity with flexibility, since any triangle can be mapped onto any other by such a mapping.

As expected, this process does produce sets which are as complicated as Julia sets, and display self-similarity properties, at least when the set of affine mappings is not too expanding overall. But now there is a bonus. It is possible to start with a given set that displays self-similarity properties and, by analysing these properties, create an iterated function system which generates the given set.

The by-now-classical example is the fern shown in Figure 10. This set can be partitioned into four subsets, each of which is the image of the fern itself under the affine mapping indicated (strictly speaking an affine mapping should be invertible, but this requirement is relaxed here). Running the iterated function system with the given probabilities, gives a rendering of the fern.

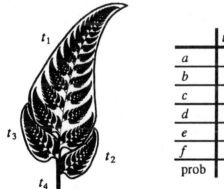

	t_1	t_2	t_3	t_4
a	0.85	–0.15	0.2	0
b	0.05	0.25	–0.25	0
c	–0.05	0.25	0.25	0
d	0.85	0.15	0.2	0.17
e	0	0	0	0
f	1.5	0.75	1.5	0
prob	0.8	0.09	0.09	0.02

Figure 10 Partitioning the fern

Such iterated function systems are extremely effective for plotting sets like the fern which display obvious 'affine self-similarity'. Much more ingenuity is needed to plot images whose affine structure is less transparent. Nevertheless, Barnsley and others have devoted very considerable efforts to investigating applications of iterated function systems to the important area of 'image compression', and have developed startlingly efficient methods of rendering real-life images as a result; see Barnsley and Hurd, 1993.

5. Affine firework display

The iterated function system discussed above produces a sequence which is in a certain sense obtained by mixing up the constituent affine mappings. Another way to produce a sequence from a mixture of such mappings is to make the choice of mapping depend on the current term of the sequence. A rather extreme example is the complex sequence $z_{n+1} = z_n^2 + c$, which can be expressed in 'affine form' as

$$\begin{pmatrix} x_{n+1} \\ y_{n+1} \end{pmatrix} = \begin{pmatrix} x_n & -y_n \\ y_n & x_n \end{pmatrix} \begin{pmatrix} x_n \\ y_n \end{pmatrix} + \begin{pmatrix} a \\ b \end{pmatrix},$$

where $z_n = x_n + iy_n$ and $c = a + ib$!

A more moderate example arises in connection with an intriguing problem posed some years ago by Morton Brown (Brown, 1983 and 1985). The problem was to prove that the real sequence x_n defined by

$$x_{n+2} = |x_{n+1}| - x_n, \qquad n = 0, 1, 2, \ldots,$$

is periodic, with period 9, for all pairs of initial terms x_0, x_1 (not both 0). Perhaps the simplest (certainly the most memorable) way to solve the problem is to recast the recurrence relation in the form of an iteration sequence in $\mathbb{R}^2$:

$$\begin{pmatrix} x_{n+2} \\ x_{n+1} \end{pmatrix} = \Phi \begin{pmatrix} x_{n+1} \\ x_n \end{pmatrix}, \qquad n = 0, 1, 2, \ldots.$$

Here Φ is a mixture of two linear transformations, as follows:

$$\Phi = \begin{cases} \begin{pmatrix} 1 & -1 \\ 1 & 0 \end{pmatrix}, & \text{on the right half-plane,} \\[3mm] \begin{pmatrix} -1 & -1 \\ 1 & 0 \end{pmatrix}, & \text{on the left half-plane.} \end{cases}$$

It is now easy to check that Φ^9 is the identity on the orbit of rays shown in Figure 11. Since this orbit includes the y-axis, which is where the two linear mappings join together (continuously, but not smoothly), it follows that Φ^9 must be linear on each of the sectors between adjacent rays in Figure 11, and so Φ^9 is the identity everywhere, as required.

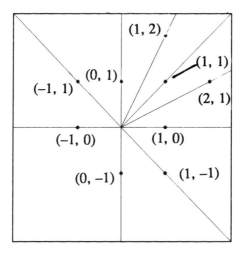

Figure 11 The orbit of (1, 0)

Actually, this problem involves just one special case of a general family, which has been investigated recently (Crampin, 1992, and Beardon, Bullett and Rippon, 1993). A simple extension is to consider sequences of the form

$$x_{n+2} = \lambda \, | \, x_{n+1} \, | - x_n, \qquad n = 0, 1, 2, \ldots ,$$

where λ is a real parameter. It turns out that if $\lambda = 2\cos(\pi/p)$, where $p = 2, 3, 4, \ldots$, then the sequence is always periodic with period p^2, a nice generalisation of the Morton Brown problem (where $p=3$). For other values of λ, however, this periodic phenomenon seems not to occur. As λ increases from 0 to 2, the corresponding orbit in $\mathbb{R}^2$ of a point such as (1,0) repeatedly pulsates inwards and outwards, producing the firework display in Figure 1(b). Other such firework displays can be arranged by choosing different families of piecewise affine mappings - the reader is urged to experiment!

References

Baker, I.N. and Rippon, P.J. 1985, 'A note on complex iteration', *American Math. Monthly*, 92, 501-504.

Beardon, A.F., Bullet, S.R. and Rippon, P.J. 1993, Periodic orbits of difference equations, to appear in *Royal Society of Edinburgh Proceedings A (Mathematics)*, 125.

Brown, M. 1983 and 1985, *American Math. Monthly*, 90, 569, and 92, 219.

Bruce, J.W., Giblin, P.J. and Rippon, P.J. 1990, *Mathematics and the Microcomputer*, Cambridge: Cambridge University Press.

Crampin, M. 1992, Piecewise linear recurrence relations, *Mathematical Gazette*, 76, 355-359.

Peitgen, H.-O., Jürgens, H. and Saupe, D. 1992, *Fractals for the Classroom*, *Part Two*, New York: Springer-Verlag.

2 Designing Tasks for Learning Geometry in a Computer-based Environment

The Case of *Cabri-géomètre*

Colette Laborde

One of the reasons for teachers resisting the use of computers in the mathematics classrooms is the change that this introduction implies for the kind of problems which might foster the acquisition of knowledge by the pupil. Some old problems become uninteresting in the new environment. Conversely computers allow the design of some new problems that are impossible in a paper and pencil environment. In this chapter I consider the computer as part of the"milieu didactique": as such it offers specific representations of knowledge, specific facilities and specific feedback to pupils. I discuss the role of these characteristics i) on the way mathematical objects may be perceived by the pupils at the interface ii) on the solving processes of the pupils when they are confronted with problem situations on the computer. The context of the chapter is the geometry program Cabri-géomètre .

One of the aims of research in mathematics education is to investigate the relationships between teaching and learning. It has often been observed that pupils do not learn what teachers expected they would: they may learn something else which was not part of the teaching content; they may even learn something false. Several pieces of research have given evidence that listening

35

to the discourse of the teacher, to a clear presentation of mathematical content does not guarantee the learning which was expected (Burton 1994). According to a constructivist hypothesis which is widely shared among mathematics educators, knowledge is actively built up by the cognizing subject when interacting with mathematical learning environments (Steffe & Wood 1990, Davis et al. 1990). This notion of environment must be taken in a broad sense, the environment is of a material nature as well as an intellectual nature.

Problems are part of such environments and may play an important role in the construction of mathematical knowledge by the learner in so far as they offer the opportunity to involve his/her own ideas, to test their efficiency and validity, when trying to find a solution. Brousseau (1986) considered the process of elaborating a solution by a learner as the interaction between the learner and a "milieu" which enables the learner to perform some given kinds of actions in order to solve the problem and offers feedback to his/her actions. In terms of systems, the "milieu" is the system with which the learner is communicating. A classical example of milieu is provided by the "jigsaw" situation. Pupils have to enlarge a jigsaw made of polygonal shapes (triangles, squares, rectangles, trapezoids,...) so that 4cm of the original jigsaw become 7 cm in the new one. The first most frequent strategy consists of adding 3 cm to every length while preserving the nature of the shapes. But when putting together the pieces, they do not fit together. The milieu provides perceptive feedback which is interpreted by the pupils as a sign of the erroneous character of their strategy through their cultural knowledge of a jigsaw according to which all pieces must fit together.

Computers are environments in which problem solving may take place. The graphical and computing possibilities of some software now allow a reification of abstract objects and in particular of mathematical objects as well as numerous possible operations on these objects and various feedback. But as Pea (1987) claimed, the mediation by the computer provides new tools for operating on these objects and therefore changes the objects themselves. The feedback provided can also be of a different nature. It may be very sophisticated in comparison with a paper and pencil situation. All these points imply that designing a problem situation in a computer based environment requires a new analysis of the mathematical objects, operations and didactical feedback involved in the situation. This perhaps explains why teachers can be reluctant to use computers in their classrooms; although such an analysis about changes brought by the computer is done in some research, it is not really part of the education of teachers.

Another relation to geometrical objects through computer

It has been often stated that the nature of geometry is dual: problems related to geometry may be of a practical nature as well as theoretical. According to the claim of Herodotus about the origin of measurement of fields due to the floods of the Nile in Egypt, geometry was originally built as a way of controlling relations with physical space. But geometry has also been developed as a theoretical field dealing with abstract objects. The situation became clearer when non Euclidean geometries were developed, since when Euclidean geometry appears as one possible theory modelling real space.

A usual mediation of the theoretical objects of geometry is offered by graphical representations called "figures". The relation of mathematicians with these representations is complex and in a sense reflects the dual nature of geometry; mathematicians draw them, they act on them as if they were material objects (by means of a kind of experimentation) but their reasoning actually does not deal with them but with theoretical objects. This is why Parzysz (1988) proposed to distinguish the "figure" which is the theoretical referent (attached to a given geometrical theory: Euclidean geometry, projective geometry,...) from the "drawing" which is the material entity. But the introduction of geometry software enabling the drawing of figures on the screen of computers leads us to refine this distinction in order to account for this new kind of mediation.

From drawing to figure

As a material entity, a drawing is imperfect: the straight lines are not really straight. But mathematicians ignore these imperfections and work on an "idealized drawing". Questions which up to now have not been taken into account are becoming crucial for designers of graphical representations on the screen of a computer:

- To what extent are the imperfections of a drawing considered as noise by the users and do not prevent them from having access to the idealized drawing?
- To what extent is a sequence of small segments accepted as representing a straight line? Between what limits may the pixels of a line vary in order to give the visual impression of a straight line?
- What is the upper limit of the length of the common segment of a circle and one of its tangent lines?

The computer indeed reveals a phenomenon which is of importance for the pupils in a paper and pencil situation. For the latter the process of eliminating the imperfections of a real drawing is not so spontaneous as for mathematicians because it actually requires mathematical knowledge. Mathematicians know that a circle and a tangent line do not have a segment but only a point in common; they know that the tangent is perpendicular to the radius and they are able to infer this from the drawing even if the angle is not exactly right. So for the pupils the move (which is of a conceptual nature) from the actual drawing to an idealized drawing is not spontaneous.

An idealized drawing may give rise to several figures depending on the features which are relevant for the problem to be solved. The referent attached to a drawing cannot be inferred only from the drawing but must be given by a text in a discursive way. Two reasons may explain why.

i) Irrelevance of some properties of the drawing.
    Some relations which are apparent on the drawing may not be
    part of the figure.
*   The size of the sides of a triangle may not be relevant for the
    problem being solved, The position of the drawing on a sheet
    of paper is usually without any importance to a geometrical
    problem. Only some features of the drawing might be relevant
    to the solution of the problem. A drawing is a model of a figure
    and as for any model, all relations expressed on the drawing
    cannot be interpreted as representing relations in the referent.
    The interpretation of the drawing as a signifier is determined by
    the theory in which the user of the drawing is working. In
    projective geometry parallelism on the drawing is not relevant
    since all straight lines intersect. In 2D perspective drawings of
    3D figures, the intersection of two straight lines does not refer
    to the intersection of the 3D corresponding lines.

ii) Variability of elements of a figure.
    An important feature of geometrical figures is that they involve
    elements varying in subsets of the plane (considered as a set of
    points). A drawing cannot account for the variability of its
    elements.
*   Does a point of a segment on the drawing belong to the
    segment or to the straight line supporting the segment?
*   Must secant circles on the drawing necessarily be considered as
    secant for the problem?
    A description of the figure in natural language, or whatever
    symbolic language, is needed for interpreting a drawing.

The geometrical figure behind the screen

Computers have been used to design programs enabling the
multiplicity of drawings attached to a given geometrical figure to
be manipulated. It has been done by several means such as
programming language or repeat facilities. The improvement of
interface facilities now allows direct manipulation and the drawing
on the screen can be manipulated by means of the mouse and

dragged while all geometrical properties used to construct the drawing are preserved. A common feature of these programs is their use of an explicit description of the figures: a drawing produced on the screen is the result of a process performed by the user in which he/she makes explicit the definition of the referent. Such programs differ from drawing tools like MacPaint in which the process of construction of the drawing involves only action on the physical screen and does not require a description of the referent i.e. of relations between elements. Nevertheless because the description for geometry software presents some specificities, the referent of a drawing itself cannot be completely identified with the usual referent in Euclidean geometry. In software geometry the figure is determined by a construction process made of primitives and by the operations which it is possible to perform.

A new kind of referent is created by such geometry programs but also new problems may arise because of the novelty of the objects and operations that are possible on these objects. But referents and problems actually differ from one software to another according to the software specificities. Let us discuss this question through the example of the geometry program *Cabri-géomètre* in the learning of 2D geometry by 11 to 15 year old pupils.

The Cabri-figure and the Cabri-drawing

Because the drawing made with *Cabri-géomètre* on the screen has specific features, it will be denoted by the term *Cabri-drawing* and the object of the theory to which it refers by *Cabri-figure*.

Two kinds of primitives are available to make a drawing in *Cabri-géomètre* :
Primitives of pure drawing (similar to those in MacPaint) enable the user
i) to mark a point anywhere on the screen, just at the location shown by means of the cursor (which takes the form of a pen); we call this kind of point a *basis point* ;
ii) to draw a straight line while pointing on the screen at the

position of two points but these points are not created as geometrical objects; we call this straight line a *basis line;*
iii) to draw a circular line called a *basis circle* ;

Primitives based on geometrical properties enable the user to draw objects not on a perceptual basis but on a geometric basis

* for example, the user can draw a perpendicular bisector of a segment in selecting the item *perpendicular bisector* in the menu *Construction* and using the cursor to show the given segment.

When an element of a Cabri-drawing is dragged, the only constituants of the drawing which move are those that depend geometrically on this element. For example, if a segment AB is drawn and the user decides to mark its midpoint I only in showing a point correctly located from a visual point of view by means of the primitive *basis point,* this point I does not move when A or B is dragged (fig.l &2). In contrast if the midpoint is drawn by using the menu item *midpoint* its property of being the midpoint of AB is preserved by the drag mode.

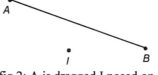

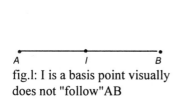

fig.l: I is a basis point visually does not "follow"AB

fig.2: A is dragged I posed on A

The combination of these two kinds of primitives and of the drag mode gives meaning to the notion of figure versus the notion of drawing. The dragging of an element of a drawing generates an infinity of different drawings on the screen while a geometrical figure is the set of geometrical properties and relations attached to a drawing that are invariant through the drag mode. In a sense, *Cabri-géomètre* materializes the notion of geometrical figure.

But the theoretical object to which the Cabri-drawing refers presents some features which differ from those attached to the usual theoretical referent. The Cabri-figure is a result of a sequential descriptive process, which introduces an order among the elements of the figure. So in a Cabri-figure some elements are basis elements or free elements on which the other elements are constructed. These free elements can be grasped by the mouse and dragged, they have a degree of freedom equal to 2, whereas a constructed element like the midpoint of a segment is completely dependent and cannot be grasped, its degree of freedom is equal to 0. Intermediary elements of degree equal to 1 can also be created: these are points on objects, like a point on a circle or on a straight line which remain on this object when the Cabri-drawing is dragged. This notion of freedom due to the drag mode and to the necessity of a construction process is not included in the theory of Euclidean geometry.

It introduces new questions like: what is the trajectory of a point on an object when the Cabri-drawing is dragged? This is dependent upon a decision of the designers of the software which is not completely arbitrary. The behaviour of the point on an object cannot be too remote from the expectations of the user. Because a figure in Euclidean geometry is invariant through a similarity, it has been decided to keep the ratio IA/IB constant for a point I on a segment AB. As a consequence, it would not be relevant to infer properties of a point on an object from observing its trajectory in the drag mode.

The same geometrical figure can give rise to different Cabri-figures. A right triangle ABC right in A can be constructed by choosing A and B as free points and C on the line perpendicular to AB at A (fig. 3) or by choosing B and C as free points and A as a point on the circle with diameter BC (fig. 4).

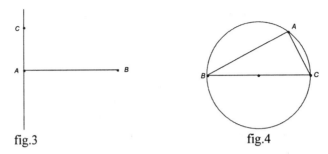

fig.3   fig.4

Construction aspects and dependence relations play an important role in the way in which the Cabri-figure may be handled. For example these influence the way functional aspects of the elements of a figure are emphasized : a perpendicular bisector must be seen as depending on one variable (a segment) or two variables (two points), a perpendicular line must be seen as depending on two variables, a point and a line.

Learning geometry in a Cabri-environment : some hypotheses

The dual nature of geometry has been indicated above. It is a theoretical body of knowledge modeling spatial and graphical phenomena but it is also developed for its own sake. We believe that the traditional teaching of geometry emphasizes the role of theoretical knowledge and ignores or underestimates the relations between drawing and geometrical theory. Geometry is not presented as knowledge allowing the interpretation of visual phenomena or even to control and predict them.

The role of drawings is more likely to be seen as an auxiliary illustration of geometrical concepts. Pupils are not taught how to cope with a drawing. They are not taught how to interpret a drawing in geometrical terms, how to distinguish the spatial properties which are pertinent from a geometrical point of view from those which are only attached to the drawing, or, in other words, they are not taught how to distinguish spatial properties which are necessary from a geometrical point of view from properties which are only contingent in the drawing.

That is the reason why there may be some geometrical misunderstandings between pupils and teacher. When the pupils are given a construction task, the teacher often believes the task involves the use of geometry whereas for the pupils it may be a task of drawing. For example, when asked to draw a tangent line to a circle passing through a point P, the pupils often rotate a straight edge around P so that it touches the circle. This action is not based on geometry but on perception. The task intended by the teacher is not the task performed by the pupils who see the task as a drawing task and not as a geometrical task. For the same reasons when pupils are asked to prove a spatial property, it may be difficult for them to understand why they are not allowed to infer properties directly from the drawing. These behaviours of pupils have been described by researchers in different countries (Fischbein 1993, Hillel & Kieran 1988, Mariotti 1991, Schoenfeld 1986).

We make the hypothesis that by designing specific tasks in a software environment like *Cabri-géomètre* it is possible to promote the learning of geometrical knowledge as a tool for interpreting some visual phenomena, explaining, producing and predicting them. According to the hypotheses given at the beginning of the paper, the software environment plays a double role. It is the source of tasks and it offers ways of actions and feedback to the pupils solving the task. In the following sections, we present these two aspects :
- how the environment can give rise to new kinds of problems in geometry and in geometry teaching concerning these relations between drawing and geometry;
- to what extent the interaction with the software environment may promote an evolution of pupils when solving the tasks.

New problems raised by the computer environment

On the one hand, the dynamic treatment of the Cabri-drawings allows for the raising of new categories of problems in geometry itself; on the other hand, the environment through its specific possibilities permits problems which could not be proposed in a paper and pencil environment to be given to learners.

New kinds of problems in geometry
Generic constructions

Some constructions depend on the mutual spatial position of elements of the figure which can change under the effect of the drag mode. For example, constructing the tangent lines to a circle from a point P outside it is not the same process as when the point is on the circle. So that when dragging the point P until it is on the the circle, the construction process does not provide the tangent line for the endpoint of the trajectory of P. Another example is provided in construction of the inverse of a point P with respect to a circle (C) with centre O; the usual process for a point outside the circle is based on the construction of the tangent lines i.e. on the use of the intersecting points of the circle (with diameter OP) with the circle (C) (fig.5), whereas the usual process for inside points is based on the use of the intersecting points of the perpendicular line to OP with the circle (C) (fig.6). There actually exist generic solutions (fig.7).

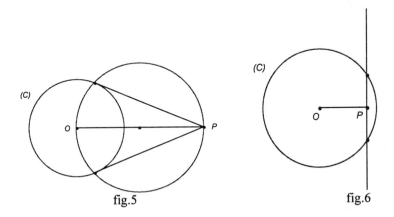

fig.5   fig.6

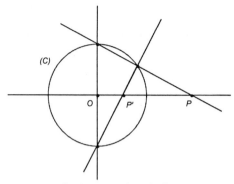

fig.7 a generic solution

Order of points

When coping only with static drawings, it is difficult to give meaning to a simple question such as how to determine by means of a geometrical construction what point among three points on straight line D is between the two others when these points are moving, on D. It is easy to answer perceptively but the perception neither provides a generic solution nor a geometrical process.

Conditional objects

The dynamic feature of a Cabri-drawing may also raise a kind of question which until recently was asked only in a formal setting: how to produce a Cabri-drawing which only exists if the elements defining it belong to a subset of the plane. For example by means of what geometrical construction process is it possible to obtain a segment AB which is visible only when A is on one of the two sides of the plane determined by a given line (BC) (on the side which does not contain a given point D)? (fig. **8 & 9**)

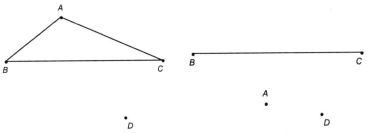

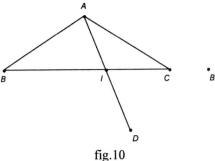

fig.8 the triangle ABC is visible

fig.9 after having dragged A onto the other side of BC the sides AB and AC are invisible

An idea underlying a solution consists of using the difference between a geometrical object and its drawing. The drawing of the segment AB is constructed as a segment joining A and a point B" whose drawing is superimposed with B when B" exists but B" as a geometrical object differs from B because it exists if and only if A is on one side of the line (BC). An intermediate point I existing only and only if A exists is constructed. It is the intersecting point of the segment AD and of the line (BC). Then B' is constructed as depending on I : B' is the image of B when applying the point symmetry around I and B" is the image of B' through the same transformation (fig. 10): $B' = S(B)$ and $B" = S(B')$.

fig.10

When dragging A, if the segment AD crosses the line (BC), I exists and therefore B' does too and the drawing of the segment AB' superimposed on AB is visible. If the segment AD does not cross the line (BC), the segment AB' no longer exists. The same construction is done for the side AC. This kind of method enables the drawing of 2D Cabri-drawings which represent 3D objects with hidden lines continuously redrawn when the drawing is dragged.

The three categories of problems presented here emerge from the fact that the Cabri-drawing is a new kind of representation of geometrical object with specific behaviour produced and controlled by geometrical knowledge:

- it can be moved by direct manipulation (and not by means of a symbolic language) and it keeps its geometrical properties in the movement;
- it is produced by a geometrical algorithm based on geometrical primitives.

This clearly shows that new kinds of representations may give rise to new kinds of problems requiring geometrical knowledge. The same idea may be used when trying to set up tasks for pupils, designing tasks specific to the Cabri-environment (i.e. the features of the environment create the task) which can only be solved by using geometrical knowledge. We assume according to our hypothesis concerning the role of problems (expressed in the introduction to this paper) that solving this kind of task may promote pupils' learning. What is of interest is to design tasks, impossible in traditional environments, which call for the use of this knowledge.

New problems for the learners

With regard to the relation between drawing and geometry, several kinds of tasks can be distinguished:

i) moving from a verbal description of a geometrical figure to a drawing ; in a paper and pencil environment this would be the classical construction tasks in which pupils have to produce the drawings of geometrical objects given by a verbal description;

ii) explaining the behaviour of drawings by means of geometry which corresponds to moving from drawing to verbal description and explanation:

interpreting drawings in geometrical terms; this occurs in tasks in which pupils have to prove why a spatial property giving rise to visual evidence is verified by a drawing;

predicting a visual phenomenon such as in problems involving a locus of points;

iii) reproducing a drawing or transforming a drawing by using geometry.

We assume that these three kinds of tasks become different in *Cabri-géomètre* because of the specificities of a Cabri-drawing. Let us examine *a priori* how each kind of task receives specific meaning in a *Cabri-géomètre* environment.

Tasks (i)

Producing a Cabri-drawing satisfying spatial properties when it is dragged requires the use of geometrical properties for construction and disqualifies the trial and error processes only visually controlled. A process such as the construction of a tangent line mentioned above provides a Cabri-drawing which obviously does not satisfy the conditions as soon as a basis point is dragged. The use of geometry is required by the task itself and not as in a paper and pencil environment by the teacher.

It is well known in a paper and pencil environment that altering the available instruments may change the conceptual tools (i.e. the geometrical properties) to be used to perform the construction of the drawing. In a software environment it is possible not only to modify the primitives available to the pupils but also to create new primitives which in turn create complex objects. An example of this kind of task follows.

The use of transformations as tools for obtaining geometrical properties.

The possibility of creating any macro-construction, i.e. a genuine geometrical function which produces as output a geometrical object when the input objects are given, allows the teacher to design his/her own menus and to ask the pupils to construct geometrical objects by means of tools decided by the teacher. For example, grade 8 pupils have been asked to construct a line parallel to a given straight line (D) and passing through a given point P by using only items among *angle, bisector, reflection, point symmetry* (Capponi 1993). Two solutions are given below (fig. 11 & 12).

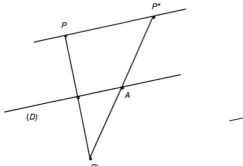
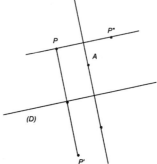

fig.11: composition of reflection and   fig.12: use of a common
point symmetry   perpendicular line

What is new is the use of a sequence of transformations for visually obtaining a property. Even the fact that reflection may be a way to obtain a line perpendicular to a direction is not always known by pupils because it is not used as such in a paper and pencil environment. Guillerault (1991) reports that pupils working with *Cabri-géomètre* were surprised to observe that the line joining a point and its image through a reflection is perpendicular to the symmetry line : "*it gives us the perpendicular line, it must be by chance*". The use of a composition of point symmetries is usually the result of a theorem; it can be used in a proof but not as a tool for producing a drawing of a figure. Using composition of transformations in a proof is at a high conceptual level and in this case might seem too complicated for pupils of this age in comparison with the simplicity of the middle line theorem in a triangle.

It seems that here the computer provides opportunity:
- not only to work in the domain of theoretical objects as it does with a proof activity but also to relate the theoretical concepts to the visual effects
- to link the visual and theoretical aspects of this notion of composition of transformations, not only in a passive way (like in a film) but in an operative way (the theoretical objects are used as solution tools of a problem to be solved).

Construction of Cabri-drawings with imposed trajectories

Because the drawing to be produced in *Cabri-géomètre* is of a dynamic nature, the satisfaction of some conditions of its behaviour can be demanded in the task. For example, draw an equilateral triangle rotating around its centre. Fulfilling this kind of condition requires the analysis of the Cabri-drawing in terms of degrees of freedom differing from a traditional construction task. The conditions on fixed and moving points may also require a construction in a different order and thus call for the use of different geometrical properties. This is the case in the equilateral triangle task in which very unusally one has to start from the center of the triangle.

Tasks (ii)

Dynamic drawings offer stronger visual phenomena than one unique static drawing does. A spatial property may emerge as an invariant in the movement whereas this would not be noticeable in one static drawing. In this latter case, it might not be observed, for example, that a straight line always passes through a given point. We assume that in tasks (ii) the software environment gives more importance to the visual observation and therefore may compel the pupils to explain why they obtain such and such visual remarkable phenomena. The software environment may facilitate the appropriation of the explanation or proof task by the child, or in Brousseau's (1986) terms it may facilitate the *devolution* process of the problem; the pupil acquires ownership of the task whereas in a traditional environment a proving task can be viewed as a school task without any relation to visual phenomena.

Such tasks have been used with *Cabri-géomètre*. Grade 9 pupils, for example, were asked to produce the locus of points from which it is possible to view a segment AB subtending a right angle and to try to explain why it is a circle (it is possible experimentally to verify with *Cabri-géomètre* that it is a circle by various ways, using facilities like measuring segments or drawing a circle with diameter AB). The necessity of devising an explanation may be even stronger if pupils are asked to produce phenomena which are impossible to display visually. Questions like "is it possible to construct a triangle with two perpendicular angle bisectors ?" or "is it possible to construct a triangle with two obtuse angles?" (Bergue, 1992) require a proof because it is not possible to obtain a triangle satisfying the required constraints on the screen. At the beginning of the task pupils were convinced of the possibility of obtaining such a triangle and they tried hard to do so. But after several unsuccessful trials they became aware that it might be impossible and the question 'why ?' was then meaningful for them because it was based on their own experience.

Tasks (iii)

Tasks (iii) when dealing with Cabri-drawings involve both the interpretation of Cabri-drawings in geometrical terms and their reconstruction by means of the geometrical primitives of the software. We call these *black box* tasks. A Cabri-drawing is given to the pupils; they do not know how it was constructed and the facilities of the software giving access to the construction process are removed. The task for the pupils is to reconstruct the same Cabri drawing, i.e. a drawing on the screen behaving in the same way as the given Cabri-drawing when it is dragged. These black box situations cannot be given in a paper and pencil environment since a drawing on its own does not convey information about respective relationship of its parts. A discursive description of the figure should be given in addition to the drawing.

An example of such a black box experiment in a grade 9 class is presented below (fig. 13) (Boury, 1993). The black box is made of a parallelogram ABCD and a central figure EFGH. The pupils must be able to construct a similar central figure attached to another parallelogram PQRS so that the whole has the same behavior in the drag mode .

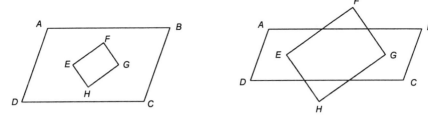

fig.13 : some snapshots of the aspects of the Cabri-drawing to be reproduced

This kind of situation can also be used for learning about a transformation. Two geometrical figures, one the image of the other by an unknown transformation, are given as Cabri-drawings which can be moved; the task of the pupils is to find the hidden geometrical transformation, to reconstruct the image of the first figure and to check whether they obtain the second one. Such black box situations require several operations :

• the exploration of the unknown figure and its behaviour when it is dragged, an analysis of the geometrical properties which remain invariant by dragging and in particular the distinction between what is contingent from what is necessary in the visual appearance of the Cabri-drawing under the drag mode.

• checking whether the supposed geometrical properties of the figure are satisfied by using other facilities of the software. For example, if three points seem to be on the same straight line, the pupils can draw a straight line passing through two of them and check whether this line goes through the third point when the drawing is dragged.

These two operations are based on an interaction between visualization and conceptualization, the first being an interpretation of visual aspects by means of theoretical notions, the second one consisting of validating theoretical interpretations by checking if two drawings remain visually superimposed while the drawing is dragged. This kind of task seems to contribute, in a very appropriate way, to pupils' construction of relations between spatial and visual properties and geometrical knowledge.

In summary, tasks (i) and (iii) should require the a priori use of geometrical knowledge and tasks (ii) seem to motivate the need for a proof.

Pupils' solving behaviours in the interaction with the software environment

It is not because theoretical objects are embodied in a material environment that the learner will immediately have access to the meaning intended by the designer of the environment or the teacher. It has often been stated about the LOGO turtle that the pupils did not conceptualize the notion of angle simply because they were faced with a task including the construction of angles (Hoyles & Sutherland, 1990, Hillel & Kieran, 1988). The learners construct a representation of the functioning of the software and of the tasks which are given to them which may differ from the expected representation.

Numerous observations of grade 8 and 9 pupils working with *Cabri-géomètre* have given evidence of the pupils' solving processes in this environment, about the use of their primitives and of the drag mode.

From purely visual to geometrical strategies

The drag mode disqualifies drawing strategies using only the primitives of pure drawing which produce traces of point, line, or circle visually positioned on the screen but which actually are not interrelated objects. This feedback turned out to be effective in that very often pupils start with a strategy of drawing by eye and evolve after they have observed that their Cabri-drawing is not resistant to the drag mode.

Let us give in detail a representative example of pupils' behaviours in the task of the construction of a parallel line given above (III.2): pupils worked in pairs without the help of their teacher. They had to draw a line parallel to a given straight line (D) and passing through a given point P; the available primitives were: basis point, basis line, basis circle (pure drawing primitives); segment, straight line passing through two points, triangle (which can be used visually as pure drawing primitives or related to existing points); point on object, intersection of objects, symmetrical point, reflected point, angle bisector (which are geometrical primitives). Julien and Daniel were two boys working together.

First phase : erratic use of primitives (fig. 14).
They constructed P0, a reflection of P with respect to (D), then the line (PP0). They drew a basis point P1 and constructed P2, its reflected point with respect to the line (PP0). They drew a basis point P3 and constructed P4 its reflected point with respect to the line (P1P2). They drew a basis point P5 and its reflected point P6 with respect to the line (P3P4).

Second phase : by eye strategy (fig. 15)
They drew P7 a basis point by eye at the same distance from (D) as P and they constructed the straight line passing through P and P7. This line (PP7) was apparently parallel to (D).

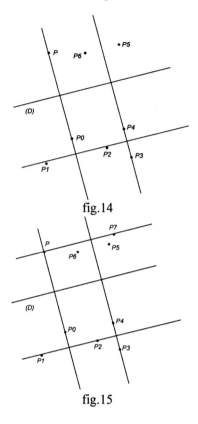

fig.14

fig.15

They tried to validate their construction by dragging P. It was then obvious that the line (PP7) did not remain parallel to (D) (fig.16).

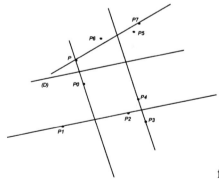

fig.16

They deleted all points except P0 and started again.

Third phase: geometrical construction without aim (fig.17).

They put a point P8 on (D). They constructed P9 the reflected point of P8 with respect to the line (PP0) then P10 and P11 symmetrical points of P around P8 and P9 and the line (P10P11). They dragged P and observed that the line (P10P11) remained parallel to (D). But it was not a line passing through P.

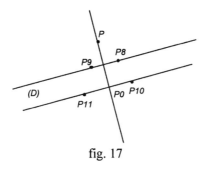

fig. 17

They were then faced with the task of reproducing the same procedure of drawing a parallel but at the right place (i.e. passing through P). They analyzed their drawing, erased it completely and began a third strategy.

Fourth phase: aim directed geometrical construction (fig. 18)

They constructed P1 the reflected point of P with respect to (D), P2 any point on (D), its symmetrical point P3 around the intersecting point of (D) and the line (PP1). They constructed P4 and P5 the symmetrical points of P1 around P2 and P3. The line (P4P5) remained parallel to (D) and passing through P when P is dragged.

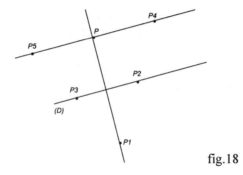

fig.18

Julien and Daniel wrote an explanation of their construction: "*Our straight line passing through P is parallel to (D). When one does a line passing through the midpoints of two sides, the line is parallel to the third side and is half of the line*"
This example shows several strategies. The strategy by eye was not the first one and it probably occurred after a lot of erratic trials of construction of reflected points because in desperation the pupils did not know how to obtain a satisfying drawing on the screen.

The first strategy was a systematic use of the available geometrical primitives on available objects. We call this strategy an *empirical combinatorial strategy ;* it is not aimed at a definite target but guided by the hope of discovering something from the drawing on the screen. Again this is the strategy used in the third phase and the pupils came up with a line parallel to (D) but not satisfying the second condition. The drag mode enabled them to notice this spatial invariant. And it is only at that moment that Julien and Daniel were *really faced with a geometrical task*: how to reproduce this spatial invariant with the additional constraint that the line must pass through P. To manage this, they had to analyze their procedure ; this analysis may lead them to understand why this procedure provides a parallel line. This actually happened with Julien and Daniel as shown by their written explanation. The same kind of evolution of move from empirical *combinatorial strategy* to a geometrical strategy was observed several times in another case.

The pupils constructed the reflected point P' of P with respect to (D) then the straight line (PP'), then they drew a basis point P0 and constructed P1 the reflected point of P0 with respect to the line (PP'); they dragged P and observed that the line (P0P1) remained parallel to (D) (fig.19).

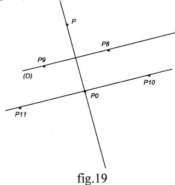

fig.19

Their problem was then to produce such a line but passing through P. It could be done if P takes the place of P0 and if there is a line perpendicular to (D) playing the role of (PP'). They drew P'0 the reflected point of P0 with respect to (D) and obtained thus a perpendicular line (P0P'0) to (D). They constructed the reflected point P" of P with respect to the line (P0P'0). The line (PP") is parallel to (D) (fig.20).

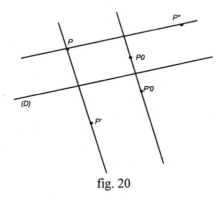

fig. 20

59

Again the same process occurred. By systematic use of the primitive reflection on the available objects, the pupils created a spatial invariant they attempted to reproduce with an additional constraint. This led them to analyze the geometry underlying their procedure.

From these examples, we observe the occurrence of three strategies :
* purely visual strategies;
* combinatorial use of geometrical primitives without definite intention;
* geometrical strategies aimed at a definite result.

What must be stressed is that the features of the software play an important role in the move from one strategy to the next one:
* the drag mode disqualified the purely visual strategies;
* but it also enabled the pupils to notice a spatial invariant;
* because the pupils knew that the spatial phenomenon they produced by chance was the result of the use of geometrical primitives, they were sure that it could be reproduced using geometry. The fact that the software "knows geometry" triggered the pupils to the search for a geometrical solution.

In addition, to these, strategies have also been observed mixing visual and geometrical steps. This behavior is to be related to the existence in *Cabri-géomètre* of pure drawing primitives providing a circle or a straight line at the position indicated with the mouse on the screen. The pupils sometimes consider the drawn circle or the straight line as geometrically depending on the other elements of the drawing. We assume that this confusion originates from the conjunction of an incomplete mastery of the software and of the habits of the paper and pencil environment in which it is impossible to distinguish from the external appearance what is drawn from what is geometrically constructed. But again the drag mode disqualified these strategies and prompted the pupils to find another solving strategy.

Three such strategies occurred :

i) The common perpendicular line

It consisted of drawing a basis line (D') apparently perpendicular to (D) and constructing the reflected point of P with respect to (D') (fig. 21). Such a strategy is based on two geometrical properties; the line joining a point and its reflected point is perpendicular to the axis; two lines perpendicular to the same line are parallel but one of the elements is not constructed but only drawn.

ii) The parallel line passing through P is visually drawn as a tangent line to a basis circle which is visually centered on the intersection of (D) and the line (PP0) and passing through P (fig. 22).

iii) A square is drawn using; the drag mode so that the displayed measures of the sides are equal.

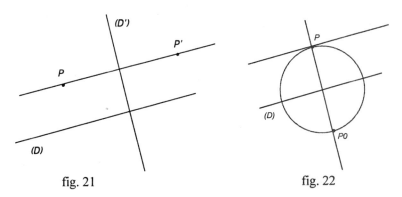

fig. 21   fig. 22

In the following diagram the evolution of the 21 pairs of pupils observed on this task is described.

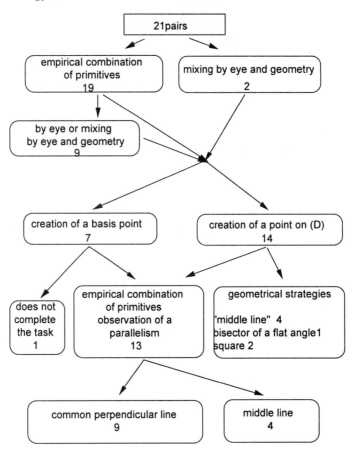

It is important to note that all pairs except one finally solved the task with an appropriate geometrical strategy but that this final strategy occurred after the pupils did other trials empirically or partially by eye. The software through its feedback allowed the evolution of the pupils. What is also important to note is that these pupils were already familiar with the software (about 20 hours of practice); however when faced with a new task posing them some problems they reverted to visual strategies although the inefficiency of these strategies had been demonstrated.

Finally a new aspect of the *didactical contract* (Brousseau 1986) due to the software must be stressed. The pupils systematically tried every possible menu item on every object with the hope of discovering the solution or at least a hint. It is as if these menu items must be used because they are made available by the teacher. This last aspect must not to be underestimated by teachers when they design tasks in a software environment. There should not be a high probability that a solution can be found through an empirical combination of menu items. The importance of the design of tasks and problem situations given to pupils is convincing when they can be observed evolving from a use of simple drawing to a use involving geometrical knowledge, from a "problématique" of the practice to a "problématique" of geometry (Berthelot & Salin 1992).

How pupils use the drag mode

The drag mode is used in two ways by the pupils when solving a geometrical task: as a validity criterion of their constructions (tasks i and iii) and as a means for exploration (tasks ii and iii). Several ways of using the drag mode as a validity criterion have been observed and can be classified with respect to the relative importance of visualization and geometrical tools.

First level : "*all must move together*"
Pupils universally recognize that a geometrical figure given in the statement of a problem must give rise to a Cabri-drawing all of whose parts are connected together or, in other words, if a Cabri-drawing has no moving parts by dragging, it is not the drawing of a geometrical figure. "*All must move together*" is often said by pupils when verifying their construction. The pupils have even developed a specific vocabulary with a more or less mechanical flavour : "*we must attach the point to the straight line*", "*the two straight lines must be linked*". An intersecting point of two straight lines is viewed as a linkage between them. So it can be claimed that the pupils have constructed a contextualized representation of the notion of geometrical relation.

Second level : visual recognition of a familiar shape

At that level the drag mode may be used more or less extensively. Three ways of use are possible. In the construction involving familiar shapes like rhombi, rectangles or parallelograms, the pupils drag their Cabri-drawing in a prototypical position in order visually to recognize the expected shape: a rhombus is constructed with horizontal and vertical diagonals, a parallelogram with a pair of horizontal sides. The validity criterion here is only of a perceptual nature and very limited. The drag mode may be used with a wider scope of action: the Cabri-drawing is moved in all positions in order to check if the familiar shape is preserved in all positions. A third possibility could consist of putting the drawing in critical positions (for example in the case of two points superimposed). We did not observe this behavior by pupils.

Third level : checking of geometrical properties

Visual aspects and geometrical knowledge are jointly used to check the validity of the obtained Cabri-drawing. Through the drag mode, pupils test the geometrical properties that they know about the figures to be constructed. For example, to be sure to have drawn a rhombus, the pupils verify if the diagonals remain perpendicular when the drawing is dragged although they did not construct the rhombus by means of this property. Checking the validity of the construction like this can be done in two ways

- purely visually: i.e. only by eye
- using the facilities of Cabri, measuring facilities or geometrical primitives: to check whether the diagonals of the rhombus are perpendicular, one can measure the angle (measuring facility) or draw a perpendicular line to one of the diagonals and check whether it coincides with the other diagonal when dragged.

The use of the drag mode depends on the Cabri-drawing to be produced; some shapes or some properties are more easily visually recognized than others. The equality of lengths requires more measuring than perpendicularity or parallelism. But a midpoint is a case of equality of lengths where perception is relatively more efficient than in the case of comparison of two non parallel segments. From our observations, it appears that the use by pupils of geometrical primitives to check a property is unusual unlike their spontaneous use of measuring facilities. One may assume that specific situations must be set up to teach the students how to use the drag mode in this way.

Conclusion

Software provides rich environments modelling or embodying mathematical objects. The study of the example of *Cabri-géomètre* in the case of geometry stresses the necessity of analyzing the new kind of relation to knowledge which is constructed by the pupils through a software environment. This analysis must also deal with the tasks which can be used by the teacher in order to promote learning. New tasks are possible. But what is their meaning for the pupils? It seems that a software environment providing feedback may facilitate empirical trials made only with the intention to discover the expected solution by chance. The designed tasks must avoid the possibility that such trials enable a successful solution. It appears that visual feedback based on geometry plays an important role in the evolution of pupils, both by showing the inadequacy of their strategies and by giving evidence for some visual phenomena. But it also appears that the pupils could more extensively use the conjunction of drag mode and geometrical primitives in order to receive more sophisticated feedback. This leads us to emphasize the relation between visual and geometrical phenomena with the existence of new interface possibilities.

In Cabri-like software the scope of visual phenomena is dramatically enlarged and at the same time these phenomena are controlled and produced by theory. They could be used more in the teaching and learning of geometry which could give a stronger meaning to geometrical tasks for pupils which, up until the present have often been viewed by pupils as school tasks without any relation to visual phenomena. Learning geometry could be viewed as learning to control these relations.

References

Bergue, D. (1992) Une utilisation du logiciel Géomètre" en 5ème, *Petit x* IREM de Grenoble, n°29, pp. 5-13

Berthelot, R. & Salin, M.H. (1992) *L'enseignement de l'espace et de la géométrie dans la scolarité obligntoire,*Thèse de l'Université Bordeaux1

Boury, V. (1993) *La distinction entre figure et dessin en géométrie : étude d'une "boîte noire" sous Cabri-géomètre,* Rapport de stage du DEA de sciences cognitives, Equipe DidaTech, LSD2IMAG, Université Joseph Fourier, Grenoble

Brousseau, G. (1986) Fondements et méthodes de la didactique des mathematiques, *Recherches en didactique des mathématiques* Vol. 7, n°2, pp. 33-115

Burton, L. (1994) Whose culture includes mathematics ? In: *Cultural Perpectives on the Mathematics Classroom* S. Lerman (ed.) (pp. 69-83) Dordrecht, Boston, London: Kluwer Academic Publishers

Capponi, B. (1993) Modifications des menus dans *Cabri-géomètre.* Des symétries comme outils de construction, *Petit x,* n° 33, pp.37-68

Davis, R.B., Maher C.A. and Noddings, N. (eds) (1990) Constructivist Views on the Teaching and Learning of Mathematics, *Journal for Research in Mathematics Education,* Monograph n°4, Reston, Va: NCTM

Fischbein, E. (1993) The theory of figural concepts, *Educational Studies in Mathematics,* Vol.24, n°2, 139-162

Guillerault, M. (1991) *La gestion des menus dans Cabri-Géomètre, étude d'une variable didactique* Mémoire de DEA de didactique des disciplines scientifiques, Laboratoire LSD2-IMAG, Université Joseph Fourier, Grenoble

Hillel, J. & Kieran, C. (1988) Schemas used by 12-year-olds in solving selected turtle geometry tasks, *Recherches en didactique des mathématiques,* Vol.8, n°1.2, pp. 61-102

Hoyles, C. & Sutherland R. (1990) Pupil collaboration and teacher intervention in the LOGO environment, *Journal für Didaktik der Matematik,* 11 (4), pp. 323-43

Mariotti, A. (1991) Age variant and invariant elements in the solution of unfolding problems, *Proceedings of PME XV,* Furinghetti F. (ed.), (Vol. II, pp.389-396), Assisi, Italy

Parzysz, B. (1988) Knowing vs Seeing. Problems of the plane representation of space geometry figures, *Educational Studies in Mathematics,* Vol.19, n°1, pp. 79-92

Pea, R. (1987) Cognitive Technologies for Mathematics Education. In Schoenfeld, A. (ed.), *Cognitive Science and Mathematical Education*, Hillsdale, N.J., LEA publishers, 89-122

Schoenfeld, A. (1986) Students' beliefs about Geometry and their effects on the students' geometric performance. *Paper presented at the tenth international Conference of the group Psychology of Mathematics Education, London*

Steffe, L.P. & Wood T. (eds) *Transforming children's mathematics education: International Perspectives*, 498pp., Hillsdale, NJ: Lawrence Erlbaum

3 Collaborating via Boxer

Andrea A. diSessa

Collaboration, as an educational strategy in mathematics or science teaching, is usually thought of as arranged by creating the proper social organization and spirit in the classroom. In this chapter I examine how collaboration may be supported by material means. I present several case studies of the use of the computer system, Boxer, in collaborative modes, and I identify the reasons Boxer seems especially good at supporting collaborative work.

Introduction

Collaborative learning is a watchword in contemporary educational reform, especially in the United States of America Socially and collaboratively oriented images like "a community of learners" (Ann Brown and colleagues) or "cognitive apprenticeship" (J. S. Brown, 1989) have spread like wildfire. Vygotsky and other social theorists have a strong beachhead in thinking about classroom learning.

In this chapter I will not argue for or against collaborative learning. Instead, I accept collaborative strategies as part of a balanced repertoire that we need to understand and enhance in the pursuit of educational goals.

In seeking to enhance collaborative modes of instruction, the most obvious parameter at our disposal is the social organization of the classroom and of classroom activities. I focus, instead, on a less obvious but perhaps no less important issue, the nature of the material basis for mediating collaboration. Written language, of course, may be the best example of this, but technology is a more malleable and redesignable material substrate, hence worthy of particular consideration. Even given a focus on technology, there are more or less obvious things to which to attend. Network communications systems and explicitly collaborative software are among the more obvious foci. Again, my attention is in a less obvious, but arguably not less important direction.

Boxer is the name of a flexible, general purpose computational system designed to serve the needs of students, teachers and educational materials developers. Boxer has all-the-time accessible resources for text and hypertext editing, for dynamic and interactive graphics, for complex data handling and, centrally, for programming. Although I shall show and talk about some of Boxer's features, readers are referred to other articles for more detailed descriptions (e.g., diSessa and Abelson, 1986; diSessa, Abelson and Ploger, 1991).

At the top level, Boxer's goals are overtly social. We want to create a genuinely new and powerful medium—like written text only extended by essentially new capabilities computers can add. We would like to see this medium adopted by as broad a group as possible as the basis of a new literacy that can better foster, especially, mathematical and scientific thinking (diSessa, 1990).

At the next level, however, Boxer does not have on its surface any particularly collaborative features. Instead, Boxer was designed with an eye toward (1) learnability and comprehensibility, and (2) general expressiveness. Yet we have discovered, somewhat to our surprise, that some of our best successes have come from collaborations mediated by Boxer. This chapter presents and seeks to understand some of our early collaborative successes.

In retrospect, Boxer's success as a collaborative medium should not have surprised us. Our analysis here suggests that the same characteristics of Boxer that account for its comprehensibility and expressiveness also account for many of the ways in which it is a good collaborative medium. As with natural language, written or oral, collaborative structure need not be explicitly visible to be effective. You cannot see the fundamentally collaborative aspects of language in syntax, grammar or lexicon

A series of case studies follows. After the first case history, I present an analysis of Boxer's characteristics that help explain what happened. Further examples use and extend this analysis.

First course

The first organized Boxer course (a summer course for students age 13-16 on statistics) was, not unexpectedly, somewhat ill-prepared (Picciotto and Ploger, 1991). The software implementation at the time was incomplete and buggy. Machines crashed frequently and their 16 MHz 68020 processors (less than half the speed of entry-level Macintoshes now) fairly crawled. The teacher we recruited was superb and experienced with programming in BASIC and Logo, but he had not taught a statistics course before with any technological help, and he had received almost no Boxer training. There was no Boxer documentation, and we had had almost no experience teaching Boxer to anyone. To make matters worse the course was only six weeks long (only half the time each day devoted to Boxer), so we had little time to recover from false steps; and students in the course had an almost unmanageable range of expertise, from "hackers" to essentially computer-naive individuals.

Planning for the course initially was faulty as well. Buoyed by perhaps overoptimistic expectations with little counteracting experience, the Boxer post-doctoral researcher helping organize the course convinced the teacher to have the students implement every tool they used, from scratch.

In the third week, the course bordered on collapse. Even the students who were experienced in programming had failed to implement a workable version of the very first statistical tool assigned by the teacher. It was time to re-think.

Jumping ahead, by the end of the course each of four teams of two or three students had produced a fine programming project that illustrated and elaborated statistical principles from the course. Two of the four project groups contained computer-experienced students who, not unexpectedly, created fairly impressive products. For example, one group created a tool that generated "bar and whisker" charts showing confidence intervals on samples drawn from specified populations. The program did not rely on numerical charts or other "tricks," but used an internal simulation of sampling to generate reliable approximations to rather fancy statistical formulas.

Yet even the least sophisticated students also produced excellent working program-projects. One group of three, who, among them, had had only the barest prior experience programming BASIC, developed an excellent pedagogical simulation. They developed a generalization of the standard science museum display of the normal curve emerging from balls dropped into a triangular array of pegs, collecting in bins at the bottom. Their simulation represented sequential samples of binary values (yes, no) from a selectable distribution (say, 55% of the population say "yes") on some polled issue. A "yes" response corresponded to a move downward to the right, and a "no" corresponded to downward to the left. The simulation could be single stepped, allowing process explanations "in slow motion." It could also be run one batch (say, 16) samples at a time (say, producing one "ball" in the "7 yes" bin), or in multiple batches to show the growing histogram distribution. This program collected and sorted numerical data internally as well as showing it graphically.

Although a sample of four projects is scarcely definitive data, our prior experience with summer programs of this sort using Logo were rather different. Many groups failed to produce the quality of projects demonstrated by this class. Even worse, projects combining hackers with less computer sophisticated students were typically dominated by the hackers to the point that other students completely lost ownership of the computer aspects of the project. Only hackers could run and explain their group's program.

Although we were not prepared with the kind of data collection one might have hoped for, we had both the expert observations of the teacher and videotaped exit interviews of students concerning their projects. The teacher also believed these achievements were clearly beyond what he had experienced before. The video tapes, including one narrated by a student who was nearly computer-naive beginning the course, showed excellent mastery of the project programs, including debugging on the fly and expert perusal and explanation of the program's internal structure.

How had the course been turned around from near disaster to unusual success? In a word, code-sharing saved the day. After bailing out from "every programmer for him/herself," the teacher had begun bringing some central program-tools into the class to give to students. For example, he programmed a simple statistical calculator that had several essential statistical functions ready at hand. Graduate student observers dropped in other tools, e.g., a MEMBER? primitive (is X a MEMBER of set Y?) and a simple sorter of numerical data. Students too began contributing to this pool of tools, which other students borrowed.

The transition to a tool-building and sharing culture was not as simple as project members and teacher priming the pump. Some students initially thought borrowing code was cheating. Others kept code they had received private, as if it were theirs to hoard, until this inconsistency was publicly pointed out. Still other students persisted in writing deliberately inscrutable code with great pride to the end of the course.

So, clearly, sharing was a cultural construction and achievement, at least in some measure. But, did Boxer have a role in fostering it? We believe Boxer did have an important role.

In the first instance, after five good years of further experience, we have found code sharing consistently to be a powerful influence in Boxer classes. It is, for example, much more prominent than in any of our many years working with Logo.[1] Continuing examples below help make this point. More profoundly, when we look to see why sharing in Boxer works well, it seems some of Boxer's structural innovations are at the root. Here we list four, which we elaborate and extend in other cases, below.

(1) Visibility—Boxer is designed to allow students to see all aspects of the system as directly as possible. A good example is variables, which are visible and manipulable boxes containing a value. When a variable changes, the result automatically appears on the screen. In contrast, no wide-spread language has a notation for the *fact* of a variable having a certain value. Instead, if variables are shown at all, they are shown in statements that command their existence (declarations) or change a value ("set" or "make"). Not only can you see more structure in Boxer (see also below), but you can dynamically see how programs change their environment. In terms of learnability, the idea of variable, a well-documented problem in children's learning of programming, ceases being a barrier. In terms of sharing, seeing what another's code does and how it does it is much easier. Figure 1 illustrates some aspects of Boxer visibility.

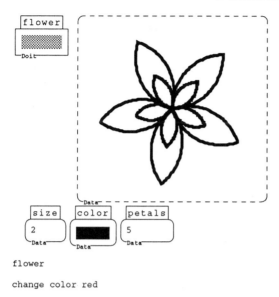

flower

change color red

Figure 1. All aspects of Boxer are rendered visible on the computer screen. In this case a program, flower *(shrunken, to hide its contents), and three variables that it uses are concretely present on the screen. A user may change a variable (e.g. ,* size*) by simply editing its contents (e.g.,* size *was changed from 1 to 2 and a new larger flower is produced by* flower*). If* color *is changed by a programming statement, the* color *variable will visibly reflect the change.*

(2) Pokability—Variables can be set by hand simply by editing their contents, as well as by program control. More importantly, programs themselves can be poked in bits and pieces to see what they do. Concrete stepping (executing bits of a program one at a time) is trivial and frequently used in Boxer to understand a piece of code. All a user has to do is point to each line in sequence and execute it with a keystroke or mouse click. Figure 2 shows the result of opening flower and stepping one of its parts, petal.

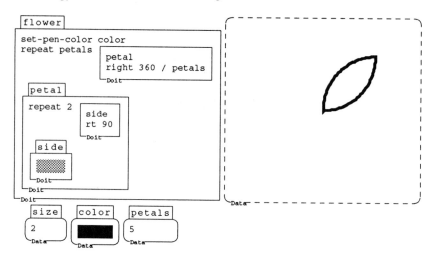

Figure 2. Opening flower *reveals its code. Pointing to* petal *and double-clicking the mouse shows* petal*'s effect. Note we have also opened the subprocedure* petal *to inspect its code.*

Poking and interpreting the results of a poke is made even easier in Boxer because of another of its characteristics. It is easy to arrange that both code and the output of that code are simultaneously visible on the screen. This makes for very easy association between, say, a program step and the effect it produces. One clicks the former and simultaneously watches the latter. In contrast, for example, it is unusual for code (script) and its effect to be co-present in Hypercard, and this is impossible, in fact, in most Logos, where code only appears in a separate editor or on the "flip side" of the graphical locus of effects of that code. Figure 2 illustrates also how easy the co-presence of procedures, variables and graphics can make interpreting how a procedure works.

(3) Structure—Boxer provides a natural and powerful visual structure to organize code. One uses boxes inside boxes (inside boxes) to organize hierarchical code. Subprocedures may be defined locally in any box, just where they belong, at any level of the

hierarchy. This provides a much more expressive presentation for program structure than the typical list of procedures. Programmers don't *have* to program so as to communicate with better organized structure, but Boxer provides resources that allow this. We have some reasonably well-studied cases of student programmers achieving breakthroughs in programming complexity by mastering the expressive possibilities inherent in Boxer organizational resources (Ploger and Lay, 1992).

Boxer also provides a natural and intuitive way of *perusing* complex code—shifting focus, suppressing or displaying detail as needed. Boxes can be expanded or shrunk, or they can be "entered" by expanding to full screen. This makes for an easy inspection of program structure, allowing students to keep the screen as simple as possible, but exposing as much context as desired (within limits). Figure 2 shows both hierarchical structuring of flower and easy perusal via successively revealing substructure. The transition from Figure 1 to 2 involved opening shrunken boxes flower and petal; side remains shrunken, to be opened at need.

(4) "Chunking" into visible, manipulable units—Because of Boxer's structuring, it is fairly easy to arrange meaningful chunks that appear as complete, manipulable units on the screen. So a complex program with many subprocedures, may appear simply as a small black box that can be cut out as a unit, transported and pasted in an appropriate place in another program. Well-designed units "travel" more easily in Boxer than in less concretely organized kindred languages. Figure 3 illustrates that flower's variables may also be encapsulated inside flower, and the whole unit simply cut and pasted into a new context.

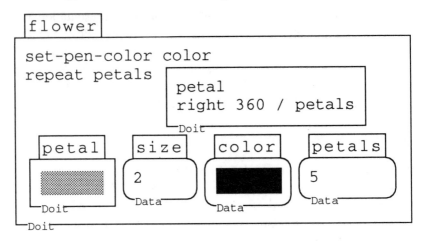

Figure 3. Flower *and all its subprocedures and variables constitute a visible and functional chunk that may be cut from an old context and pasted into a new one.*

"Travel" is a very apt metaphor for moving things around in Boxer. Boxer's very spatial presence of boxes and text inside of other boxes is not limited to code, but, in fact, organizes every "place" in which users interact in Boxer. So, a very concrete first step to borrowing code is literally to cut it from its old place of use and paste it into its new place, never opening the old code that is borrowed nor boxes containing code in the new context. That latter step is only necessary if processes from both contexts need to be interleaved. Our next case study provides an excellent example of this stepwise process.

We were not in a position to document in detail how sharing worked in this first Boxer class. But the class's success first alerted us to the power of sharing in Boxer and began the analysis that pinpointed areas such as those listed directly above. We could see more directly how things worked in examples below.

Steve's Graphing Adventure Game

Four years ago we gave a full year-long class on physics for sixth graders, age 11 and 12, based on Boxer (diSessa, in press a). Two of the students, Steve and Bill (pseudonyms) produced an impressive program as a final project. It was an elaborate graphing adventure game in which players were invited along on a fantasy journey in outer space. The hero of the fantasy was confronted by a series of challenges involving motion, and the player had to select from among a set of 5 position, velocity and acceleration graphs the one that indicated a motion that would extricate the hero from his current predicament. If you selected the correct graph, you advanced to the next level of the adventure. If not, you would be scolded for selecting the wrong graph (with an insulting but enlightening description of the motion represented by your selection), and you would be sent back to level one to start again. Level five of the challenges was, itself, an independent game of "space invaders," shooting invading aliens with three levels of difficulty.

The program itself was stunning in size and complexity for sixth grade students. It consisted of approximately 500 boxes, with file size over 100 kilobytes. The program stored and displayed dozens of graphs, contained a textual introduction and help with reading graphs, produced interactive text and dynamically changing menus according to the current game context, and it had a cumulative scoring subsystem. The story-line and interactive text was marvelously inventive, filled with typical sixth grade humor. (See Figure 4.)

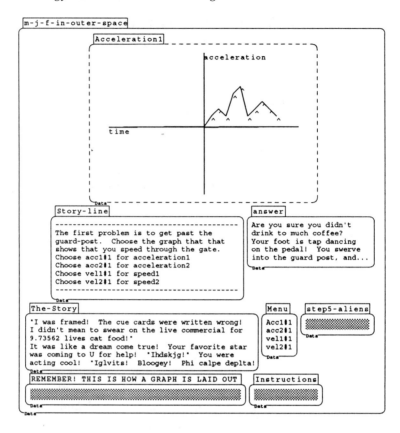

Figure 4. Part of Steve and Bill's graphing adventure game (only one of five challenge graphs are shown). Story-line *poses a challenge.* Answer *provides feedback (in this case, from selecting an irregular graph for a smooth motion).* The-Story *sets the scene and* step5-aliens *is an "arcade" game one can enter and play after solving four graphing challenges.*

How did Steve and Bill manage such an elaborate production? First, Steve was an intent and good "hacker." He was the most dedicated programmer in the physics course, but not by a very wide margin. (In the first half of the course, several other students, including one girl seemed as competent and interested.) Steve did, in fact, do almost all the programming for the project, but it seems clear various modes of collaboration and code sharing helped tremendously.

Bill produced essentially all the interactive text for the game and, we believe, he provided the basic form as well. Excellent synergistic collaborations between students with complimentary talents cannot be claimed as Boxer successes, except to this extent: Boxer's visibility and inspectability seem to help ordinarily less involved students keep a general understanding of, and feelings of ownership toward, Boxer programs. Indeed, there are many aspects of producing a typical Boxer program, including box layout (user interface design) and hypertext editing, that involve little more than the skills one learns in the first few days of Boxer experience. These can be manifestly important contributions while still not requiring exotic programming skill. We mentioned the surprisingly good collaboration between expert and less expert programmers in the above case study, and we add others below.

There were other collaborative factors. On inspecting the game's code, I came across some telltale signs. (See Figure 5.) I found, for example, some documentation and instructions for a graphing tool that was obviously unfinished (it contained notes to the author about features to be added and other unfinished and undesigned aspects of the tool). This had been brought into the classroom by a graduate student. Steve saw its value and "borrowed" it.

What Steve took from the tool was quite interesting. First, he reused the basic spatial organization of having a separately named graphing "turtle" in each graph box so as easily to manage multiple graphs. Steve obviously had added new graphs using the same paradigm as the tool he borrowed. Second, Steve used the axis drawing part of the tool. Third, he took the idea of storing graphical data as lists of numbers that was the basis of the graphing tool. On the other hand, Steve left a lot of the tool behind. Much of the working structure was simply deleted from his world, including the basic routine to plot points. Instead, he wrote his own, rather inelegant graphing routine. Either he felt he did not need to borrow that, or he felt he needed something slightly different, or both.

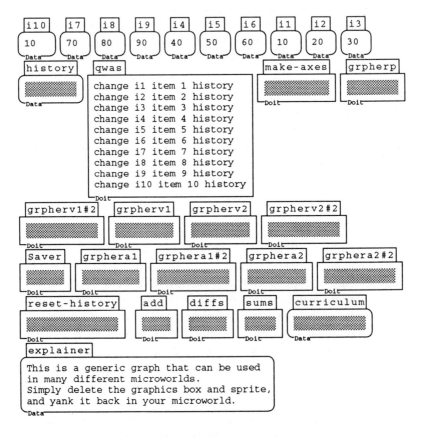

Figure 5. A small part of the code from Steve and Bill's program. It mixes original code with that from a "professionally" written Boxer graphing program at fine-grain scale. The variables at top and qwas *are part of Steve's (inelegant) rewrite of the actual graph drawing function. On the other hand, Steve borrowed the idea of storing graphs as lists of numbers*

(history), *and several other ideas and functions (e.g.,* make-axes). *The bottom two rows contain pieces from the pirated graphing program that Steve did not use, but neglected to delete.*

All in all, Steve's borrowed and new code and their articulation show that he rummaged selectively for ideas as well as code components and had little difficulty joining his own ideas with those of others. Visibility, pokability, structuring resources and clear unit boundaries, we believe, played essential roles in allowing the construction of this complex program.

There is one additional collaborative aspect of this construction that is easy to overlook. Almost all the features of Boxer that allow excellent collaboration and code sharing in general apply to individuals, as well. Good modularity and structure allow easy combining of past work with present. All the inspectability of Boxer helps programmers regenerate and extend an understanding of their "old" code. Thus, many of the good collaborative properties should show up in individual programmers' combining and extending their own work into very complex creations.

Steve and Bill's graphing adventure game had an excellent example of this. Level five of the game, as we mentioned, was an independent subgame that Steve had, in fact, finished as an earlier, independent project. To first approximation, Steve had simply and literally cut this game out of its original context and pasted it in the middle of the graphing adventure game. (It is step5-aliens in Figure 4.) All the working structure of the game—menus, graphical display and hierarchical code—came, intact, in a working visible unit. Steve and Bill only needed to find an appropriate place in the adventure game to add the old game. Contrast this with systems that require linking code by splicing it into the thread of activation, within the loops and eddies of the "main program."

Trivial to perform and easy to conceptualize joining of old and new code is a vital first step. But then customizing and refining are equally important. Given that the old game was played squarely in the midst of the program structure of the new game, interlinking is made easy. For example, Steve and Bill "locked" the invaders game box by adding a command that instantly reshrank the box if someone tried to enter it before successfully completing the first four levels. (Boxer contains easy hooks to activate code on a user's entering, leaving, or changing a box.) This relied on a hidden global variable that kept track of the player's progress. Indeed, on exiting the game, Steve and Bill had a small segment of code to neaten up the space invaders game for a new player and to pass on the score the current player had achieved, upon which some of his/her future fortunes would depend.

Steve put most of his linking code exactly where it belonged. For example, the "triggers" that activated an entry and exit from his aliens box were placed in that box. Strongly associating code with place makes an excellent modularity principle in Boxer's overall structure. It is especially powerful on modifying and extending for new and more elaborate contexts.

The People Mover

Among the nice pieces of video data we have from our year in sixth grade with Boxer is one that shows, from start to finish, a two hour programming project undertaken collaboratively by three students. The project was to program a "people mover"—a moving walkway such as found at airports—so that one could experiment with different speeds and directions of walking and walkway motion. Our intent was to create an engaging and programming-mediated encounter with some of the basic ideas of relative motion. Students did learn a fair amount about relative motion from this exercise, in interesting ways. The point I make here is only to note some of the things we learned about Boxer-mediated collaboration from this video.

One of the students in the group was an exceptional Boxer programmer who had been using Boxer for the better part of a year. He, naturally, contributed substantially to the success of the group. But so did the others. In fact, about a third of the time the other two collaborators worked without him, and it is notable that the work continued nearly seamlessly through his departure and return. This was excellent indication of joint ownership.

Of the two non-experts, one had had Boxer experience. The other was a complete computer neophyte. One substantial point about collaboration is that even the neophyte contributed significantly. Some of his contributions, in fact, were critical to the project; they concerned how to computerize (in arithmetic) some of the group's intuitive ideas about relative motion. The important point about Boxer is mainly that the student understood enough of what was being programmed, and how, that he could make productive suggestions without prior programming experience.

The more general point, which we could see in all of the students' interactions, is that Boxer provides an extraordinarily rich and direct visual presence. Students can not only see a lot on the screen in terms of program structure (e.g., where the pieces are that accomplish various tasks) but also they can see how things operate. As noted previously, Boxer is nearly unique in allowing one to see at the same time the effect of a program and the programming structure that causes those effects.

Visible variables and pokable code allow easy access. This is familiar from "long distance collaborations" (participants not working at the same computer) described in the case studies above. The novel thing we could see in this context was how much the screen served as a place to point and explain, like a super (interactive) blackboard for students to coordinate their ideas and actions, make visible and articulate their contributions. We have noticed in other videos of joint programming how much a contribution the Boxer screen makes to a wide and effective communicative channel (diSessa, in press b). Talking, alone, is less effective. Boxer expressiveness thus manifests itself also in real-time interaction effectiveness as well as in "asynchronous" collaboration.

Vectors

For our sixth grade class and subsequent high school courses on physics, we developed an extension to basic Boxer. In effect, we added vectors as a basic data type. With the addition, one makes a vector by pressing a key, like ordinary boxes. Vectors appear as interactive graphics boxes containing an arrow that one can resize and reorient using the mouse. Vector operations were also added. One can add two vectors and see the resultant, one can tell a sprite (Boxer's mobile graphical objects—like a Logo turtle) to move along a particular vector, and one can name and change a vector like other Boxer variables. In addition, all graphics boxes in Boxer can be "flipped" to reveal a non-graphical presentation. In the case of vectors we chose to show the vector's numerical components on the flip side. These could be manipulated directly or by programming commands as an alternative to direct mouse manipulation of the vector arrow in the graphical presentation.

The sample microworld in Figure 6 is a simple construction using vectors. Vel (velocity) is shown in graphics presentation, and acc (acceleration) is flipped.

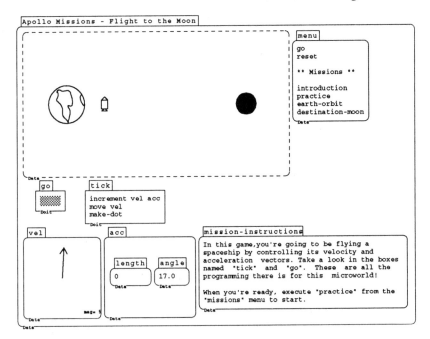

Figure 6. An exercise microworld using Boxer vectors. Vel *shows a vector in graphical presentation so that the vector arrow can be directly manipulated in real time while the simulation is operating.* Acc *is a vector that has been "flipped" to show coordinates.* Tick *is the complete program that moves the spaceship "each tick of the clock" according to velocity* vel *and acceleration* acc, *leaving dots along the way.*

Vectors provided extraordinary resources for our whole teaching and learning community. We developed many tutorials and exercise microworlds out of vectors. The flexibility they afforded us is illustrated by one occasion in our high school course. We had had students take stroboscopic pictures of balls tossed and dropped in the air, scanned them in, and were prepared for students to study these motions analytically by driving sprites around over the images. Unfortunately, the lessons on two dimensional kinematics were not going as well as we had expected. We decided to scaffold the students' analysis with a tool that suggested how one should think about the motion and what results one might get. Using vectors we spent about an hour as a group designing and implementing the tool, which was used successfully the next day in class. To emphasize the flexibility and ease of modification Boxer-plus-vectors provided, we note that one student suggested yet another, different type of analysis; with the teacher's help, he succeeded in modifying the supplied tool.

Vectors infused the course more generally. We used simple vector programs to define and illustrate basic kinematics terms, a role that would ordinarily have been taken by algebra (diSessa, in press a). In addition to exercise and tutorial microworlds, vectors spread to students as a central part of personal projects. It is rare that intellectual tools, like abstract vectors, can be concretized to the point that they provide such obviously helpful support for student-initiated activities.

Along with the general properties of Boxer programming that made vectors successful contributions to this community, this example illustrates one key property that, as far as we know, is unique to Boxer. Namely, Boxer vectors were complete, self- contained interactive objects, with their own mouse interface, and yet they were as well first-class programming data objects that could be used in student or teacher extensions in the same way other built-in Boxer objects could be used. Vectors could be named, changed by programming statements, supplied as inputs and returned as outputs of programs. This dual citizenship—as simple screen-interactive objects and also as full-fledged, extensible computational entities—meant easy learnability and near-complete adaptability for students' and teachers' particular needs. As such, they are powerful tools that can grow effectively in particular collaborative communities to serve local purposes.

A Class Box

The last case is a very simple but potentially powerful collaborative product. It relies on Boxer's simple, perusable spatial structure, easy composition of complete working units, and the ease with which text and hypertext annotation may be added. The teacher of a high school class using Boxer to teach about infinity and fractals decided to produce a class box. It contained:

a) all the tutorial materials the teachers had developed, in Boxer, to teach both Boxer and the mathematics of the course, including the exercises he assigned;

b) teacher comments on how students did with various materials in the course, including successes and difficult points;

c) all of the students' final projects, each in a mutually agreed standard form, which had working programs, textual explanation of mathematical analyses, and demonstrations. The students worked in groups of two and three, and the final projects of this particular class were visually impressive, artistically presented and generally mathematically cogent.

Figure 7 shows the top level of the class box, and Figure 8 shows an example of student work.

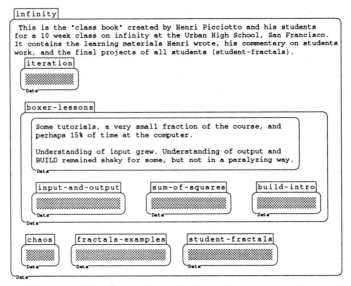

Figure 7. A "class box" containing all of a teacher's prepared materials and student projects from a class on infinity. One box is opened to show some teacher commentary on how his tutorials worked.

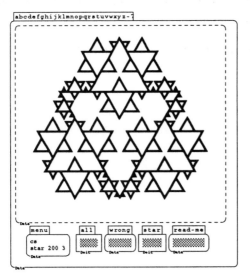

Figure 8. A sample student project. Typically, students included a complex graphic and some mathematical analysis, say, computing its fractal

dimension (not necessarily an integer), total area or path length. In the
wrong *box, the student here explains a prior failed attempt at this design.*

In addition to providing a motivating and informative class product, such a hypertext document provides excellent dissemination of instructional ideas to other teachers. Seeing student work is an excellent source of ideas and calibration of expectations.

The Future—Community Boxer

Boxer was not designed with collaboration explicitly in mind. However, we have seen after the fact that many of Boxer's learnability and expressiveness properties contribute to excellent code sharing, and to asynchronous and synchronous collaborations. I have traced Boxer's success in supporting collaboration to properties such as:

- A high degree of visibility of all Boxer structure, which allows collaborators to see what each other has done and is doing.

- The easy capability to activate any part of a Boxer program and observe what it does.

- An expressive spatial and hierarchical structure, which allows one to "put things where they belong" and peruse complex products top-down, successively revealing details at need.

- The possibility to pack complete functional systems into a visible unit that can be easily transported to other contexts.

At present we are considering extending Boxer to better allow multiple machine collaborations. Our goals are straightforward. We want to provide generic multi-machine collaborative support in the same way Boxer now provides general facilities for programming, personal or small group work, microworld and tool building. We want users to design and modify their own collaborative structures in the way Boxer users now develop, share and modify traditional Boxer materials. The means toward this end are also relatively clear. We intend to expand Boxer's current spatial metaphor and its sharing and hypertext structures across network connections. Users may "inhabit" the same Boxer space or share substructures arranged to support many forms of collaboration.

Notes

---

[1] It is worth noting that the structure and presentation of Logo programs, per se, have not changed essentially at all through the years, even while the programming environment has been enriched with text processing (LogoWriter), better file and other organization, and (especially with Microworlds) ready tools such as a paint tool and prefabricated buttons. It is the program structure and presentation, however, that we argue is the root of Boxer's collaborative spirit.

Acknowledgment

This paper is based on an earlier version that appeared in P. Georgiadis, G. Gyftodimos, Y. Kotsanis, and C. Kynigos (eds.), *Logo-like learning environments: Reflection and prospects*, Proceedings of the fourth European Logo conference, Athens, Greece: Doukas School, 351-357. The work described was supported, in part, by the National Science Foundation, grant numbers NSF-MDR 88-50363 and NSF-RED-92-52725. The opinions expressed are those of the author and not necessarily those of the NSF.

References

Brown, J.S., Collins, A. and Duguid, P. (1989), 'Situated Cognition and the Culture of Learning', *Educational Researcher*, January-February, 32-42

diSessa, A. A. (1990), 'Social niches for future software', in M. Gardner, J. Greeno, F. Reif, A. Schoenfeld, A. diSessa and E. Stage (eds.) *Toward a Scientific Practice of Science Education*, Hillsdale, NJ: Lawrence Erlbaum, 301-322.

diSessa, A. A., (in press a), 'The many faces of a computational medium', in A. diSessa, C. Hoyles, R. Noss, with L. Edwards (eds.), *Computers for Exploratory Learning*, Berlin: Springer-Verlag.

diSessa, A. A. (in press b), 'Designing Newton's laws: Patterns of social and representational feedback in a learning task', in R.-J. Beun, M. Baker, and M. Reiner (eds.), *Dialogue and Interaction*, Berlin: Springer-Verlag.

diSessa, A. A. and Abelson, H. (1986), 'Boxer: A reconstructible computational medium', *Communications of the ACM*, **29** (9), 859 - 868.

diSessa, A. A., Abelson, H., and Ploger, D. (1991), 'An overview of Boxer', *Journal of Mathematical Behavior*, **10**(1), 3 - 15.

Picciotto, H. and Ploger, D. (1991), 'Learning about sampling in Boxer', *Journal of Mathematical Behavior*, **10** (1), 91-101.

Ploger, D. and Lay, Ed. (1992), 'The structure of programs and molecules', *Journal of Educational Computing Research*, **8** (3), 347-364.

Section Two:
Why Technology?

Five authors, in Section Two, bring different answers to bear on the questions provoked when technology enters the mathematics classroom, raising issues that are central to using technology as a bridge between teaching and learning. From an overview of the calculator scene, we move through chapters raising many different pedagogical issues about what happens to learners and teachers when technology enters the classroom.

Adrian Oldknow begins, in Chapter 4, *Personal Technology and New Horizons*, by exploring the implications of an anticipated injection of hardware, particularly by graphic calculators, into secondary schools in the United Kingdom. He describes an agenda necessary to ensure that experiences can be generalised and made more likely to impact upon practices. Although written with respect to the current state of the art in the U.K., we believe that this agenda is appropriate much more widely. He defines the agenda in terms of evaluation and research, a curriculum review and an international exchange of ideas. Perhaps we are now reaping the harvest of the failure to implement such a necessary programme when computers first became available to schools? As a result, we now find fragmentary software development, uncertainty among teachers and a lack of consistency between teaching and learning styles and available software.

In Chapter 5, *Software for Mathematics Education*, Eric Love shifts the focus to a consideration of the metaphors which are chosen to describe software use for the teaching and learning of mathematics. He points out that many software tools have been created to address "the discourse of achieved competence of the professionals and (not) the discourse of learning". He questions just how appropriate is the metaphor of 'tool' and, indeed, whether the software designed for use by professional mathematicians is necessarily the best environment for learning mathematics.

John Mason, in Chapter 6, carries the problematics further by pointing out that "*Less may be more on a Screen*". Far from

95

computer images making the learning of mathematics more straightforward, he suggests that, in enhancing learning, electronic screens may not only demand more time but, more importantly, more personal space. The construal of mathematics is, he indicates, a complex task of making sense of what is seen on the screen, through working on the mental images thereby provoked.

Hilary Povey, the author of Chapter 7, *Working for Change in Teacher Education*, is a mathematics teacher educator. She describes why and the ways in which Logo serves as an effective vehicle to stimulate a re-thinking of the relationship between teacher and learner, and learner and learning, when development of a critical pedagogy is the aim. Learning to use Logo, and therefore learning geometry through Logo, becomes only one purpose of the course she describes. More important is the acquisition of a reflective approach to mathematics itself, as well as a questioning of its learning and teaching.

The final chapter in this Section, Chapter 8, *Looking At, Through, and Back At: Useful Ways of Viewing Mathematical Software*, is written by Liddy Nevile. She asks us to consider computer software not only by looking *at* the software and its potential uses, but also by looking *through* the software to the mathematics learning. which it provokes. She offers this technique as a way of gaining useful teaching strategies to support teaching mathematics through software. However, she also looks *back,* reflecting upon the degrees of control offered to students by different pieces of software. Amongst others, she discusses the spreadsheet and dynamic geometric construction software such as Cabri-Geometre.

All five authors raise issues which bear upon technology as facilitator (or block) to the learning of mathematics as a feature of empowerment. They address the implications such a view holds for the mathematics classroom from the perspective of teacher and students.

4 Personal Technology and New Horizons

Adrian J. Oldknow

Recent government initiatives in the U.K. have given a stimulus to the use of hand-held computer technology in secondary school mathematics. It is quite possible that a `critical mass' of schools will be using such technology within the next three years. If it is to be put to best effect in enhancing students' learning then a program of associated evaluation, research and development is urgently needed. The chapter describes what has taken place and suggests an agenda for associated action.

1. The tools - personal technology

There is now a variety of electronic computing products which can fit comfortably in a pocket or bag and which can be used almost anywhere. These range from graphic calculators of a variety of sophistication, via `personal organisers', to very small computers with built-in software. For convenience this range of products is given the collective title `personal technology' - where `computing' is understood.

At the time of writing there are about seven distinctly different types of graphic calculators from the four main manufacturers: Casio, Hewlett Packard, Sharp and Texas Instruments. Some of these come in more than one model, offering differing amounts of memory and connectivity at differing prices. New models are continually being developed.

Personal organisers usually include facilities for storing information (data-base), entering text (word-processor) and scheduling appointments (diary). Some have a built-in spreadsheet, and some have the facility to insert additional software on small forms of magnetic media.

There is also a variety of so-called `palm-top' computers which are compatible with standard personal computer formats and which can run software designed for them, including some powerful mathematics packages, such as *Derive*.

2. Mathematical facilities

Within the current range of graphic calculators and software available for small computers we can take for granted the ability to do tasks such as the following:

- Plotting graphs of the form $y = f(x)$ and reading off y-values for corresponding x-values ("tracing"). By adjusting the limits on the axes ("re-scaling" or "zooming") to locate points of interest: roots, intersections, turning points etc. to a high degree of accuracy.

- Generating a sequence of values from a one-term recurrence relation of the form: $u_n = f(u_{n-1})$ and hence investigating conditions under which such sequences converge, diverge, oscillate and exhibit chaos.

- Calculating statistics for single and paired variable data, and producing graphic output in the form of histograms, scattergrams and regression lines.

- Entering, saving, running and editing programs in a programming language. Storing data in variables of different types (numeric, statistic, matrix, vector, complex etc.) and manipulating them.

It is clear that once a facility has been implemented in software for a computer there is no intrinsic reason why it cannot be built-in to the system software of a hand-held computer or graphic calculator. Whether it will, or not, depends on the marketing judgement of the manufacturer. It is in their interest to keep developing new products to maintain a marketing edge over their competition. Obviously they keep a close eye on the educational market in judging both price and educational thresholds.

The following exemplify just what is currently available on at least one of the current range of graphic calculators.

Arithmetic with exact fractions.
Conversion of a decimal to a fraction.
Conversion between units e.g. cm : yds.
Output of a table of values from a function.
Plotting in parametric form.
Plotting in polar form: $r = f(\theta)$.
Plotting conics in implicit form.
Plotting surfaces of the form: $z = f(x,y)$.
Plotting solutions to O.D.E.s (ordinary differential equations).
Autoscaling the y-axis.
Zooming in on a user-defined box.
Plotting a numerical derivative.
Plotting a numerical integral.
Plotting an exact derivative.
Automatically finding roots, extrema and intersections of graphs.
Matrix: determinant, inverse, eigenvalues and eigenvectors.
Arithmetic and functions of complex numbers.
Generating and plotting terms in sequences.
Drawing cobweb and staircase diagrams.
Sorting statistical data in lists.
Calculating quartiles and drawing box plots.
Graphing normal distribution functions.
Graphing cumulative frequencies.
Solving sets of simultaneous linear equations.
Solving polynomial equations.
Solving for an unknown in an implicit equation.
Solving a set of first order O.D.E.s.
Finding integrals and derivatives symbolically.
Expanding and factorising algebraic expressions.

Performing vector calculations - norm, dot and cross products.
Manipulating data lists and character strings.
Structured programming commands such as: For, Repeat, While, If..
An object-oriented functional programming language.

Not only are the manufacturers of graphic calculators continually developing their products, but manufacturers of personal technology such as pocket-book or palm-top computers are developing products which are closing the distinction between graphic calculators and computers.

3. Technical features

The most important differences which effect the educational deployment of personal technology are:

Quality of the display

Some screens are much sharper and easier to be shared with several students than others. Most have adjustable contrast and can be used in normal lighting conditions. Some have rather short battery life when used in high contrast.

Printing

None of the current graphic calculators can plug directly into a standard parallel printer. Some connect to their own model of printer. Many now allow printing to be done via a personal computer. Many of the personal organisers and pocket computers will print directly, though some have only a serial interface fitted.

Connection with each other

Some of the current graphic calculators now have sockets for cables (usually supplied) which allow programs, data and pictures to be interchanged. Some personal organisers, pocket computers and graphic calculators communicate using infra-red signals.

Connection with a PC

Some graphic calculators provide host software for computers (PC or

Mac) which allows data, screens etc. to be exchanged between the calculator and the computer, and hence to be stored, manipulated and printed. Most organisers and pocket PCs have this facility.

Large output

Most graphic calculators have special versions to display on an overhead projector (OHP). Some sit directly on the OHP with a "see through" LCD display. Others `echo' the calculator screen via a cable to a second, larger, LCD which is placed on the OHP. In some cases that second screen can be plugged into any calculator, rather than a dedicated one - which enables students' work to be shared with others. Few of the pocket computers allow connection to an external monitor or OHP pad.

4. The experience base in the U.K.

The development of relatively cheap personal technology came just too late to take advantage of funding available through the earlier national government initiatives to promote IT in schools. A number of schools and colleges were able to acquire sets of graphic calculators through the government funding scheme - the Technical and Vocational Educational Initiative (TVEI) - for developments in the curriculum for 14-18 year olds. These have mainly been used with 16-18 year old students taking advanced (A/AS-level) mathematics courses. There has been a curriculum development project in this field supported by the National Council for Educational Technology (NCET) - (see Ruthven, 1992).

A number of the newer A-level schemes, such as those developed by the Schools' Mathematics Project (SMP) and the Nuffield Advanced Mathematics Project, assume that all students will have access to graphic calculators - though most centres expect students to buy their own. Most public examinations in mathematics at ages 14 (Key stage 3), 16 (General Certificate of Secondary Education (GCSE)) and 18 (A/AS) allow the use of programmable graphic calculators, but not personal organisers or computers.

With the availability of such technology it is possible to take quite different approaches, for example to statistics and matrices, which concentrate on understanding and applying, rather than learning further manipulative techniques - (see Oldknow, 1993).

Although the use of graphic calculators at A/AS-level is now well established it would be wrong to give the impression that their use is yet the norm. The School Curriculum and Assessment Authority (SCAA) is responsible for approving A/AS syllabuses and examinations. In July 1993 it redefined the `common core' of material which must be included in any such syllabus to include newer approaches in knowledge and understanding, such as in mathematical modelling. Students will be taking examinations based on the new syllabuses for the first time in 1996. Among the skills and assessment objectives contained in the core SCAA have also introduced the following:

> Candidates should be able to demonstrate that they can
> ....(v) appreciate how to use appropriate technology, such as
> computers and calculators, as a mathematical tool, and have
> awareness of its limitations. (SCAA, 1993)

At the moment this assessment objective is optional, but it may have the effect of encouraging more examination boards to promote and assess the critical use of personal technology and IT.

It is also not yet the norm for mathematics departments in higher education either to encourage the use of personal technology or to insist on students providing their own - in contrast, say, with practice in USA. In fact the current trend seems to be towards using ever more powerful software on ever more sophisticated hardware.

Although external examinations in mathematics, such as the GCSE which is usually taken at age 16 and the Standardised Assessment Tasks (SATs) at 14, generally permit the use of graphic calculators, their use in the 11-16 secondary school curriculum has, until very recently, been quite limited.

Between April 1993 and March 1994 the government, through £2.55m funding provided to NCET, supported around 130 projects making use of personal technology in different subjects and at different stages of the 5-16 curriculum. Within this scheme there were a number of schools using such technology in the 11-16 mathematics curriculum. The project supported a variety of approaches. For example some schools were able to equip all their 1994 entry (Year 7) and their relevant teachers with TI-81 graphic calculators for use in mathematics and other subjects, such as science. In another school a class of 15 year-olds were given Hewlett Packard HP-95 palm-top computers which have built-in software, such as the *Lotus 1-2-3* spreadsheet and a graphing calculator, and access to the *Derive* software package on ROM-cards. That scheme provided the first widespread introduction of personal technology into U.K. mathematics teaching 11-16. A limited evaluation was carried out by the National Foundation for Educational Research (NFER). This was completed in March 1994 and considered the full range of areas of the curriculum, key stages of the pupils and types of technology. However its timescale was such that it could only satisfactorily deal with technical and managerial aspects of technological change rather than the evaluation of learning outcomes. (Stradling, 1994)

The NCET pilot project only provided the items of personal technology (albeit in large quantities) and a limited amount of software. There was no funding provided for external support for the participating schools. Fortunately other agencies, such as the Nuffield Foundation and higher education institutions, reacted quickly, but in a limited way, to enable (a) the provision of support for the schools, (b) the generation, testing and exchange of new ideas and approaches, and (c) research into pupils', teachers' and parents' attitudes to the use of personal technology in mathematics.

5. The U.K. launch pad

For the past five years the U.K. government has made available a sum of around £30m annually to support the development of IT in schools. In February 1993 it organised a conference with the mathematics professional associations (The Association of Teachers of Mathematics (ATM) and the Mathematical Association (MA)) to review the potential impact of IT on the mathematics curriculum and to propose strategies. In the light of this conference, and similar ones held for science, technology and geography, the government has radically changed the focus of this IT funding for schools.

From April 1994 funding to secondary schools was no longer provided for `IT across the curriculum' but was directed towards developing the use of IT to support the teaching and learning of mathematics, science, technology and geography. Government funding has also been made available for development work in support of the introduction of that refocused scheme. (MA/ATM 1994)

Through these Grants for Educational Support and Training (GEST) around 20% of secondary schools' mathematics departments have received funding for hardware, software and professional development for teachers in the first year of the scheme: 1994/5.

The typical amount of funding provided for each participating mathematics department is in the region of £3000-£6000 and discussions with LEAs reveal that most participating schools have acquired class sets of personal technology (very largely graphic calculators). The focus of the GEST IT funding in secondary schools in the next two years is to stay on mathematics, science, technology and geography. Within the next three years, then, it is realistic to expect that the majority of mathematics departments secondary schools in the U.K. will have sizeable amounts of personal technology to deploy in the 11-16 curriculum.

Independently of such developments within schools, pupils are acquiring and using their own personal technology - and so there is also an element of `bottom-up' pressure on the education system for change in the curriculum.

6. Fuelling the take-off

The injection of hardware on the scale anticipated can not, of itself, be expected to have major benefits for the learning and teaching of mathematics. As may be expected there is currently much well-intentioned, but uncoordinated, activity rapidly to provide curriculum materials for the use of personal technology in schools. This is coming from groups of teachers, through the professional associations, through publishers, through the manufacturers, through NCET and through funded projects. Through this process there will undoubtedly be made available many examples of good practice, and exciting and motivating activities for pupils, on which schools introducing personal technology can draw. It has been common experience with such developments in the past that materials alone cannot be relied upon to bring about change in practice - more especially so when the diversity of sources may convey no clear common messages.

It remains to be seen how the professional development of teachers within the schools using GEST resources for mathematics will be managed across the country. In some areas there are already plans to involve teachers from the NCET personal technology project schools in this development, but it appears that others may not involve any specialist mathematics input.

7. Steering the course

How can we ensure that this unexpectedly rapid improvement in the IT provision for mathematics does have the intended beneficial effect on the quality of mathematics education? The still very recent experience of the NCET project is clearly valuable - but it has considerable limitations. It is all too easy to be carried away on a wave of enthusiasm. The teachers involved in that project are largely enthusiasts, and have, by and large, received a considerable measure of external support. However there are many of those schools in which staff assimilation of the technology, the means of access by the pupils, and the style of use in the curriculum have developed at a much slower pace than anticipated.

At this stage it appears that we can make some judgements about different ways in which the technology can be deployed to support mathematical understanding. This needs putting on a firmer footing so that the results can inform practice. There is then a need for an evaluation of the ways that personal technology is currently being used in the U.K. (mainly in the NCET project schools) to support the mathematics curriculum. This can only have a limited scope but such information is needed urgently. It is not yet clear from what source funding for such an evaluation might be forthcoming. This needs to be followed by a more thorough research programme into both the quality of learning with, and the attitudes of pupils, parents and teachers towards, personal technology. We need, but cannot have, the results of such research now - but we must ensure that it is initiated as soon as possible. The pressures on the main research funding agencies are intense and again it is not clear where support for such a research programme may be found. This is an international issue - the lessons learned from such use of personal technology need to be available to inform all concerned with mathematics education.

There needs to be a curriculum review. We have many examples of isolated activities for the use of personal technology at different stages in the curriculum. We need to take a more systematic view of how progression may be sustained throughout the curriculum. In particular we need to take a close look at how changing practices in the use of IT in industry, commerce and research are altering the nature of mathematical skills required in young people. It should be possible to take an innovative approach to many aspects of the mathematics curriculum which builds upon the power of personal technology to aid understanding and to promote challenges. Again this needs to be undertaken from an informed basis - and emphasises the parallel needs for evaluation and research.

There also needs to be a good forum for the exchange of ideas both nationally and internationally. The conference, Technology in Mathematics Teaching, provided a very valuable meeting ground for such an exchange of ideas and experiences. This kind of conference is vital for the spread of good practice in the use of personal technology.

Recent changes in the fabric of the education system in the UK have had a detrimental effect on the provision of support for mathematics teachers. Ten years ago U.K. teachers experienced the problem of the introduction of an element of assessed coursework in mathematics for pupils taking GCSE. Very few teachers had themselves any experience of working at sustained problem-solving in mathematics, and communicating their findings, in the ways that they were expected to promote with their students. It was not just a matter of a technical change of structure - it had to do with a view of the nature of mathematics and mathematical activity. We would be naive to expect that there will not be a similar problem with the widespread introduction of personal technology in secondary school mathematics. Few teachers will have had any experience of learning, or doing, mathematics with the aid of technology. There is a need for teacher support to encourage and develop their own perceptions of mathematics if they are to be active agents in the promotion of change.

This chapter has concentrated on the impact of personal technology in schools. The organisation, funding and practices of the school, further education (FE) and higher education (HE) sectors are very different. But are their purposes so different? Why do students need to undergo such discontinuities in educational experience? We need better liaison between these sectors so that the skills which learners acquire are drawn upon and developed as they progress through their education.

We do not know what lies over the horizon, but the journey there should certainly be interesting and exciting. We have the vehicle but we must make sure that it stays pointed it in the right direction.

References

MA/ATM 1994, `*The IT Maths Pack*', Coventry: National Council for Educational Technology.

Oldknow, A.J. 1993, `Designing new Advanced-level Courses with Modelling at the Core', in de Lange, J. et al. (eds.) *Innovation in Maths Education*, Chichester: Ellis Horwood.

Ruthven, K.R. 1992, `*Graphic Calculators in Advanced Mathematics*', Coventry: National Council for Educational Technology.

SCAA 1993, `*GCE Advanced and Advanced Supplementary Examinations - Subject Core for Mathematics*', London: Schools Curriculum and Assessment Authority.

Stradling, R. et al. 1994, `*Portable Computers Pilot Evaluation Report*', Coventry: National Council for Educational Technology.

5 Software for Mathematics Education

Eric Love

The ways in which software is conceptualised in educational settings will affect its use and place in the teaching-learning process. For example, the metaphor of 'software tools' for doing mathematics evokes particular relationships in the use of such software for mathematics learners and users. This paper examines this metaphor, where not only the notion of 'tool' is problematic, but so are those of 'mathematics', 'learner' and 'user'.

There is a recurring debate amongst mathematics educators in the UK. concerning the relative usefulness in teaching mathematics of various categories of computer software such as 'small specific software', short BASIC programs, large 'generic' tools (see, for example, Ball et al, 1991, Mackinnon, 1993). While some of these categories are descriptive, others employ metaphors which indicate a particular relationship of the use of the software to teaching and learning. Problems with the metaphor of 'software tools for mathematics' are discussed here. Prior to this, it is necessary to examine the use of other relevant terms such as 'mathematics', 'learner' and 'user'[1] .

Mathematics and Its Contexts

It has long been observed that 'school mathematics' is a set of practices related to, but far from identical to, the 'mathematics' envisaged by professional, academic mathematicians. Learners do not learn some pre-existing mathematics embedded in a variety of contexts; they learn about how to operate in that context. Similarly, the work of 'users of mathematics' is not usefully seen as such people taking over known mathematics and applying it in their contexts. Dowling (1991), in an analysis of a number of studies of mathematics in the workplace, in everyday practices and in non-Western cultural settings remarks:

> [The researchers who] set out to illustrate how mathematics is or can be used as a tool in everyday or working life, actually succeeded in demonstrating the existence of a discontinuity in signification of the practices they observe. The result of this discontinuity is that the utilitarian notion of mathematics as a set of tools can no longer be supported. It actually represents a pathologising of everyday and working practitioners who do not see the toolbox or its contexts. (p103)

In Dowling's phrase 'mathematics' is multiply signified. While *professional mathematicians* define (in their cultural setting) what is known as mathematics, *learners* are those who are not so much learning 'the mathematics' of the professionals, as those engaged in practices which are recognised as 'school (or post-school) mathematics'; and *users* are those whose interest is elsewhere and in the course of their work have to carry out operations which can be read by professionals as having a mathematical basis or appear to use mathematical ideas.

There are deep differences in the ways in which 'mathematics' is practised, thought of and embodied in the work of these three different groups. Whereas the studies discussed by Dowling related school mathematics to work practices, even where a group of 'users' have been taught advanced mathematics it is equally misleading to think of their work as being 'mathematical'. For example, Bissell and Dillon (1993), examined the practices of control engineers who, a mathematician might say, use second order differential equations. They found that these 'users' do not actually work in terms of mathematical techniques for solving such equations, but adopt a language which connects the patterns of graphical representations or formulae directly with features in the physical system, which they then describe and manipulate in terms of the patterns. In comparison with the apparent use of 'school mathematics' where the workers did not see themselves as using mathematics, the practices of the engineers incorporated elements of their mathematical studies, but transformed them and embedded them in their context. They were not using mathematics as 'a tool'.

One way of characterising the distinctiveness of these three practices – of mathematicians, learners and users – is to describe them as inhabiting different discourses. Because for each group 'mathematics' signifies something different, there are confusions when the discourses of one group are applied to those of the others. These discourses include particular kinds of computer software.

One further complicating factor needs to be mentioned. Practices of doing mathematics by all three groups are in the process of a profound shift generated by the increasing embeddedness in computers of mathematical processes to which, hitherto, direct attention has had be given. This continuing shift in the characterisation of mathematics for each of these groups makes highly problematic the notion of 'software for mathematics'.

Computer Software For Mathematics

'Mathematical software' embraces that devised for various purposes:
- specifically for use by professional mathematicians
- for 'users' of mathematics
- for learning purposes
- for purposes not primarily recognised as 'mathematical'.

Of course, design intentions do not confine the software to one category of use. Over the last twenty years, the emphasis on mathematical processes has encouraged mathematics educators to provide learners with a wide range of tasks and activities in order to develop their mathematical thinking. Thus, Logo is valued not only for its geometric aspects, but as an opportunity to learn about the use of variables, or the modular construction of solutions to a problem. Spreadsheet-based tasks for learning algebra (Healy and Sutherland, 1991) and an object-oriented drawing package for exploring geometrical situations (Mackinnon, 1993) are recent examples of the use of software not specifically designed for mathematical work.

There is no doubt that possibilities for teaching mathematics can be recognised in software, whether it is designed for that purpose or not. But is the software then in some sense 'mathematical'? Noss and Hoyles (1992) appear to be suggesting it is when they claim:

> It is not difficult ... to ... design the set of software that
> provides the pupil with a representational system of some
> mathematical structures and relationships. (p 442)

However, it is possible to interpret this claim as the mathematics being *read into* the software as it can be read into the practices of users. This is supported by Noss and Hoyles when, in accounting for the lack of use of mathematical aspects of Logo by children, they suggest that Logo activity is in some sense similar to the "informal mathematics" used in practical situations. Such 'mathematics' is in the eye of the beholder and not the practitioner. While the mathematician may see 'a representational system' for the mathematics being built into the software, what is worked on by the software users are the contexts, not the mathematical representations. As Noss and Hoyles conclude,

> ... a microworld cannot be defined by the software alone
> in isolation from its other components: the learner, the
> teacher, or the context within which it operates.... we
> now see the contextual component as a central and
> pervasive element ... it is the very medium we are
> investigating. (ibid. p. 465)

Software Tools

Mason (1992) has developed the idea that some computer software might comprise a set of 'tools' which can be used for doing mathematics – both for solving mathematical problems and for more general exploration of the mathematical aspects of a situation. His attractive vision is that

> Through exploration, using a device such as compasses, protractor or software, pupils work towards the mental-tool state in which the device is an extension of their own thinking. … [with such software] pupils will be able to focus on thinking creatively, on using tools for a purpose, not just in the approximate and resistant material world, but also in the exact and pliable mathematical world of the imagination.

This vision indicates a seamless use of tools – an aid in learning mathematics, used for exploring some aspects of mathematics, and in solving problems outside of mathematics. Nevertheless, there are potential difficulties with any implementation of software tools; three of these are now considered:

- the mathematics curriculum is altered by the existence of tools;
- tools have no independent existence outside of their contexts;
- computer tools are subject to ever greater redundancy.

Tools and The Mathematics Curriculum
One scenario is as follows. A large part of pupils' time in learning mathematics in school is spent learning how to perform various techniques – whether these are methods to perform calculations, solve equations, find areas of shapes, calculate statistics, draw pie charts, find derivatives or plot graphs. Since all of these can be carried out by calculators and computer software, part of the mathematics curriculum in schools would be for pupils to learn how and when to use such software tools.

There are several difficulties with this. Firstly, although some time may be needed for students to learn how to use software that will, for example, carry out differentiation, draw pie charts, or multiply matrices, this learning is not likely to occupy the whole of the mathematics curriculum time in schools. But there is a deeper issue. The school mathematics curriculum would have to be radically reconceptualised because learning how to *carry out* techniques,

113

rather than *using* them, *is*, substantially, the school mathematics curriculum. For example, the key focus of the topic of 'quadratic equations' in school mathematics is being able to solve such equations. This fact so colours the view of quadratic equations in school mathematics that it is hard to imagine any other focus being considered as plausible. When a tool that will solve quadratic equations is available, what is its role?

In the school curriculum, the reason for learning such techniques is almost always for them to be a basis for learning some other technique. Moreover, as has been indicated, school mathematics is relatively autonomous, and the techniques learned are not used by most people, in any direct sense, in the world outside. Thus it is not simply that learning these tools would take less time, but more importantly, the content and the structuring of the mathematics curriculum would have to be almost totally rethought.

Tools and Their Contexts

It is a mistake to think, especially in the teaching-learning situation, of tools as having some kind of existence outside of the contexts in which they are used. As Noss and Hoyles (1992) put it:

> Although the idea of a tool is a common metaphor, we run the risk of implying that there exists a set of decontextualised tools that can be "applied" across a range of contexts without changing them. We would still like to stay with the idea of using or avoiding tools, but we want to stress here that pupils are not entirely free to choose what tools they use any more than people are free to use any language they like. (p. 465)

While this seems to me to be true, it does not go far enough. Expert users are able to think of such aspects as 'tools' because they project their previous experiences of paper-and-pencil mathematics on to the situations in the computer software and use these tools as surrogates for their previous manual techniques. Learners, of course, do not have this previous experience and thus have the double handicap of

114

knowing neither under what circumstances they might use the tool, nor how it works.

It is worth recalling that there are two discourses here: the discourse of achieved competence of the professionals and the discourse of learning. Software such as Maple and Mathematica, has been created within the former. The use by learners of such software tools presents considerable difficulties. Goldenberg (1991), in an analysis of the difficulties encountered by pupils in working with graphing software, argues that the needs of 'learners' are different to those of 'users' because:

> When engineers and scientists use graphers, they are
> often interested primarily in the behaviour of a particular
> function. Although students, too, must deal with
> particular functions, most of the educational value is in
> the generalisations they abstract from the particulars. The
> shape of $-2x^2+30x-108$ is of no educational
> consequence, but it may serve as a data point about any
> of several broad classes: a particular family of quadratics
> (e.g. ones that differ only in the constant term); more
> generally, all quadratics; still more generally, all
> polynomials or even all functions.

Goldenberg draws upon his observations of the misconceptions created by student learners with graphing software, to specify a range of attributes that need to be incorporated into versions for learners. But if this happened, the software would now be designed with specific teaching purposes in mind; it thus would become part of 'school mathematics'. Many of the problems learners have with such software are reminiscent of those described by Noss and Hoyles (1992) in their work on Logo. In general, pupils fail to attend to the mathematical ideas built-in and use aspects of the software that avoid the mathematical ideas intended to be the focus of their study. This appears to be an instance of the didactic transposition (Chevellard, 1985), where learners focus not on the possibilities of the software as a tool, but on learning how to execute it. It is quite likely that learners will not see the software as a tool they can use, especially since, being specially created for them, it will later have to be discarded as competence is achieved.

Functions Embodied in Computer Software

The techniques view of mathematics has become a tools view of computer software. But the opportunities for carrying out these techniques are diminishing rapidly exactly *because of* the use of computers. Those techniques needed to perform calculations, draw graphs or whatever are incorporated into the computerised version of the application. A shop assistant does not have to learn how to use a calculator in order to add up bills or calculate change; nor learn about numbers to enter amounts of money: these features are embodied in automated checkout tills. In the language of Chevellard (1989) the mathematics is becoming 'frozen into' the technology.

This is happening not only with things 'in the outside world', but with software which is thought to be useful for learning mathematics. So, with spreadsheets, aspects of which can be regarded as mathematical (e.g. cell references as variables) are becoming buried (cells can be referred to by picking them up with the mouse). This 'concealing the mathematics' is an advantage to users of spreadsheets, although of course it might make spreadsheets less convincing as dcvice for learning about algebra. It is reasonable to suppose that much other software not specifically designed for educational use will develop in similar fashion. This implies that the use of such software for teaching mathematics may be a temporary phenomenon and as the mathematics gets ever more buried the software will be seen as less suitable.

Chevellard's well-known observation that as society become more and more mathematised, there is less and less need for anyone to know this mathematics (Chevellard, 1989) might be adapted: as more and more mathematical techniques are able to be carried out by software there is less and less need for anyone to learn to be able to carry them out, even with aid of the software.

In Conclusion

'Doing mathematics', whether by professionals, users or learners is changed because of the existence of calculators, graph plotters, symbol manipulators, and other software. There is a dynamic interaction between the work that people do, computer software to carry out operations hitherto taught as part of a mathematics curriculum, and of what 'mathematics' is thought to consist. Trying to pin down parts of this dynamic by speaking of "the mathematics embedded in software", or "producing software for teaching mathematics" or "tools for doing mathematics" will inevitably fail adequately to characterise this continually changing interaction.

School mathematics is part of this dynamic, so there is no certainty of arriving at a stable position where a set of mathematics appropriate for operating in particular environments which have traditionally used mathematics can be identified. For, if some process can be embedded in software, and if users find it easier, or more economical, or more efficient (and no less effective) to carry out their work with the software, then explicit attention to the process is likely to wither except amongst those whose enterprise it is to be interested in such processes. Such people will include mathematicians, but perhaps not mathematics educators.

Notes

[1]It is hoped that the use of quotation marks around terms will not be too irritating to the reader; it is essential to the argument of this paper that such terms are not merely assumed to have conventional and unproblematic meanings.

References

Ball, D. et al 1991 'Debate' in *Micromath* 7(2) pp20-30.

Bissell, C. and Dillon, C. (1993) 'Back to the backs of envelopes.' *Times Higher Education Supplement*, 1993, p. 16.

Chevellard, Y. 1985, *La Transposition Didactique*, Grenoble: Pensee Sauvage

Chevellard, Y. 1989, 'Implicit Mathematics: Their Impact on Societal Needs and Demands' in Malone, J Burkhardt, H. and Keitel, C *The Mathematics Curriculum: Towards the Year 2000* Perth: Curtin University of Technology.

Dowling, P. 1991 'The Contextualising of Mathematics: Towards a Theoretical Map' in Harris, M (Ed), *Schools, Mathematics and Work* London: Falmer Press.

Goldenberg, E.P. 1991 'The Difference Between Graphing Software and Educational Graphing Software' in Zimmermann, W. and Cunningham, S. (Eds) *Visualisation in Teaching and Learning Mathematics,* Mathematical Association of America.

Healy L. and Sutherland, R. *Exploring Mathematics with Spreadsheets* Oxford: Blackwell

MacKinnon, N. 1993,'Friends in youth' in *The Mathematical Gazette*, No 478 pp 2–25.

Mason, J. 1992 'Geometrical Tools' in *Micromath* **8**(3) pp 24-7

Noss, R. and Hoyles, C. 1992 'Looking Back and Looking Forward' in Hoyles, C. and Noss, R. *Learning Mathematics with Logo*, Cambridge: MIT Press.

6 Less may be more on a Screen

John Mason

One possibility for computers in education is the opportunity they provide for virtual construction and exploration in impossible as well as possible worlds. What are the conditions for success? Will students necessarily learn more from electronic apparatus? To make the most of electronic screens requires exploiting not replacing the mental screen, which in turn involves seeing screens as things to look through rather than at, in order to contact potential generality through particular screen images.

Introduction

The potential offered by electronic storage and display of static and dynamic images promises a transformation in the teaching of mathematics. Such a transformation is far from inevitable, but rather will require careful attention to lessons learned from text, television, and videotape, if we are not to waste time and energy in repeating mistakes from the past.

Just as manipulation of physical apparatus guarantees neither construal nor abstraction of mathematical ideas from that experience, so experience of screen images guarantees neither construal of nor abstraction from those images. Indeed, screen images usually require even more work than text to see through the particulars to the general.

As we move from a world dominated by passively 'dumb' screens (text on paper, posters on walls, images projected on screens, television), to a world dominated by actively 'intelligent' screens (manipulable text, images, and animations, with the power to process these and to express one's thinking in mixed and multi-media), it might be worthwhile to try to learn from experience with presentations of mathematics on mental and currently available physical screens, so that we can take maximal advantage of what mixed-media and computational environments offer.

A Language for Discussing Screens
Screens as a Metaphor

I use the term 'mental screen' to refer to the internal 'locus' through which we access an inner world, not as something 'watched', but as an all-encompassing, fully participatory experience that is so hard to describe. It is like a window through which we enter various worlds, and I am stressing the parallel with electronic screens, OHP-screens, or even texts as screen-windows.

The term 'screen' in the context of mental experience is *not* intended to summon an image of a homunculus watching an inner projection screen; indeed one of the features of the mental 'screen' lies precisely in the all-encompassing, fully participatory experience that is so hard to describe metaphorically. But the metaphor of an inner screen as a window through which we enter mental worlds, fits well with electronic screens, and by extension, OHP-screens and even texts as screens, through which we enter worlds.

Looking At and Looking Through

Symbols and diagrams on a screen (or page, or board) act as agents for the person who produces them. They stabilise mental imagery and support extended thought. Despite their particularity, they act as windows for the producer to look through to generality (Griffin & Mason, 1990). The student often sees only the particular, looking at the screen as if at a wall of particularity.

The purpose of *looking through* a screen is to shift from *seeing that* something particular is the case, to *seeing* the screen image *as* an instance of a more general phenomenon. The expressions *looking at, looking through, seeing that,* and *seeing as* can act as reminders in the moment to trigger a fresh response when working with or planning for students.

Working On and Working Through

The main problem in presenting mathematics to others, whether through exploration or exposition (Love & Mason, 1992) is to attract or induce students to *work on* rather than simply *work through*, to contact the inner task as well as completing the outer task (elaborated later). Having students work through numerous exercises may not actually lead them to exercise the required mental processes, because their attention is focussed on completing each exercise in turn rather than making sense of the whole set of exercises. Getting students to make connections, for example, is not achieved merely by the teacher telling them the connections. Connecting is something that students do, stimulated by a need to organise, synthesise, and simplify, possibly stimulated by images, text, and tasks prepared by a teacher.

Contacting I-You rather than I-It

Kang & Kilpatrick (1992) draw attention to the difference (elaborated particularly by Buber, 1970) between the I-It relationship, and the I-Thou relationship. The former involves distinction and separation: the I contemplates the It, employs It, expresses and teaches It. The latter signals union; knower and known are fused, integral, often pre-articulate. Teaching has come to mean exclusively the transformation (Chevellard, 1985, called it the *"transposition didactic'*) from I-Thou to I-It, as if an I-Thou relationship were unreachable in working with students (see Mason 1994a, 1994b for development of this notion). Language games with the word *ownership* (making an idea one's own, and the related play-on-words *'owning* your own maths' are attempts to reach the I-Thou dimension in the context of an investigative approach to teaching mathematics.

The hope of many for electronic screens is that somehow that I-Thou relationship can be engendered. I am suggesting that it may be harder rather than easier to reach, through images. The full range of energies of the triad of student-teacher-content are necessary to realise (literally, to make real) Buber's I-Thou.

Particular and General

Words are, by their very nature, generalising media, in the sense that words signal abstractions which have to be re-particularised in context and it is necessary to work to make them refer to the specific. By contrast, screens are particularising media, in the sense that they attract the physical senses of sight and sound, and have to take special steps to indicate that some more general notion may be being exemplified (Mason & Pimm, 1984). Students are, by their very studenthood, caught in trying to construe what examples are exemplifying.

The *transposition didactique* of Chevallard mentioned earlier, draws attention to the transformation that takes place between an expert's knowing, and instructional materials for students to come to know. The expert has highly developed sensitivities and awarenesses which provide the basis for informed and expert action, whether in a cognitive, academic domain, or a practical, material domain. By seeing the world through those awarenesses, experts cope with particular instances. The student is unaware of generalities perceived by the expert, and comes to that awareness through re-generalisation for themselves, guided by the expert's articulations and participation in contrived and spontaneous educational experiences.

The effect of the *transposition didactique* is to induce the *didactic tension* (Mason, 1986) which frames the teacher-student interaction:

> the more precisely the teacher specifies the behaviour which results from the awarenesses to be developed, the easier it is for the students to manifest the behaviour, without recourse to the awareness from which it is supposed to be derived.

The didactic tension induces education to turn into training, and the *transposition* reaches its apotheosis. The didactic tension is a source of energy, and not something to try to avoid or nullify. Its ramifications are particularly visible in software that attempts to direct and channel students attention to the author's exposition (however presented as animation or visual metaphor).

Teachers dwell in the general, but manifest particulars (through worked examples and case studies). They want their students to see the general through the particular, and to be able to specialise, to see the particular in the general. Formulae and theorems are the principal manifestation of mathematical generality, but students spend most of their time working through particulars, doing exercises.

Integration through Musculature

The graphics screen, and the computer mouse/trackerball are significant developments for education, and soon we will have access to the glove, and to a variety of other tactile inputs. The expression *mouse mathematics* (Mason, 1989) signals the effect of experiencing generality through muscular movement. Geometry programs have been among the first, naturally, to exploit mouse movement to enable the user to try out in rapid succession a finite, but apparently infinite, range of particular cases, but I predict that there will be increasing use of the mouse for this purpose, precisely because it offers direct participation in generality by the user. The click-and-drag feature enables full use to be made of the spatial metaphor which computer screens are able to exploit. Other examples include click-and-drag text editing, and fill-right or fill-down on a spreadsheet.

The musculature provides direct access to a principal seat of memory and awareness that is little used in traditional education beyond the primary years. Another sense available is sound, with the potential for using clicks, for example, to indicate passage of units as the mouse moves.

Consider the difference between the following presentations:

Diagram:

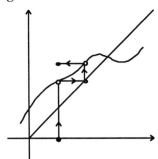

Animation:

Imagine the diagram animated so that the sequence of line segments builds up, going vertically from the point on the x-axis to the curve, then horizontally to the line, then vertically to the curve again, then horizontally to a position over the starting point; then the point on the x-axis moves uniformly along the axis with the construction following it; then the animation repeats with the track of the final point being left behind.

Mouse-driven: imagine a computer program in which movement of the mouse drives a point along the x-axis, and the construction depicted/described above follows. Pressing the mouse-button leaves a trace where the end point of the construction has been.

The diagram is static, it must be *looked through* rather than *at*, so that it is seen as but a single frame from a film, with the rest of the film imagined or sensed; the animation offers a one-dimensional film, the dimension of time providing a focus of attention to indicate firstly that a sequential construction is involved, and secondly that the construction can be applied to any point on the x-axis, and thirdly that the final point traces out a curve or function; the mouse-driven program connects physical movement with change and variation on the screen, while the eyes can be directed to what is invariant in the midst of the change.

Tasks

Getting students active is relatively easy. Arranging that students learn from activity is quite another matter. They usually learn something, but what they learn is liable to be quite different from what the teacher intends. Distinctions between task-as-envisaged and task-as-set by the teacher, and between the task-as-perceived and the activity in which students actually engage, have been made by many authors, for example Christiansen & Walther (1986, p243-307), and Love & Mason (1992). For me it all comes down to the fundamental questions:

> What are the students attending to? Where is student attention in the presence of 'screened mathematics'? What is the role of their mental screen?

Questions about what students construe as their task become even more potent when students have available a computationally expressive medium such as Boxer, or more familiar examples such as LOGO, or symbol manipulation with text processing as in Maple, Math CAD and Mathematica, in which to explore special cases, variations and generalisations. Even when students work on assigned tasks supported by carefully established educational contexts and by corresponding teacher-actions, learning as intended does not follow automatically from their activity on tasks. The student constructs a task which they *can* do, not wilfully, but as a natural part of sense-making. And while they are doing what they do do, their attention may not be on the aspects intended to attract attention, so the effect of working on the task may be quite other than intended. The more control is exercised by the student, the more problematic is the construction of student tasks.

Hart (1985), among many other authors, has challenged the orthodoxy which maintains that learning is enhanced simply by the presence and use of apparatus in mathematics at all ages. Apparatus alone does not produce or enhance learning: it is what the student does with and while using the apparatus that matters. The same applies to suites of exercises and problems, and most especially to electronic screens. Indications of effective ways of working are available from those who have exploited mental imagery in the classroom, for example Beeney et al (1982), and Stevick (1986).

Whereas excitement at the possibility of screening electronic images may induce authors to fill screens with details of multiple representations, the result is likely to render users as passive viewers, despite attempts at generating interaction through menus and questions. Experience in other media suggests that a more fruitful approach may be to learn from a Gestalt perspective, to provide a minimum of screen detail which will sufficiently intrigue the viewer to try to make sense of what is seen, through filling in of details, and so initiate a sense-making action which will continue into accounting-for that succession of images.

It is so tempting to tell-show all that you know when you are producing materials for others, rather than locating the minimum needed to invoke the sense-making powers of the student.

Aspects of Tasks

Tahta (1980, 1981), distinguished *inner* and *outer* meaning of tasks, and to these I add *meta* meaning. *Outer* meanings have to do with explicit aims and activities. If students are set a list of exercises, then the outer meaning (purpose) is to complete each individual question, and students may choose to do different ones at different times, or otherwise to work through the questions. Outer meaning includes the explicit apparent mathematical content of the task, described using mathematical terms such as adding fractions, factoring quadratics or differentiating comnposite functions. Such labels purport to summarise a mathematical story providing meaning for the task, but are not necessarily the way students conceptualise the task. To exploit the outer meaning of a task, it makes sense to engage students in story telling by reconstructing what they have seen and done, because they do not often appear to generate such global accounts spontaneously .

Inner meaning of tasks refers to the mathematical themes and processes, such as multiplicity of definitions or of perspective, the

126

linking together of previously disparate elements into one continuous family, the perception of an infinite class of elements as a single entity, a shift from process as process to process as object for further processing, the choosing of constraints and the effect of those constraints, the role of processes like specialising and generalising, or the notion of an inverse operation. If inner meaning or purpose is explicitly mentioned as part of the task, it becomes outer, and so changes the nature of students' encounter with those ideas. Inner meaning is very unlikely to emerge explicitly without an established classroom practice of reflection and story-telling, nor even implictly if the teacher is also unaware of or not attending to them.

Meta meaning of tasks refers to awareness of more global themes or perspectives, often arising through metaphoric resonance, such as the stressing of different points of view that yield an invariant result, the simultaneous holding of several points of view by means of one invariant, as well as the role and use of the topic in society, who benefits from its application, and so on.

Finally, there is always a personal dimension to any task, in the opportunities to observe one's own reactions and propensities, and to struggle against these, thereby perhaps to increase sensitivity so as to inform action in the future.

Imagery

The power of imagery has been recognised for thousands of years. One source of its power lies in its often pre-articulate nature, directing and formatting attention without conscious awareness. Another is in the way it speaks to us metaphorically, and triggers new thoughts metonymically.

It is tempting, once you become convinced of the endemic role of imagery in education, to start wanting to 'give students a useful image'. No matter how carefully you provide imperative descriptions, no matter how carefully you display an image and point out features which you are stressing, images are like text: people see what they are attuned to see, and not what you are seeing. Text is bad enough for permitting students to construe what they read, in a variety of ways; images are even more open to ambiguity, to alternative stressing and ignoring (Gattegno, 1990). Yet at the same time, images can be remarkably effective.

Computers have been described as *conception stretchers*, (Dennett, 1990) and one of the most significant ways in which conception stretching takes place is in the access to images and to image processing provided by computational devices. Dennett also suggested that

> 'conceiving of something complex . . . is a matter of learning your way around in a 'space' you must construct in your mind.'

How that mental space is related to the screen is of increasing interest in the educational community generally, as evidenced by the rapid increase in imagery-related papers at international meetings, particularly at PME (Psychology of Mathematics Education) since 1992, and a sudden increase in the number of books about imagery in a variety of disciplines.

In attending to mental imagery, it is tempting to think that a visual metaphor or image *is* what the student sees, and *is* how they see the mathematics. For me, symbols are not the mathematics, nor do they signify some Platonic or other mathematical referent; rather, they are an expression of seeing, of awareness, an integral part of thought experience. Diagrams are not the mathematics, nor do they signify some Platonic form; rather, they are an integral part of thought experience, an artificial snapshot of time-evolving sense-making. Words are not the mathematics, nor do they signify some 'thing' which is the mathematics; rather, they are an expression of a fleetingly one-dimensional stressing of certain features of thinking in process. Words and images which purport to signify, need not necessarily signify a signified, because words and images are at best fuzzy projections from a complexly experienced high dimensional process space into at most one, sometimes two, and occasionally three physical dimensions (Mason 1992).

Sketches and Oils

A sketch is often used as an aide memoir, literally to re-member and re-construct a complex of ideas. It serves to stabilise mental images so that the powers of mental imagery can be employed to fill in details, to augment, and to connect (Fish & Scrivener, 1990). Examples include notes for a lecturer, diagrams in texts, sketches on a page, snatches of ideas, opinions and beliefs. Sketches used in this way are often idiosyncratic, depending on stimulated recollection of the artist. By contrast, I am using the term *oil* to refer to a completed oil painting made up of layer upon layer of richly combined pigments.

The temptation of diagrams is that somehow a diagram 'conveys more' (worth a thousand words) than text, and similarly, that an animation is worth a thousand diagrams. But the very richness and complexity of diagram and animation create correspondingly greater potential for ambiguity and multiplicity of interpretation. In the presence of ambiguity of images, it is important to work at flexibility and multiple interpretations, so as not to be trapped in a single interpretation.

There is a spectrum of image-types, from images which are economical and sketch-like, which invite active construal, to images which are so rich and complex like an oil-painting that the non-specialist viewer can do little but immerse themselves in the experience. While the latter, like good exposition, have their place, an unbalanced diet can lead to learned dependency on the teacher to explicate. Somewhere along that spectrum is a threshold, which may be person and context dependent, but which marks a shift in construal from active Gestalt-completion of detail, to passive saturated viewing.

When an author prepares a presentation (as lecture or text or experience), details are filled in for the audience, and the sketch itself is lost. The playful, eagle-soaring insights of the presenter become memorised mouse-scrabbling 'learning' by the audience. Perhaps students need to go through mouse-scrabbling details in order to experience and appreciate the eagle-soaring insights and over-view; perhaps not always. There is an endemic pleasure in sorting things out, and the presence of an audience lends energy and meaning to such an enterprise for the presenter. But the audience hears the crafted product, and does not experience the in-fill. To make sense they have to masticate before digesting (re-constructing) for themselves.

Screen Images

Goodman (1990) distinguished 'images of memory and imagination' from 'optical and sensory images', though he emphasises that neither need be pictorial. But such a distinction seems to blur when it comes to mathematical images. Imagined and constructed generalised images sit side by side sensory images of mathematical apparatus and diagrams. Lacan (1985) links imagination not based on sensation with the symbolic:

> '...there is nothing in the nature of the wheel that will describe the pattern of marks that any one of its points makes on each turn. There is no cycloid in the imaginary.
> The cycloid is a discovery of the symbolic' (p208)

With graphics screens, the notions of symbolic and imagistic begin to intertwine. Sometimes it is hard to distinguish between the mental experience and the physical screen; the latter acts as a window, and the mental screen becomes the world experienced through that window. Animation is particularly engrossing: what is seen is not simply the screen image, but something more substantial, more general, more personal; a world beyond the window is contacted (*looking through* as opposed to merely *looking at*). Just as a picture-frame is essential for focussing attention but often goes unnoticed, so in the merging of the physical and mental screens, the consequent framing focuses attention, but often goes undetected.

The very pre-articulate, all-pervasive nature of mental imagery makes it critically important in human life generally and in education particularly. S. Arieti (quoted in Neville, 1989), even coined the term *endocept* to refer to 'that vague sense, that dim and faint form of knowing' often associated with imagery. Electronic screens permit us to throw a variety of images in front of students; students can become very involved, even entranced in watching. Television images, like pop music, form a background tapestry to their lives.

From a constructivist perspective, what are we to make of children assimilating words to pop songs and television advertisements without any apparent effort, and yet struggling to remember a few words in some foreign language, a few trigonometric relationships, or a mathematical diagram? Can we really get away with saying that some images are necessarily easily construed and recalled? As a society, we can now present images to other people in a variety of sophisticated and engaging ways, and we are doing it comprehensively and ubiquitously. But what are we doing? Have we really made any advance? Or are we producing electronic versions of the 19th century lecture format, which formed the core of Victorian movements such as the Mechanics Institutes (whose buildings are still to be found in some English and many Australian towns) but which has been displaced by forms developed for television?

An image alone is likely to be inadequate as stimulus to learning, especially if it does not challenge or surprise, or otherwise activate sense-making. Although I accept the constructivist description of learning and teaching in which people construe for themselves, I am never-the-less not at all sanguine about the power of visual images to influence people's sense-making. To construct meaning requires 'building-materials', which are fragments of past and present experience. Mental images form a major part of those materials, and mental imagery plays a significant role in determining the overall direction of thought as meaning is made, often subconsciously. So although each person admits and builds on the images that have meaning for them, images have a way of infiltrating the sense making mechanism.

In Conclusion
I am convinced that our visually–dynamic image-dominated culture,

far from facilitating education, makes even more demands for teacher and researcher time to be devoted to making use of static and dynamic images, and to processing, analysing, and re-expressing those images with a variety of mental and electronic tools. Animated images in particular may decrease the time needed to encounter an idea, but there may be a corresponding increase in the time needed for construal. In other words, electronic screens may enhance learning, but may not make it less time consuming. The attraction, the entrancement of moving images, is potentially a force for real development in education, but it is also potentially a force working against the type of inner and outer task construal which we associate with learning mathematics.

References

Beeney, R., Jarvis, M., Tahta, D., Warwick, J. & White, D. ,1982, *Geometric Imagery*, Derby: Association of Teachers of Mathematics.

Buber, M. 1923, *Ich und Du*, translated Kaufman, 1970, *I and Thou*. Edinburgh: Clark.

Chevellard, Y. 1985, *La Transposition Didactique*. Grenoble: La Pensée Sauvage

Christiansen, B. & Walther, G. 1986, Task and Activity, in Christiansen, B. Howson A. & Otte M. , *Perspectives in Mathematics Education*, Dordrecht: Reidel.

Dennett, D. 1990, Thinking With a Computer, in Barlow, H. Blakemore, C. & Weston-Smith, M. (ed), *Images and Understanding*. Cambridge: Cambridge University Press.

Fish, J., & Scrivener, S., 1990. Amplifying the Mind's Eye: Sketching and Visual Cognition, *Leonardo*, 23 (1) p117-226.

Gattegno, C. 1990. *The Science of Education.* New York:: Educational Solutions.

Goodman, N., 1990. Pictures in the mind?, in Barlow, H., Blakemore, C. & Weston-Smith, M. (ed), *Images and Understanding*. Cambridge: Cambridge University Press.

Griffin, P. & Mason, J. 1990, Walls and Windows: A study of seeing using Routh's theorem and associated results, *Maths Gazette* 74 (469) p 260-269.

Hart, K. 1985, Children Use Their Own Methods, *Proceedings of Edinburgh Mathematics Teaching Conference* 1985, University of Edinburgh, p24-28.

Kang, W. & Kilpatrick, J. 1992, Didactic Transposition in Mathematics Textbooks, *For the Learning of Mathematics* 12 (1) p2-7.

Kieran, T 1989, personal communication.

Lacan, J. 1985, Sign, Symbol and Imagery, in Blonsky M. (ed), *On Signs*. Oxford: Blackwell.

Love, E. & Mason, J. 1992, *Teaching Mathematics: Action and Awareness*, Milton Keynes: Open University.

Mason, J. & Pimm, D., 1984, Generic Examples: Seeing the General in the Particular, Educational Studies in Mathematics 15(3) p277-290.

Mason, J., Burton, L., Stacey, K., 1984. *Thinking Mathematically.* London: Addison Wesley.

Mason, J. 1994a, Researching From the Inside in Mathematics Education: locating an I-You relationship, in Ponte, J. & Matos J. (eds), Proceedings of PME XVIII, Lisbon, Portugal, p176-194.

Mason, J. 1994b, *Researching From the Inside in Mathematics Education: locating an I-You relationship*, Extended version, IP5, Milton Keynes: Centre for Mathematics Education, Open University

Mason, J. 1992, Images and Imagery in a Computing Environment, in *Computing the Clever Country*, Proceedings of ACEC, Melbourne.

Mason, J. 1989, Change, *Teaching Mathematics and Its Applications*, 8 (4), p162-168.

Mason, J. 1986, Tensions, *Mathematics Teaching* 114, p 28-31.

Neville, B. 1989, *Educating Psyche: emotion, imagination, and the unconscious in learning.* Melbourne: Collins Dove

Stevick, E., 1986, *Images and Options in the Language Classroom*, Cambridge Language Teaching Library, Cambridge: Cambridge University Press

Tahta, D., 1980. About Geometry, *For the Learning of Mathematics* [1] p2-9.

Tahta, D., 1981. Some Thoughts Arising From The New Nicolet Films, *Mathematics Teaching* 94, p 25-29.

7 Working for Change in Teacher Education

Some First Steps for Monday Morning[1]

Hilary Povey

Logo has been found to be a powerful tool for prompting teachers of school mathematics to experiment with new ways of teaching and learning. This chapter is concerned with initial teacher education and looks at the ways in which the students' learning of Logo may help them to rethink their understanding of the nature of mathematics and to develop a critical, emancipatory pedagogy.

At the University where I work all our first year mathematics students who are following a course in initial teacher education are introduced to the Logo programming language. This unit is part of their mathematical studies and our brief is therefore to equip them with a programming language which is powerful enough for them to use as a tool in their own mathematical enquiries and one which encourages the development of their own mathematical thinking. Logo meets this brief.

However, when we, that is the two colleagues who are responsible for teaching the unit and I, chose Logo we had an agenda additional to the mathematics. We wanted to address a number of other concerns, matters to do with pedagogy and learning, to do with the preparation of our students for their role as teachers. Logo has been found to be 'a remarkable language with which to challenge teachers to think about the mathematics with which they are familiar in new ways' (Fletcher, 1988, p1) and as such it has prompted many teachers of school mathematics to experiment with new ways of teaching and learning. This is because it is a mathematically rich and powerful environment which lends itself to learner control: it is possible from the beginning for learners to devise worthwhile projects for themselves where success or failure is not absolute but judged by themselves. We thought that learning Logo might prompt our students to rethink some of their preconceptions both about mathematics and also about teaching and learning, since our concerns lie not just with our students as learners of mathematics but also as intending teachers.

Within my own pedagogy I want to take this further. My aim is to help our students become teachers who want to empower the learners with whom they work, both *qua* learners and also as critical citizens within a democracy, where empowerment is taken to mean developing democratic social forms that enlarge and enhance those individual capacities which lend themselves to individual autonomy and collective responsibility and freedom. (Giroux and Freire, 1988, pxiv)

I start from the assumption that an essential prerequisite in this process is that I work with the students in such a way as to encourage their own empowerment in the first place. It is with these issues, rather than with the mathematics that the students have undertaken, that this article is concerned.

Pedagogical aims

Logo was devised by Seymour Papert and his colleagues at MIT and vividly described in his book *Mindstorms - Children, Computers and Powerful Ideas* (Papert, 1980). From its inception, the introduction of Logo programming into the mathematics classroom has been seen as opening up the possibility for radical change.

> It is not true that the image of a child's relationship with a computer [...] goes far beyond what is common in today's schools. My image does not go beyond: it goes in the opposite direction [...] In my vision, the child programs the

computer and, in doing so, both acquires a sense of mastery over a piece of the most modern and powerful technology and establishes an intimate contact with some of the deepest ideas from science, from mathematics, and from the art of intellectual model building. (Papert, 1980, p5)

He writes of mathematics being learned in a new way and of the emergence of a humanistic mathematics that is not separated from the study of people and culture. He writes also about the power of Logo for developing thinking in general and for offering a new perspective on the process of learning itself. So here we have the two, interconnected, elements in our other agenda: how mathematics is to be learned and what is the nature of mathematics.

This early thinking centred on the *individual* as knower and fitted within the educational ideology of the progressive educator[2]. Within a progressive ideology, the learner has responsibility for her own learning, the learning is involving and participatory, the teacher becomes a facilitator, the teaching is designed not only to enhance knowledge but also to change the learner and affective as well as cognitive aspects of learning are respected (Brandes and Ginnis, 1986). The first cluster of attributes, then, which we want our students to develop are autonomy, independence, personal authority, choice, responsibility. The emphasis is on the *processes* involved in the work that they do; we endeavour to create a context in which they feel a sense of ownership of and control over their mathematics; the mathematics is personal in the sense of being personally directed and chosen; and both thinking and feeling are recognised as contributing to or hindering learning.

But there is more to be said than this. Firstly, the three of us teaching the course want to engage the students in reflection about the nature of mathematics. Unsurprisingly, most of our students come to us espousing an absolutist view of mathematics, corresponding to a dualist position (Perry, 1970) or, using an adapted model of ways of knowing, to being a received knower (Belenky et al, 1986). As well as being charged with the students' ethical and personal development implicit in the fact of our being their teachers, we also want to challenge the conceptions of the nature of mathematics that the students bring with them because such conceptions will inform and structure their eventual work in mathematics classrooms[3]. We believe that mathematical knowledge is socially constructed and that mathematical meaning is negotiated, that mathematical truths are not absolute nor do they pre-exist before they are known, that mathematics is personal in the sense that it reflects the positioning of the knower, and that mathematics, like reason, is not innocent. This entails a pedagogy which emphasises questioning, decision making, negotiation, discussion, recognition of difference, respect for multiple ways of knowing, and the location of authority with the knower. We want the students to understand the contingent and constructed nature of ways of knowing. We want to challenge

> the "banking" concept of education, in which the scope of action allowed to the students extends only as far as receiving, filing and storing deposits [... where] knowledge is a gift bestowed by those who consider themselves knowledgeable upon those whom they consider know nothing (Freire, 1972, p46)

and to make them aware that who is considered knowledgeable and who is not reflects the social positions and power of the knower.

As well as an epistemology that permits an understanding of how power relations shape what it is possible to say and think, there are also aspects of pedagogical practice which can foster or deny the development of critical teachers who are, and who see themselves as being able to be, transformative individuals. In order to help them see beyond the taken-for-granted, we need to give them a range of experiences which challenge their preconceptions about teaching and learning and then enable them to reflect on those experiences. We need to take this one step further so that the category of student experience is not limited pedagogically to students exercising self-reflection but opened up as a race, gender and class specific construct to include the diverse ways in which students' experiences and identities have been constituted in different historical and social formations [... offering] students the opportunity to read the world differently, resist the abuse of power and privilege, and construct alternative democratic communities (Giroux, 1992, p75)

Beginning to understand *themselves* as located in and by a variety of positions is the first step in coming to terms with the fact that when they are teaching they too will be engaged in the production of ideologies. If the students are going to become teachers who prepare their students for critical democracy, they also need to experience a pedagogy which develops the critical capacity to challenge, to find their own voice, to take risks, which allows for the possibility of hope. Each of these encourages the capacity to see beyond the world as it is and is an essential component in devising a plan for what the world might be. Such a pedagogy will also encourage the students to respect the affective side of the learning process: they will be involved in the production of and have an investment in passion, will feel and come to respect joy, frustration and pain.

Another prerequisite is that the students become able to sustain the paradox of understanding the constructed and contingent nature of knowledge on the one hand with the capacity to be committed on the other (Perry, 1970). Belenky et al (1986) in their study of women found a pattern of such development being dependent upon and flourishing as a result of sustained discussion over time with sympathetic peers. I conjecture that much of the same pattern of development will be true of other learners who are not part of the hegemonic group and who have had to struggle to find an authentic voice. Such students are well represented on our courses of initial teacher education, which recruit large numbers of women and working class students particularly among the mature students. Fostering their sense of themselves as being a learning *group* where the development of the individual enhances the development of the group, where ideas are exchanged and respected, where expertise is gradually developed and experienced as a group attribute, will contribute to their capacity to become authoritative and connected knowers (Belenky et al, 1986). It is my expectation that this in turn will allow their development as critical, democratic teachers.

How we try to achieve our aims

Some of these pedagogical aims are addressed by the nature of the Logo language itself. The fact that it is structural and procedural and allows experimentation with objects, especially visual ones, *permits*, although it does not *require*, an approach to programming which was termed 'soft mastery' by Turkle (1984). It is possible to program in Logo by arranging and rearranging objects, adopting a 'negotiating style', without working to a pre-determined plan. Some individuals, when solving problems,

> prefer to work by *identifying with* objects and contexualising knowledge through this identification. These people prefer to break down the barriers between self and object in order to work inside the situation rather than 'plan, freeze the plan and then execute the plan'. (Sutherland and Hoyles, 1988, p42).

It is possible to program successfully in Logo without adopting a learning style which emphasises a separated mastery (*sic*) over formal knowledge, that is without adopting the learning style currently cultivated by the hegemonic group.

Some other aims are addressed by the pedagogic style. Our approach to the twenty hour unit is to invite the students to initiate their own projects, to devise their own questions to explore, acquiring the knowledge they need as and when required by their own enquiry. (The only limit we impose is that they begin work within the field of geometry because prior experience indicated that we were less effective in supporting the students' learning when they explored initially without this restriction.) On the walls of the classroom in which they work, there is a good range of visually exciting and mathematically appropriate materials for inspiration and they are also encouraged to use a well supplied reference collection elsewhere in the building when searching for a source of inspiration. We have put together the beginnings of a collection of materials[4] to support their independent learning so that they have sensible access to the information that they need when they need it. As soon as possible, students are encouraged to understand that whilst the tutor is more experienced in Logo than the students and has better access to information about Logo than they do initially, since the students are setting the problems, it will be an everyday occurrence that students will know more than their teachers about aspects of their work. An atmosphere is sought in which some pairs quickly become experts about particular things in which others are likely to be interested as tools in their, different, enquiries. Knowledge and authority are shared.

We ask the students to work in pairs but each to keep a diary of the work done. This diary is to include a description of how they got started; what goals they set themselves, how they decided upon them, how they modified them and how they extended them; planning and final results; some reflection on the experience of working in pairs; an account of the mathematical problems they encountered and how they overcame them; an account of the learning problems they encountered and how they overcame them; and reflections on what they learnt about learning. This diary provides the sole means of assessing the unit.

What evidence is there of success?

I have used one main source for the evaluation of the work so far. This is the students' diaries[5] which provide a wealth of written information about what the students have done and what their reflections on the experience are. I have used the diaries of two cohorts of students on the two year Postgraduate Certificate of Education (Mathematics Conversion) course. There were very few students in either of these cohorts who ended up with a negative view overall, although difficulties were highlighted and real pain expressed. Nevertheless, most found the experience at least worthwhile. In the extracts from their writing below, I have indicated how I believe some of our aims for the course have been met and where we have failed. My claim is not that all or most of the students achieved our aims all or most of the time but that success occurred some of the time with some of them. My belief is that by listening to these voices we will become more likely to achieve our aims in the future.

There was considerable evidence of students grappling with the issue of being **in control of their own learning** and having some success at doing so. A considerable amount of fear is expressed about 'falling behind other pairs in the class' at the beginning of the course and many students have great difficulty initially in devising projects for themselves.

> I have found this method of learning a new way of "going about things". Traditionally I have been "taught" as opposed to this method of setting my own goals and finding out the answers to the questions myself. At times I found this frustrating not knowing where I was going or not being able to do something, but [...] I found it in some respects a more beneficial way of learning and importantly understanding the work I was doing because it was where I wanted the learning to lead me.

Even though the students in these two cohorts are all postgraduates, many of them found great difficulty in distinguishing between, on the one hand, devising independent lines of enquiry, posing their own problems and so on where they need to act autonomously using the teacher at most as a guide and, on the other, acquiring information for their work where they are using the teacher or the text as an authority. Some have felt embarrassed about obtaining information (for example, how to stop a recursive procedure running), feeling that they should have somehow 'discovered' this information for themselves. They also feel that asking for help is 'cheating'. The experience of the Logo unit gives them an initial opportunity to rework some of these naiveties and misunderstandings, essential in developing a personal pedagogy which will in turn allow school students in their charge to explore and articulate the relationships between power, knowledge and authority.

Students frequently make connections between the difficulties they experience and their own previous education.

> I want something new but find it difficult to decide or set my own task. Maybe because I am use[d] to teacher telling me do this and do that [...] and I'm becoming good at following and tackling teacher's question that I don't know how to find/create my own questions. I asked myself, "What do I want to achieve?"

This second student ended up devising and seeing through a very successful and difficult project and wrote at the end 'I would love to continue my work'. She, like many others, also had to grapple with the issue of choosing something safe with a high probability of success or **taking a risk** of failure. The assessment does not particularly penalise or reward this choice so the student only has their own thoughts and feelings to consider. The ability to take risks and not to surrender in the face of failure are crucial qualities to be fostered by a radical pedagogy.

Some of the students find aspects of the **negotiation** with partners difficult but the comment

> this decision making bit was really good we threw ideas back and forth looking at the pros and cons of each way of doing things and eventually decided [on a joint plan]

is typical of the success many achieve. Even though **the personal** nature of the diary is emphasised from the start, the writing in the first few weeks is often anonymous in tone and frequently no mention whatsoever is made of the student's partner even though they have of necessity been working together throughout. It is rare for the students initially to write in a direct and personal way: very few of them name their partner even if they are allowed to exist! Challenging this and pointing out that this impersonality is not producing a true account enables many students to acknowledge both themselves and others as real in the learning process. What we are not currently enabling our students to do is to see those selves as, at least in part, constructed. There is an implicit acknowledgement of the relevance of the personal but little awareness of the personal as political. I intend next year to ask for an introductory 'chapter' to the diary, perhaps like most introductions written last, which gives an autobiographical account of the writer and of their previous learning. As trust within the group develops it may be possible and productive to compare, contrast and analyse these accounts.

Moving beyond simply working with a partner, a number of students have commented on **their discovery of the rest of the class** and their experience of the power and support of their peers.

> [I am] getting used to being able and encouraged to asking other pairs around for help and/or discussion of a problem. This is also a different approach to learning to that adopted when I was at school.

> It is good how the students all help each other. For example [J] and [A] spent a lot of time explaining how recursion works and then again in trying to help us work out the IF command. There is a good atmosphere to learn in. It is what I would call a form of cooperative learning. People were working in their groups but there was no sense of competition and everybody I asked was willing to share their knowledge and help me find the answers I required.

An interesting additional perspective on this was given by my unintended absence one day.

> [H]'s not here today. It was only after 1/2 hour that I realised that [H]'s not in. Everybody looked more relax without [H] but less work? No! We talk to each other and

ask problems and look at what's happening to people's terminals.

The absence of [H] although at first disconcerting was compensated by the availability of the other students and their readiness to answer problems and share their experiences [...] how important it is to set up a course such as this so that it is student [led]. In a school situation it would be good to think that if the teacher was away then the students would be in control and would be able to go some way towards solving their own problems.

It is worth emphasising here that these particular students are all graduates and yet their expectations of teaching and learning are such that working for oneself and not for the teacher and working for others and not for oneself are both sensed as novel. Again, both are central to a notion of a critical pedagogy.

For many of the students there is an **acknowledgement of passion.** The following comments are representative of many.

I could hardly explain the sense of excitement and joy when the program successfully ran

I found myself having complete mood swings while learning Logo from elation [...] to frustration and depression

I thought I understood how a recursive procedure work. But it still didn't work. What it did was drawing a circle, forever. This really makes me tired. I wish I'm not doing this [...] I am not happy at all, not with all this mess.

That was the moment of truth! I got a tree (branches really) drawn on my screen! What a joy, but I didn't kiss the screen as [C] did ...

To work through difficulties, frustration and even despair to triumphant success is itself an empowering experience. To acknowledge that success does not always come but that the struggle is still worthwhile though painful is an essential element in creating an individual who can see themselves as transformative. To recognise that impersonality, coldness, formality are not essential attributes of learning but may rather be used as the tools of the elite opens up the possibility of challenging the *status quo*.

The students' experience of mathematics changes

> one thing seems to lead to another when using Logo. Each problem solved raises new questions whose answers raise even more questions. Because of this, Logo is a good environment for investigative learning as no matter how fast or slow you are at coming to grips with Logo there are always new questions arising.

This in turn allows some to challenge their existing understanding of **the nature of mathematics** and of mathematical knowledge.

> My attitude to mathematics is showing itself. I have not thought very deeply about the way I think about maths. That is, one of my misconceptions was that if you can't do maths you can look it up in a book and see how someone else has done it. They will have been examining the problem from exactly the same perspective but they are cleverer and have managed to work out the solution. I think that this point of view is attributable to my mathematics background where most of my experience of maths is algorithmic 'quick fix' solutions to standard problems. For some reason, I had disassociated mathematical thought from eg sociological thought. I had assumed that for maths problems there is one way of solving problems leading to correct answers. By doing this course I have discovered the *obvious*. That is, problems can be approached in many different ways and conclusions

can be challenged. This is the first time I have concentrated on problems in maths for so long and I have learned that maths problems are like other problems - sometimes the solution might be a new set of questions and the reward isn't necessarily in the solution - but in finding out the *way(s)* that a problem can be tackled.

It is fair to say that most or all of the students move some way towards the vision of teaching and learning we are asking them to embrace and this comment is not untypical:

I was very sceptical [...] but the amazement and sense of satisfaction I got [...] quite astounded me [...] motivation can stem from being an active learner

However what is lacking is any sense of how their experiences in the teaching and learning of mathematics are also part of a wider agenda. Occasionally recognition of **difference** occurs

I have found working with [B] a useful experience because he looks at things in a completely different way to me

but this is seen as an unexplained difference between individuals with no awareness that such differences occur systematically and are related to the positions in which the individual is located. We have some success in enabling our students to develop the critical capacity to challenge, to create an ethic of caring and mutuality within the group, to find their own voice (it is not uncommon about half way through the course for pairs roundly to reject an intervention from me, amid laughter, as they experience separation from the 'expert') and even to take risks, but we do not currently enable them to bring these to bear effectively on understanding the social practices, subject positions and relations of power which constitute schooling. What is needed is a new set of practices that will make these articulated and explicit. I will offer two small examples as first steps in this process which we might try 'on Monday morning'.

I have mentioned above the need for more directly autobiographical writing and the need to have this writing actively appropriated by paired and group work. Similarly, structured writing incorporated into the unit and into the diary around, for example, goal setting or difference[6] will provide opportunities for

> critical attention to theorizing experience as part of a broader politics of engagement [...for] such experiences [to] be the object of theoretical and critical analyses so that they can be connected rather than severed from a (*sic*) broader notions of solidarity, struggle, and politics. (Giroux, 1992, p80)

We can also raise directly with the students the fact that, of course, we do have a very firm agenda of our own which we do not allow the students to negotiate. Whenever one person sets a task for another there is always an issue about potential disempowerment and, when the person setting the task is the one with the formal power in general and the assessor of the work in particular, this risk is certain to be present. The formal relations between the students and the tutors are not negctiable, of course, but by making them explicit and raising how much is and is not negotiable, their *visibility* makes it easier for students to remove them from the taken-for-granted. A group exercise which we have used before in another context is to ask the students to decide who has what power within the session, unit and course.

Mathematics is perhaps not the most likely subject within which to develop a critical pedagogy but teachers working for change are to be found throughout the curriculum. I believe that learning Logo, within an environment that encourages personal exploration and accompanied by a dialogue which affirms, interrogates and extends students' understandings of themselves and their world, allows the actualization of some of the theoretical notions underpinning the construction of a counter-hegemony (Weiler, 1988). For those of us engaged in initial teacher education, it can open up an environment within which we can explore resistance and work with the students to develop pedagogical practices for a critical democracy.

Notes

[1] From Willis (1977): **Monday morning and the millennium** ... practitioners have the problem of Monday morning. If we have nothing to say about what to do on Monday morning everything is yielded to a purist structuralist immobilising reductionist tautology: nothing can be done until the basic structures of society are changed but the structures prevent us making any changes ... To contract out of the messy business of day to day problems is to deny the active, contested nature of social and cultural reproduction. (p185f)

[2] I am here following Paul Ernest's use of the term 'progressive educator'. Ernest (1991) usefully separates out five educational ideologies: those espoused by the industrial trainer, the technological pragmatist, the old humanist, the progressive educator and the public educator. Elements of *both* of the last two underlie the position I am adopting here.

[3] The relationship between how one views mathematics and how one teaches it is complex and many of the connections between one's philosophy of mathematics, one's philosophy of mathematics education and one's classroom practice remain to be elaborated and explored; but that there is such a connection is both acknowledged and also seems common sense. For further discussion of these issues see, for example, Thompson (1984), Ernest (1991), Burton (1992).

[4] So far we have produced some materials ourselves but have also drawn heavily on the SMILE Logo Pack (1986) and Brian Harvey's *Computer Science Logo Style* (1985) for inspiration.

[5] A personal and informal style is encouraged in these diaries: we do not want to sacrifice immediacy to polish and technical accuracy. I have reported the writing verbatim and therefore include

idiosyncratic turns of phrases, lapses in punctuation and so on. Any insertions or deletions are shown within brackets [ ].

[6] Writing, for example, might be about *Have you found it difficult to set goals? Describe your experience of that difficulty. Why do you think that you have found it difficult?* or *How have the positions that you occupy, for example those relating to gender, 'race' and class, affected your approach to this unit?*

References

Belenky, Mary F, Clinchy, Blithe M, Goldberger, Nancy R & Tarule, Jill M (1986) *Women's Ways of Knowing: the Development of Self, Voice and Mind*, New York: Basic Books

Brandes, Donna & Ginnis, Paul (1986) *A Guide to Student-Centred Learning,* Oxford: Basil Blackwell

Burton, Leone (1992) 'Implications of constructivism for achievement in mathematics' paper presented at ICME 7, Montreal, 1992 in Malone, John A and Taylor, Peter C (eds) *Constructivist Interpretations of Teaching and Learning Mathematics*, Keynote and other contributory papers to ICME7, Topic Group 10, Quebec, August 1992, Curtin University, Perth, Western Australia.

Ernest, Paul (1991) *The Philosophy of Mathematics Education,* London: Falmer

Fletcher, Trevor J (1988) 'Logo - a catalyst for thinking' in Watson F R (1990) *Logo and Mathematics,* Keele: Keele Mathematical Education Publications

Freire, Paulo (1972) *The Pedagogy of the Oppressed,* Harmondsworth: Penguin

Giroux, Henry & Freire, Paulo (1988) Introduction to Weiler, Kathleen (1988) *Women Working for Change: Gender,*

Class and Power, South Hadley, Mass.: Bergin & Garvey

Giroux, Henry (1992) *Border Crossings*, London: Routledge

Harvey, Brian (1985) *Computer Science - Logo Style (Volume I Intermediate Programming),* Cambridge, Mass.: MIT Press

Papert, Seymour (1980) *Mindstorms - Children, Computers and Powerful Ideas,* Brighton: Harvester Press

Perry W G (1970) *Forms of Intellectual and Ethical Development in the College Years,* New York: Holt

SMILE (1986) *The Logo Pack,* London: ILEA

Sutherland R and Hoyles C (1988) 'Gender perspectives on Logo programming' in Hoyles C (ed) (1988) *Girls and Computers: General Issues and Case Studies of Logo in the Mathematics Classroom* BWP 34, London: University Institute of Education

Thompson, Alba (1984) 'The relationship of teachers' conceptions of mathematics and mathematics teaching to instructional practice' in *Educational Studies in Mathematics,* 15, p105-127, 1984

Turkle, Sherry (1984) 'Child programmers: the first generation' in Turkle, Sherry (1984) *The Second Self: Computers and the Human Spirit,* New York: Simon and Shuster, reprinted in Jones, Ann and Scrimshaw, Peter (eds) (1988) *Computers in Education 5-11*, Milton Keynes: Open University Press

Weiler, Kathleen (1988) *Women Working for Change: Gender, Class and Power,* South Hadley, Massachusetts, Bergin & Garvey

Willis, Paul (1977) *Learning to Labour: How Working Class Kids Get Working Class Jobs*, Aldershot: Gower

8 Looking At, Through, and Back At: Useful Ways of Viewing Mathematical Software

Liddy Nevile

In this chapter I draw attention to ways of making the evaluation of software for use in mathematics classes a rich and more effective process. In particular, I suggest a way of avoiding the trap of expecting software to deliver something, and of moving attention to the possibilities which can emerge if a particular piece of software is used. It is a process for teachers designed to support their practice.

Introduction

You can look at a painting, and you can look through it to the world of perceptions that the painter experienced. You can look at a mathematical diagram, and see a static particularity, and you can look through a diagram and see it as a single frame from a vast collection, a world of experience of generality. You can look at mathematical symbols, and you can look through them to the generalities they express and a range of particularities they encompass (Griffin & Mason, 1990).

Looking back is not an easy process, because it is so much more attractive to move on to the next task. Yet it is a critical process if we are to learn from experience. Polya (1962) similarly proposed looking back as an important aspect of problem solving, but it is difficult to locate sufficient energy or attention to do it thoroughly.

In this paper I look at, through, and back at some examples of mathematics education software. Looking at software, we see it as it is. Looking through it, we see what can be done with it. Looking back at it, we see what is not possible with the software, and what is missed when it is used.

Some observations

Looking at today's mathematicians, I see them spending more and more time working numerically and symbolically on tasks which seem to demand less and less practice of some traditional skills. Many of the old skills are now being handled by computers. The new skills involve the use of software as tools for thinking.

Looking through today's mathematicians to the mathematicians of the future, I see mathematicians working on mathematics which is not yet well-defined, mathematics which will emerge as higher-level computer languages appropriate mathematical techniques. I see new mathematics built on the old as old techniques are articulated and then automated by computers. I imagine a continuous change process as mathematics moves to higher levels of abstraction, from 'what is seen as' to 'what is'.

Looking *at* today's mathematics students, I see young people working in a wide range of learning environments, including using computers running

- small packages which aim to teach some particular aspect of mathematics;
- a collection of mathematically-based general-purpose programs (Maple, Mathematica, Derive, Cabri Geomètre, ...);
- a range of general-purpose software (spreadsheets, paint and drawing programs), and
- a multiplicity of programming languages (BASIC, APL, PASCAL, Prolog, LOGO).

In rarer moments, I see students working in very high-level programming environments, using screen objects to represent what was previously abstract (Williams & Nevile 1994). The artificiality of the environment allows for new forms of representation. Students work in a continuously cycling development, *manipulating* the objects, then using them to *get a sense of* relationships and processes, then *articulating* those relationships and processes in further, more comprehensive manipulation (Mason 1989).

Looking *through* today's mathematics students to their future as people using mathematics, I envisage a generation of graduates working comfortably across a range of disciplines, using a range of software to perform tasks which would otherwise require a level of mathematics well beyond their immediate competence. Already we have comprehensive stock-control systems which are operated by the mere press of a key with a hamburger icon on it, for instance. We know that some people choose to play with a computer and explore chaotic behaviour instead of reading a book or watching TV. Behind all these activities lies complex mathematics, but it is not clear whether it is the users who are doing the mathematics. Maybe the package is doing it.

Looking back at the transformation that is expected of students and workers in the way of thinking and performance in general, and mathematics in particular, there seems to be a dual movement. On the one hand, simple calculations are being automated and removed from human beings, as in supermarket checkout mechanisms. The mathematics is being embedded and hidden in machines (Keitel 1988), from washing machines to video recorders, from barcode readers to robots. On the other hand, the potential now exists for more people than ever before to have access to tools which empower them to engage in mathematical thinking and support them in that thinking. I have observed students working in collaboration with computers on mathematical aspects of problems which I suspect they could not manage without the computer's support (Williams & Nevile, 1993).

Implications

What are the implications of the changes if we look through mathematics as suggested above, and back at the practice of teachers of mathematics?

Looking *at* the implications, I see some skills which were valued in the past which I suspect will not be important in the future. I see some which have not been necessary but which are now. To look *through* current mathematical practice, I need a way of identifying new skills, and perhaps even old ones, so that I can 'see' what is being done when people are using and doing mathematics. If I see what is being done, then maybe I will be able to look through it to what I might want to teach students.

Similarly, we need a way of seeing what we are teaching our students, so we can see through that to our ideas of what we need to be teaching, thus identifying areas for improvement.

Main conjectures

Developing the technique of looking at and looking through mathematical software will help gain access to useful information about what is possible when students are using it. This process will alert us to opportunities available in the moment of use, and as teachers, to actions we may take and things we may say, when students are using it.

Working on looking back at mathematical software will take the process of evaluating the use of software in a teaching situation a useful step further towards helping identify what is missing, what cannot be done at all or better, with the software.

Looking at and through software can create lists of possibilities and looking back at the software can augment those lists. In addition, the lists of possibilities help clarify what is not possible during the process of looking back at the software. This extra dimension, the addition of the what-is-not to the what-is, enriches what is available when the what-is possible is being examined.

Looking *at* and *through* applied to software

The following four applications were generated by applying the process of looking at and then through. In each case, an analogy is drawn between conventional teaching activities and the use of a computer-based activity: a conventional teaching activity is looked at and then through to the mathematics learning which might be done. Then an example using software is offered. Again, it is the software (and its potential) which is the object of *looking at*, and the mathematics learning which is done using it which is the object of the *looking through*. It is as if there are two screens involved: first we look *at* the activity or computer screen then we look *through* it to the mental, mathematical screen. Perhaps it is as if the first screen opens and becomes a frame through which to view the second.

Application 1:

If we claim that we teach long division to give students an example of using a mathematical algorithm, we might argue that this is a valuable experience from which the students can generalise later. If we use a small, user-controlled, computer animation of graphing a particular situation such as Eureka (described later), we might argue we are giving an example of graphing as a good way of presenting a dynamic relationship.

Application 2:

If we teach conversions from metric to imperial standards, we might say we are giving our students a chance to recognise, among other things, the notion of mathematical equivalence. If we use GRID Algebra (designed by Hewitt 1992 and described later in this chapter), we might hope our students will recognise equivalence as a relationship as well as a condition; and the role of the computer in the process of determining and managing the equivalence relationship for us.

Application 3:

If we ask students to construct a set of triangles to work on Pythagoras' Theorem, we might be expecting students to recognise

157

the role of generalisation. Working with students in a computer-based geometric environment (for example Geometer's Sketchpad or Cabri-Géomètre) which maintains constructions as particular elements of them are dragged about the screen by a mouse, we might hope to give students access to the notion of generality and invariance, and of mathematical manipulation of complex data, by seeing it on the screen.

Application 4:

If we have students working on word problems, we might be hoping they will recognise that codification or symbolisation is not arbitrary and irrational, and that there is a difference between mathematical and other perspectives. For example, if students use spreadsheets for relating the number of sides of a polygon to the sum of the internal angles, they might come to recognise that rows and columns, formulae, fill down, and cut and paste and copy can be useful metaphors for some problem-solving processes.

These examples generate some questions:

Just how close is the relationship between learning mathematics and the software used in the process?

Do we use the computer to increase students' motivation, or to give them experience of computational skills which might be useful in some other context, that in using the computer they can learn mathematics, ... or that they need to use the computer to get it to do their mathematics?

Does the software have within it something which will teach the students ...or that the students need to learn ... or is it that when using the software some desirable teaching and learning opportunities become available? Are the possibilities by-products of using the software, or do they have to be worked on explicitly?

Are we teaching mathematics in a way which most appropriately exemplifies the doing of mathematics?

In order to work more directly on the ideas identified in these questions, let us look even more closely at the mathematical packages.

Digging more deeply into the software

Most people agree that using computers might increase students' willingness to work on mathematical problems or, that with the right activities, there might be more transfer, or something. It is not easy to be precise about proposed benefits. But let us work on the mathematical nature of the software being considered: what is there about using the software that offers students useful experiences of being mathematical? What mathematics will they be able to see *through* the software?

One way of looking at software is to look at how much control of the software the user has. The computer wizard Alan Kay looks for software that can do things for the user which the original programmer could never have thought of. To explore this theme in text and without dynamic screen images, it will be necessary to call upon powers of forming mental images. Mason (1992) uses imagery to awaken awareness to the way in which we interact with what is available (on a screen).

> Imagine a plane. Imagine a cube. Imagine the plane moving through the cube. Define the slicing records as the intersection of the plane with the cube. Imagine the appearance (on a computer screen) of an animated sequence which presents the slicing records of the plane moving through the cube. Imagine you don't like the colours, or cubes.

What variations are possible if you are viewing such a film or video-tape? Given that the programme (text and frames) have been pre-recorded, there is little you can do with the display itself. You only have access to control in one dimension, time. You can pause, restart, and review. You have *one* dimensional control of a film in time, and what Mason (1992) has described as a *zero* dimensional use of computing, namely no computation.

Through re-construction of the animation by several participants, using their memories and verbal descriptions as a medium, a film can become a shared common experience from which fragments can be extricated, labelled, and used to weave together an explanatory story. When it is in a medium in which changes can be made, an initial film can become higher dimensional, in computational terms. No longer a single film unfolding in time by projection, the animation of the plane and cube can generate a richly woven fabric of mathematics and mathematical thinking which can be developed and explored in many directions, and can begin to link with other mathematical topics and ideas. But this is beyond the original object, more that the animated sequence, and outside the computer.

> Imagine a computer package called Eureka presenting an image of a person sitting in a bath. Imagine the user typing a command to turn on a tap and the water level rising in the bath (if the plug has been put in the plug hole). Imagine a graph of the water-level vs time being generated and displayed on the same screen. Imagine what happens when the computer user has the plug removed by the computer.

This animation can be initiated in a variety of ways and can be used to generate many versions. Each time the animation is run, the plug can be removed sooner, or later, the taps turned off or on. The user has some control, and the package can be described as *one* dimensional. The user can, as the linear sequence is played out, change a value. But each sequence is somehow still linear, still determined: the user can choose only from what is explicitly offered by the software. Even with more sophisticated software packages which monitor student progress, provide on-line tutorial help and possibly offer suggestions, the user cannot make something else of the package.

> Imagine a GRID Algebra screen on which there are rows and columns with entries either numbers or expressions. Imagine that on this spreadsheet, as algebraic expressions are dragged across the screen they are incremented additively, but as they are dragged and down the screen from the top row, they are incremented multiplicatively .

In the GRID Algebra software, the user can handle formulae, symbols and numbers, and ways of building relationships between the contents of one cell and others. This little spreadsheet somehow does not have time constraints (like Eureka). Instead it has spatial constraints. The path from any cell to any other must be constructed of linear segments, and downward movement is only possible between compatible rows. But if these rules are adhered to, the contents of the starting and finishing cells, and all those designated as stopping places en route, will be related algebraically. The choices made by the user are not taken from a menu of possibilities listed in the software package, but from all accessible possibilities (some of the rationals). The software does the algebra which supports the user's (limited) freedom to explore–negotiate the terrain. This might be enough to suggest that it is offering what could be called a *two* dimensional control.

The video offers interaction only in the dimension of time; the animation in the dimension of time and conditions (on–off); the Eureka graphing in time but with a few choices. The Grid Algebra spreadsheet offers variation in the dimensions of time, the choice of objects and the kinds and combinations of relationships. Each added dimension, as it were, adds a dimension to the screen through which we can look at what is being done. Each added freedom being offered by the computer results from the addition of constraints in the software. The more detailed, complex and varied the constraints, the more freedom the user gains. When there is a single constraint, yes or no, there is very little freedom for the user.

> Imagine a geometric construction such as the centroid of a triangle. Imagine that with the movement of a computer mouse, the user can drag one of the vertices of the triangle by dragging it with the mouse. Imagine moving the triangle's centroid across the screen, all over the place, and watching as the constructed relationship (is–the–centroid-of) is maintained by the triangle changing.

Mouse-mathematics introduces a certain sense of spatial freedom which can be experienced metaphorically as generality. Constructions are maintained consistently throughout the system.

161

The dynamics are mouse-controlled rather than pre-determined. The user can develop a sense of a hyper-film (film in the sense of successive frames suggesting movement, hyper in the sense that the film is only one choice in a two-dimensional range of possibilities). The *extra* dimension here comes after the freedom to determine the objects and their relationship. The constructions are maintained through whatever distortions the user chooses. The constructed objects may not have been anticipated by the software programmer. Their use may not have been anticipated. The user is free to move beyond what the programmer prepared; the limitations are not content bound or construction-bound. There are some construction limitations but they come about as a result of computational limitations, not design.

Looking through the screen, the user has access to a meta-view of what is happening on the screen, the opportunity to experience particular constructions as examples of a range of more general ones. Generality becomes manipulable, it can be felt, it becomes an it, and students can gain access to the articulation of generality through hand movements before they use words or symbols.

> Imagine working on a geometric problem within a spreadsheet-like structure. Imagine that as the geometric construction is dragged down the screen, it increases in dimensionality from one to many: three, four, five ... . Imagine that as a polygon is dragged across the screen it gains more sides, or that rows and columns are user-defined in terms of their default effects, or that they are not connected but particular cells have special functions defined by the user. Imagine the range of constraints the user might choose to use to provide experiences with geometric objects and constructions.

The spreadsheet user gains spatial freedom, for instance in the way relationships are established. The user can click on a set of cells, not connected in any obvious order, and command the computer to apply a procedure in order to maintain their sum in a particular cell, invariably, even if the contents of the chosen cells are changed. The user may also have a graphical representation of what is happening.

Generic software (text processing, image processing, spreadsheets, LOGO etc) provide menus of tools for *working on* the processing of ideas and of data, while specific software may provide choices of direction but tends to be susceptible to being merely *worked through* (Mason 1992). There are no limitations to the type of relationship and the user is free to have sections of the spreadsheet which are completely inconsistent with others. Computer scientists have long sought the combination of descriptive, imperative and constraint programming to give users wider-ranging computational freedom.

The spreadsheet is not unlike a piece of paper on which one might have notes in shorthand, quotes in copperplate, squiggles in pencil and a sketch on the side. There are rows and columns but they are not all only spatially interconnected: the user might have generated three or four conflicting algorithms, for use side-by-side and simultaneously with the same data, in order to test a conjecture. This spreadsheet would not determine the type of relationship (allowing different ways of understanding functions to become available, see Mason & Nevile, 1993a) and the user would be free to have sections of the spreadsheet which are completely inconsistent with others. There is a sense in which this piece of software is of a higher dimension. Perhaps it is *three* dimensional, although Mason was not so generous!?

> Imagine a Latin scholar sitting beside a mathematician. Imagine they are both using spreadsheets. The scholar is using the spreadsheet to record verb endings appropriate to various tenses and conjugations and noun endings for the range of declensions. The scholar copies and pastes some entries from her worksheet into some blank spaces in the mathematician's spreadsheet.

The spreadsheets do not complain. The Latin scholar and the mathematician differ in the way they want to work from specific cases to generalisations because their disciplines demand these differences, because Latin grammar and mathematics are ontologically different (for instance, exceptions abound in spoken languages but within mathematics they are often evidence that something is wrong). It is as if spatial arrangement and cell-relationships of the spreadsheet invite the user to question the very nature of the forms of expression being used, the relationship between them and their utility within a context. It supports a meta-view of what the user is doing, this time from the perspective of comparison of forms of generalisation. It will be interesting to see whether a computational environment is developed in which ontological differences are explicitly recognised and explorable.

Looking Through

Let us look again *through the* software (screen) to see what mathematics is being done.

It seems that in the same way as the computer's usage might be described as increasing across a range of dimensions, so might the users' mathematics if it is being undertaken using the software described. Opportunities for awakening awarenesses of the many dimensions of mathematics will not necessarily follow the use of more multifariously structured computational environments but, they certainly make it more likely, and it is important to be awake to possibilities.

Reconsidering the examples above, we can discern in them an increasing range of mathematising available within the environments being described, from mathematical facts in the first instance to mathematical relationships, to mathematical behaviour and on, in the last, towards mathematical ontology and epistemology. There is not so much the substitution of one form of mathematising for another, but rather an increasing richness of opportunity. It is the increasing opportunity for accessing mathematical depth which characterises the movement through the range (in the same way, content breadth is provided by the movement across the many examples of each particular class within the range).

With this awareness of possibilities, which is made explicit more easily by the use of the technique of looking at and through, we reveal yet more questions.

If the depth is to continue to be extended, what might we want to see next?

If students of mathematics can benefit from using computational environments when learning their mathematics, what do they need to learn so they can use their computers better?

In approaching the latter question, we are not confined to mathematics classes. We can ask it with respect to other areas of the students' curriculum too. For many reasons we can hope students will be better off in the future if they have a broad literacy which includes such things as how to use computer manuals, and what computation offers. Recognising this, schools have for some time been working generally on enriching of students' technology (awarenesses, ways of knowing about and using computers etc.). It seems that these endeavours should continue but perhaps they can be undertaken now as part of the more specific curriculum, in much the same way as the students' learning of their language is supported by work in all subject areas.

Identifying How to Extend Depth

In the 1970s, educational researchers first used Logo in classrooms. Many teachers are still exploring what Logo offers. It provides a rich example of software which can be mined and re-mined, as we learn more about computation and the learning of mathematics.

By looking deeper into what is possible with Logo, can we learn for the future?

Logo can be seen as a computational environment in which there are potentially all the types of software so far discussed *in one package.* The versions of Logo which are emerging in the market-place today are full of handy features, generally consume more than 100 times the memory of the old versions, and most significantly, can be understood to contain (in the presence of telecommunications and good file-handling) many thousands of microworlds and tools which have been built world-wide.

Contemplate for a moment the library of LOGO worlds and procedures around the world. It spans many different natural languages. Behind the LOGO literature, we find authors who are very young children, serious research mathematicians, musicians, medical practitioners, linguists, psychologists and many others. LOGO now comprises a set of computer files *and* the language which interprets them, in which they are written.

> Charles (aged 13) said he was happy with Pythagoras. He had had to prove the theorem to himself, so he had written a Logo program. But it was all in his mind, because he was in the mountains at the time, without a computer.

This incident exemplifies the way in which students can make the computer environment part of their world and their mode of thinking. Implicit in this anecdote are the beginnings of the notion of proof, the use of formal language, some ideas about generalising, an appropriation of meaningful (to him) forms of computation to restate what has been said, and more. (We might not expect him to recognise all these things without the support of a teacher.) By typing them into the computer they would become manifest : the Logo code, the procedure running, and the screen-product.

> Imagine that Charles has written a Logo program which takes feedback from the screen. As he drags the mouse, the computer re-constructs the triangles and squares involved from wherever the turtle is located to the opposite side of the triangle.

Still we see the product, the procedures and the process of construction.

> Imagine that Charles' procedures are now replaced by a black box, a set of geometry constraints, which does everything he wanted so that all he had to do was choose triangle construction from a menu and click the mouse in three places, choose square construction from the menu and put squares on the sides of the triangle, and then he was free to drag the vertices about.

What would we, as teacher of mathematics, see and what could Charles see?

We would see, probably, the special case and the process of generalising with the use of the mouse. But it is quite likely that Charles would see only the generalisation (what has been called the function collage, see Mason and Nevile 1993b). Charles has entered the mathematics at a high level and is denied access to the lower levels by the new context. Good and bad, we might say.

> Imagine that Charles now assumes that Pythagoras' theorem applies also to pentagons and tries to test his conjecture. He constructs a pentagon with the squares and what happens? Something pretty odd. He wants to know more about triangles because he decides they must be somehow special.

Let us focus attention on the move from the special case of one instantiation of a triangle, through the process of generalising, to the generalisation which is the theorem. Can the process be reversed, can the student now take the triangle as a generalisation and work on what is being generalised in the concept of triangle? Particular examples of Pythagoras' Theorem can be developed but is that the same as specialising from the general? This is asking for a lot: being able to work on the particular case, on generalising the particular case to form the general case, and to work from a particular case back to some special cases from which generalisation is available. But it constitutes mathematical thinking.

Now we are looking back at what the software does from our position of vantage which we established by looking at, and through the software. We have some processes which can serve as useful explications of mathematical working and now we can use what we have to see what we would like in addition, or what is missing.

Conclusion

Purposeful and detailed *looking at* software can often awaken our awareness to aspects which are easily overlooked when we use the software; *looking through* the software can awaken our awareness to what is possible within the domain for which we are using the software, and *looking back at* the software can help us identify what is not supported by the software and what we might want in software of the future.

I am indebted to John Mason for many of the good ideas in this chapter.

References

Griffin, P. & Mason, J. (1990) Walls and Windows: a study of seeing using Routh's theorem and associated results, *Maths Gazette* 74 (469) pp 260-269.

Hewitt, D. (1992) *Grid Algebra*, Milton Keynes: Centre for Mathematics Education, Open University.

Keitel, C. (1988) 'Mathematics Education and Technology', *For The Learning of Mathematics*, 9 (1) pp 7-13.

Mason, J. (1989) 'Mathematical Abstraction and the Result of a delicate Shift of Attention', *For The learning of Mathem,atics*, 9 (2) p2-8.

Mason, J. (1992). 'Images and Imagery in a Computing Environment' in *Computing the Clever Country* , Proceedings of the Australian Council for Computers in Education Annual Conference. Melbourne: CEGV.

Mason and Nevile (1993a) 'Diagrams as Functions in Geometry' in *New Directions in Research in Geometry & Visual Thinking*. (Forthcoming) Brisbane, Australia: Queensland University of Technology.

Mason and Nevile (1993b). 'Geometric Diagrams as Function Collages' in Mousley & Rice (eds.) *MATHEMATICS: Of Primary Importance*, Melbourne: Mathematical Association of Victoria.

Polya, G. (1962) *Mathematical Discovery: on understanding, learning, and teaching problem solving*, Combined edition. New York: Wiley.

Williams, D & Nevile, L, (1993) 'Calculating, constructing, collaborating and communicating mathematically in Boxer', in Mousley & Rice (eds.) *MATHEMATICS: Of Primary Importance*, Melbourne: Mathematical Association of Victoria.

Williams, D & Nevile, L, (1994) 'Boxer boxes: Collaborators and mediators of student expression' in *Proceedings of APITITE '94*, Brisbane, Australia, July 1994.

Section Three:
Using Calculators

Calculators have now been widely available within the developed world for at least twenty years and the earliest research on their use in classrooms is nearly as old. Yet, generations of school children, many of whom possess calculators, are still completing their mathematical education, especially in the early years, without experiencing the impact on their classroom learning of these small, powerful devices. Kenneth Ruthven points out in Chapter 12, that "evidence suggests that the limited use of both calculators and computers in the mathematics classroom cannot simply be attributed to difficulties of access to the technology."

Simple, four-function calculators are used by adults to handle big, or awkward, numbers or to check a calculation. In projecting these two uses onto young people, the role of the calculator as a representational device through which to explore and learn about number is denied. And yet children are highly imaginative and exploratory in their use of artefacts. As Pat Perks shows in Chapter 9, *Calculators and Young Children: A bridge to number?*, they bring playful spirit of enquiry to bear on their use of this calculators to expand their world of numbers. She reports on a classroom development project in the early years of learning and lets us see the growth in understanding of the children as well as their teachers. However, she concludes with the warning that "unless teaching takes full advantage of their power calculators will remain peripheral to learning."

In Chapter 10, *Visualisation, Confidence and Magic*, Teresa Smart discusses how the graphic calculator was integrated into the mathematics work of a class of fourteen-year-old girls. She highlights three outcomes - their increased confidence, their use of visualisation as a problem solving strategy, and the disquiet felt by some pupils at the 'magic' effect of the calculator. At the same time, she makes clear that this work was undertaken by a researcher and a teacher both committed to a mathematics curriculum and classroom which was exploratory, shared power and provoked change.

Mary Margaret Shoaf-Grubbs also worked with an all-female group, in her case elementary college algebra students in a liberal arts college in the U.S.A.. In Chapter 11, *Research Results on the Effect of the Graphic Calculator on Female Students' Cognitive Levels and Visual Thinking*, she reports on the results of a traditionally structured experimental/control group study which investigated an assumed link between performance and understanding when a graphic calculator was introduced into the experimental class. Using a pre- and post-test format, she, like Teresa Smart, found that the experimental group's performance was significantly enhanced particularly in the area of spatial visualisation and that this linked positively to enhanced understanding. In addition, she developed what is believed to be a new form of scattergram data analysis in which the progress not only of each class but also of each student can easily be followed and analysed.

Kenneth Ruthven concludes this Section with his overview, in Chapter 12, *Pressing On*, of the current state of calculator usage in schools and of research into its effects. He identifies a need for a "more considered framework within which to analyse, plan and implement the use of calculators in mathematics education." He speculates that this framework will need to take account of five important facets of calculator use in the mathematics classroom. These are the role of procedure in calculator usage, the use of strategy to support calculator usage, the impact of pupils' dispositions towards calculators, the part that the calculator plays in a re-thinking of the mathematics syllabus and the pedagogical implications of such re-evaluation.

Each of these issues can be found playing its part somewhere in the three preceding chapters, as well, of course, as in many other chapters in this book.

172

9 Calculators and Young Children: A bridge to number?

Pat Perks

> *Providing young children with calculators can help them to explore many aspects of mathematics. This chapter considers some work done in infant classrooms and offers examples where the calculator can act as a bridge to the learning of number.*

The technological development of the calculator began to make itself felt in England in the early seventies. As machines became cheaper and smaller, it became possible to consider their use in schools. Enthusiasts were eager to take advantage of the technological innovation; for example in 1975 Laurie Buxton, at the time Senior Inspector for mathematics in the Inner London Education Authority, was granted money to buy 200 simple four function calculators to be distributed to a secondary school and four primary schools (Fielker, 1976). Although much of the development work was done for secondary school use of calculators, 1980 saw a Junior Calculator Pilot Study in the County of Durham, which included the creation of pupil materials.

The largest curriculum development project in the U.K. on young children using calculators was the Calculator Aware Number project (CAN), directed by Hilary Shuard from 1985-89.

Calculators were provided for pupils in Y2[1] classes. From early reports of this project there was a suggestion that even younger children would benefit from using calculators. As a result, in 1988, the City of Birmingham began its own three year project. The schools in the central third of the city received sufficient calculators for their Y1 pupils during the first year, the other administrative areas over the next two years. The provision of the calculators was supported by in-service courses, advisory teacher support in classrooms and a pilot project where Y1 teachers from four schools worked closely with an advisory teacher for a year, producing a booklet (Coates, 1989) at the end of the year, and helping on courses for the other areas of the city later. The first phase of the project was evaluated in Perks (1990) which included material based on interviews with children and teachers from the pilot schools and questionnaires to the other schools.

The enthusiasm for the use of calculators amongst the small group of teachers working on the pilot project was clear from the reports of their meetings, their booklet, and their willingness to help with the teachers' meetings in the next phases of calculator distribution. However as the questionnaires were collected from the other schools in the central area at the end of the summer term 1989 a very different picture emerged (Perks, 1990). Many calculator boxes had not been opened, many teachers expressed their concern about not knowing what to do with them. Only in a very few classrooms were calculators being integrated into the mathematics curriculum for the infant pupil.

Distribution of calculators continued however, with the south of the city receiving them the following year, and distribution in the north the year after. Teacher support groups met in these areas and advisory teacher help was available on request. In the north of the City, a group of eight teachers from six schools met on thirteen occasions during the year following the distribution of the calculators. These teachers were responsible for reception (R), Y1, Y2 and Y3 classes. Four of the eight teachers were mathematics co-ordinators in their schools, the junior class teacher (Y3) being one of these. As a result of the evaluation (ibid.), the objective of the meetings was to collect ideas, use them in classrooms and record anecdotes as evidence of changing performance. The anecdotes were collected by teachers' recording of incidents of mathematics in their classrooms or by a more formal arrangement when I, in my role as an advisory teacher, worked with the class whilst the teacher looked for and noted down the reactions of pupils. The anecdotes covered many areas of classroom behaviour, but the principal focus of the teacher discussions, which used these anecdotes, was whether they were seeing different mathematics when their pupils were using calculators and if so whether this offered better opportunities for learning.

For these teachers the major advantages offered by using calculators for teaching and learning could be considered under four headings.

1. Enjoyment and confidence: the calculator provided children with a new resource, a different way of recording which allowed them to work with enjoyment and provided ways of working on activities which improved their confidence.
2. Opening up the curriculum: the use of calculators provided opportunities to work on ideas in areas not usually in the domain of infant classrooms.
3. Enhancing the current curriculum: the use of the calculator offered different ways of working on the existing curriculum.
4. Creativity and Exploration: the calculator provided a resource for the children to experiment with, it allowed them to be creative with numbers and to explore their ideas.

Many of the activities and experiences described here will be familiar to those who know the work of the CAN project. Similar results are described by Groves (e.g. 1990, 1994) in the Australian Calculators in Primary Mathematics Project. The enthusiasm of those teachers involved and the exciting work done by the children convinced many that calculators provide a fascinating and important addition to learning in mathematics.

1. Enjoyment and Confidence.

To those working in classrooms it soon became apparent that many children enjoy using calculators. In those classrooms where calculators were readily available the children used them when they wished, either for mathematics or for play. In play situations their use as walkie-talkies or mobile telephones seemed the most frequent. The advantage of having them as part of the equipment on a table was that the children were free to choose whether or not to use them. After a year of working with calculators, most children who were asked if they enjoyed them, (Perks, 1990), said yes because:

> You can do anything on them.
> They help you to learn your numbers, do your work.
> You can do your sums and writing.
> They help you do your number work.
> You can do hard sums.

However, one child interviewed thought that calculators were too hard.

> Because you keep on doing everything. You have to write it down, it's too hard. Everyone needs the teacher.
> (Perks, 1990, p 51.)

After further probing it appeared that this child did not like writing, she enjoyed making up the sums on her calculator, if she could show them to someone without writing them down. This raises questions such as:

> Might the necessity of having to write be a limiting factor for some children, might it slow their learning of arithmetic ?
> Will the calculator offer a different learning environment for such children and enhance their progress?

Many of the activities designed by the teachers required children to use calculators without other forms of recording. For some children this encouraged them to try a result, check it and reject it quickly. It allowed them to test their ideas in a way which pencil and paper did not.

The idea of using the calculator as an answering machine became a popular activity. Activities such as "Fill your screens with ..." provided a task where teachers could work on ideas such as recognising the keys, counting cubes, the answer to sums, (e.g. the class are first shown seven cubes , five of these are then hidden without the children knowing how many, the remaining two cubes are shown to the children who are asked to fill the screen with the number of cubes which are hidden,) These activities could be used regularly, for short time slots. Teachers were able to see what children had on their displays, whether the children understood the instructions and they could also encourage the children.

> Because they could wipe their screen they would take a chance. Some of my youngest children are a bit reluctant, but they will have a go and became much braver as a result. (Reception teacher.)

2. Opening up the curriculum.

The major area where teachers felt that the calculator opened up the curriculum to their young children involved the use of large numbers, especially through counting, and recognition of negative and decimal numbers.

2.1 Counting:

> If children are to become familiar with very large numbers, such as the millions and billions which now confront people in government information, the calculator gives a way in which children can gain this experience. (Shuard et al., 1991, p 16.)

Young pupils are encouraged to count, but sometimes they do not count very far. Rarely are they offered a picture at the same time. Keying the instructions $1 + + = = = = ...$ (usually referred to as the

additive constant) onto a calculator leads to 1, 2, 3, 4, 5, 6, 7, ...
appearing on the calculator display in turn, a useful picture of the
counting process, Children can be encouraged to count to higher
numbers, to recognise relative position and numbers before and
after given numbers.

One of the popular activities was to get pupils to set their
calculators up for counting and then ask them to stop at a particular
number. When four and five year olds are asked which number
they wish to stop at, the most popular number, sometimes
considered unsuitable for work in reception classes, is 100. As you
watch you quickly notice their fingers speeding away at pressing
keys and many dash past 100 without noticing. Repetition sees
development of an awareness of the merits of care versus speed.
Those who key-press quickly begin to slow down when the
calculator reaches the nineties, 98 and 99 are treated with great
respect. The children's facility with this type of exercise indicates
that knowledge about the order of numbers is often more well-
developed that one might expect, especially when the National
Curriculum for England and Wales (Department of Education and
Science 1991, (DES)) offered in its Programme of Study (PoS) for
the average seven-year-old:
> • reading writing and ordering numbers to at least 100
> (PoS, Attainment Target 2 level 2 DES, 1991[2],

Playing with counting on the calculator allows young children to
demonstrate such skills very much earlier in their mathematics
careers.

Currently, children are assessed as being at level 3, above average
for a seven-year-old, when
> • learning and using multiplication facts up to 5x5 and
> all those in the 2, 5, and 10 multiplication tables. (PoS,
> Attainment Target 2 level 3, DES, 1991)

However, the use of the additive constant means that children can
work on the ideas of multiplication much earlier than their teachers
might expect. Adult expectations of what is suitable mathematics

for children is challenged as are the accepted hierachies of teaching when children have such a tool for learning with.

In a class of six-year-olds, the children were counting to 100 in ones and then in twos. Testing counting in threes gave 99, 102, with children explaining where 100 was and why the calculator missed it. We had just established that counting in fours gave 100 on the display when it was suggested by their classroom teacher that they would like to do some recording. I showed them a style of recording for the task with four and asked them to choose their own numbers. Some copied the example from the board before choosing another number, some went back to counting in twos or ones, most went on to five. Watching their recording it became clear that the position of 100 as more than ninety something but less than a three digit number of the type one hundred and something was firmly established. No child wrote down more than one number in the hundreds and they could all explain whether the count reached 100 or not. Five children demonstrated some interesting mathematics, by using five, then ten then fifty, with one boy also using twenty-five. How well developed was their intuition about the factors of 100? Without a calculator would they ever have had the opportunity of demonstrating such intuition?

In another school, I had been working with a small group of Y1 children to try to develop some idea of matching counting in groups with apparatus and looking at place value. Among the various activities we used, we explored counting in tens, with the calculator and towers of ten cubes. I felt that I had forced the idea and they were not ready for this. However I was to return to the school to work with the whole class on shape in two weeks and left my large bag of cubes behind for the children to play with before using them in a mathematics lesson. The favourite class activity was to count the cubes. In break and lunch times different groups would try to decide how many cubes were in the bag. They grouped them in tens, but did not put the tens into hundreds, which made counting difficult. The small group I had worked with got out the calculators and counted in tens to 970 (several times) with six cubes over, adding 6 gave 976, a number they could not read aloud!

Large numbers seem to fascinate young children. Counting them is possible with a calculator even if grouping strategies are not well developed. As Shuard and her colleagues have pointed out:

> ... the calculator gives an additional way of exploring place value by adding known numbers. (Shuard et al., 1991, p13)

One teacher was particularly interested in working on patterns in the counting numbers, points of change, and how this might help with the writing of numbers bigger than 100. With her Y2 class she often used the additive constant with instructions such as:

Instruction:	Pupils watch:
Stop at 9, think about the next number, press = and check.	9 changing to 10.
Stop at 59, think about the next number, press = and check.	59 changing to 60.
Stop at 29, think about the next number, press = and check.	29 changing to 30.
Stop at 99, think about the next number, press = and check.	99 changing to 100.

At the beginning of the work, when asked to write one hundred and nine, many pupils wrote 1009. After working on the problems above many used the idea of starting from 99 and counting onwards, giving them 109. They went on to look at 209, etc. and what the number 1009 represented.

2.2 Negative numbers

The pilot project reported that the position of the negative sign on the calculator display gave rise to some difficulty. They wrote:

> ... we found that the children tended to ignore the negative sign on the calculator screen,
> e.g. 66 - 77 = 11 (Coates, 1989, p34)

As a result, the teachers in the North area decided to work on counting backwards as one of their major activities. Keying the instructions 12 - - 1 = = = = ... gives 12, 11, 10, 9 .... Counting backwards allows the negative numbers to be seen as an extension of the number line to the other side of zero. Once the negative sign has been spotted these numbers appear to be more easily accepted by young children, than by those aged eleven, when negative numbers are more usually introduced. One teacher commented:

> It is not as if I expected them to remember, some of them are still only six, but they like playing, they really liked seeing the number line extended around the room to match their counting. At least they are being given a chance to work with these numbers before they have to be experts in them. (Y2 teacher,)

The playing with numbers seemed to fascinate many classes.

> They like the symmetry and talk about the other side of 0. At first I made sure that we always saw the 0, but they like patterns such as 15, 9, 3, -3, -9 ...as well and we are just moving onto 17, 12, 7, 2, -3, -8 ... (Y1 teacher.)

These experiences added another dimension to their "playing with sums" as in many of these classes children had in their lists of calculations problems with negative results such as 4 - 7 = -3 .

2.3 Decimal Numbers.

One of the most popular familiarisation exercises used with the calculator in many of the schools was "Fill your screen with twos", or fours or .... Filling a screen with noughts became a problem for a very short time. The decimal point came to the children's aid. One teacher commented that the decimal point appeared in many sums and in most cases was treated as something to do with money.

> 1.25+ 1.36 + 1.36 + 2.45 was one sum I saw. I was a bit surprised and asked about the 1.25 expecting it to be a

problem "It's just like one pound and twenty-five pence." was the explanation. (Y2 teacher).

One teacher was convinced that her group would not be happy with decimals, so she gave the group of pupils, she considered as middle ability, a problem on sharing 12 sweets.

> They had some cubes and shared them between two and got the calculators out to try to get the same result. 6+6=12 wrote one. While they liked this there was some worry that they were using the answer. They eventually found 12÷2=6. They moved onto 3, they shared the cubes and got the result 4, they checked this on their calculators 12÷3=4. The next one was done on the calculator first 12÷4=3, and the cubes moved to match. The next problem gave a calculator result of 2.4. Back to the cubes, 2 each to five people, but there were two cubes left over not four. "Oh its got to be the bits you get if I could cut them up", said one. Much nodding and they moved to the next problem. (Y2 teacher)

No longer are decimal and negative numbers something which has to be hidden until the children "can do them".

> In the past teachers have been able to keep negative numbers and decimals hidden from children until they thought the children would understand a full explanation. This is no longer a possible teaching strategy; children cannot be prevented from discovering negative numbers and decimals for themselves on a calculator display at an early stage. (Shuard et al., 1991, p 23)

The advantage for most of the teachers was the matter of fact way in which the children accepted these numbers. They appeared to be seen as a normal part of the number system which they had been exploring from their first days of counting.

3.   Enhancing the current curriculum.

3.1 Use of symbols

The calculator offers young children a tool for looking at the way numbers are written. Activities in many classrooms were based on either copying written strings of numbers onto a calculator display or children generating a display which they would copy. Although this sometimes leads to children writing digital numbers, this does not tend to last long if the drawing of the more usual symbols is regularly practised.

Some children show reversal problems with 2 and 5. Playing with digits as above tended to mean that pupils knew the relative key positions on the calculator, so children with difficulties could be asked, "Where is the 2?", "Next to the 1", "Where is the 5?, "Next to the 4" and they began to correct themselves, using the calculator key board as a reference. In schools, where the counting constant had been used regularly, children were encouraged to count to find the 2 and then the 5. When these two approaches were discussed by a group of teachers, they felt that the more different ways in which they could work on these problems the more chance their pupils had of resolving the difficulties. Those who worked on the reversals with reference to the calculator felt that they were successful in encouraging pupils to find ways of checking.

The calculator is a machine where the operation symbols have to be used in order for the calculation to be completed. They are not decoration - the symbols are essential. Teachers in the group talked about the way in which children are introduced to the symbols for operations. Addition is worked on for long periods before pupils are introduced to the other operations, and as a result the symbol + can become irrelevant to the pupils' work. They always add and so the symbol offers no necessary information.

One seven-year-old was working on the problem "The answer is 25, what is the question?" He brought me his calculator, the display showed "55555".

> P: I read that as fifty-five thousand, five hundred and fifty five.
>
> Pupil: "But five and five and five and five and five make twenty-five.

He had demonstrated his skills of mental arithmetic, but was not using the "+" key on the calculator. Challenged to use the fives and get the display to show 25, he entered the calculation as 5+5+5+5+5= giving the appearance of understanding the point which had been made. He returned shortly afterwards with a display showing "444445",

> because the fours make twenty and then five gives twenty-five.

The number was read to him as "Four hundred and forty-four thousand four hundred and forty-five". Asked to make the display give twenty-five he did so but on exploring other numbers he had to be prompted by the teacher reading the display in terms of the number seen. After about ten minutes of quiet working he showed us 20 + 5 = 25 and then 205 saying

> Twenty-five is twenty add five, not two hundred and five.

Children work with practical apparatus in order to establish the concepts of number, but perhaps the connections with the operations can be made more explicit, by using a calculator. A class of six-year-olds were working on a problem with seven cubes made into two towers. They were asked to make the model, draw a picture and write down the sum.

There were few problems with the practical task or with the picture, and the related numbers were written down by the picture, for example, see fig 1.

There were, however, no signs of
the sums we were expecting,
i.e. 4+3 = 7.

figure 1

By asking the children to use their calculators and to write down
how they would obtain 7 using the 4 and the 3 and then write down
the instructions used, written sums such as 4+3=7 began to appear.
There is no implication that the use of the calculator necessarily
aided their understanding, but may offer a reason to the children
why the symbol for addition is required. It also appears to offer a
bridge between the practical apparatus and pencil and paper
(Shuard,, 1988).

3.2 Concept demonstration.

One issue which arose very quickly with children using calculators
is that they find the decimal point and negative numbers long
before any syllabus expects (see section 2.2 and 2.3). At a meeting
where work on negative numbers was discussed the teachers began
to consider what they would do with the subtraction algorithm.

> If I was doing thirty-four take away twenty-seven, I
> would say four take-away seven you can't do, but my
> pupils would argue that you can! They've discovered
> that negative numbers don't just appear when you are
> using counting patterns.

This led to a discussion of the use of language - "four take-away
seven - you can't do" "but four minus seven you can"! For these
infant teachers the issue of what to do with two figure subtraction
was shelved by saying that they would let the pupils develop their

own strategies. However for those who were also co-ordinators of mathematics in their schools the problem was how to work on these algorithms appropriately. This led to a discussion of the use of the calculator for what Johnson (1978) calls "exploration for concept demonstration." (p 53)

At a later meeting, one of the teachers reported on her work on multiplying by 10 with a junior class:

> I had thought about it and the rule "add a nought" when I realised that my six-year-olds would try it with decimal numbers (they like the dot!). So I took the calculators in, gave them some numbers, told them to multiply by 10 and see what happened. I heard lots of "It gets a 0", "Its one bigger" and other ideas when I noticed that one group had chosen numbers like 2.55 and 1.23. When I asked they said they were using money and you couldn't use two pound fifty because the nought vanished. I suggested some decimals for everyone. The rule certainly wasn't "add a nought"! (Y1 teacher and mathematics co-ordinator)

3.3 Concept reinforcement.

Many examples were also offered about Johnson's "exploration for concept reinforcement" (p 53) Many of the early experiences offered to young children about number are through practical materials. Infant teachers are well practised at providing different materials to count and different sets of objects to combine to explore addition. Children are expected to abstract the generality of number and many of them appear to do this efficiently. However experience suggests that if children are having difficulty with 2+3 = 5, the provision of more equipment may only allow the children to focus on the contexts not on the 2+3. One teacher suggested to the group that there were only a limited number of ways of working on this problem in the pure number sense. A six-year old in her class had decided to set herself problems of adding two numbers, one ending in a 2, one in a 3. The child explored the problem for some time and then reported back:

> It was good, you always get five, except if you use that.
> (Pointing to the decimal point)

Two other teachers worked with this approach with their classes, using tasks such as:

The first uses an empty box for the missing digits and directs the pupil as to the number of digits to be chosen. The other uses ink-blots, covering any number of numbers. This latter is a form which some children appear to prefer. Whilst in the first class the children produced examples such as: 172+543=715, 342+343=685, ..., the other children had 2+43=45, 12+43=55, 142+53=195, 165472+24563=190035 depending on the level of their confidence.

The teachers felt that working on number bonds in a different way was valuable to all their pupils and they had several examples of children exploring other number pairs. Some tested the problem by using the 2 and the 3 in places other than the units column, such as 324 + 353 and 325+337 and:

> I found myself listening to explanations of why 2 and 3 did not always make five, except they did, it was the other numbers giving them a one and other versions of carrying. (Y1 teacher)

The task offered pupils the opportunity to explore at their own level whilst the feeling was that all the class were considering the same activity. The task allowed them to be creative, to generate their own data, to make their own conjectures and test their intuitions.

3.4 Language

Another issue for many teachers was the opportunity for working regularly on the language of mathematics. Activities such as:

> Fill your screen with the number which is one more than five.

Fill your screen with the number which is one less than five.

Count to the number which is three more than twenty-three.

allowed the teachers to work with all sorts of language including that for the operations such as using minus and subtract not just take-away, divide not just share.

It became clear that it was important to help the pupils to articulate these phrases themselves as a number of instances of events like the following were recorded.

> We were working on "Fill your screen with the number which is one less than ten," "Fill your screen with the number which is one less than nine "and so on. I had deliberately chosen to work using a pattern to help those who were a bit insecure. After we had filled the screen with sixes I asked for someone to give the next instruction, expecting, "Fill your screen with the number which is one less than six". I got "five". I asked others for the instructions, the best I got was, "Fill your screen with fives". I realised they were hooked on the answer, which I had given them by using patterns, rather than the words which I was thinking the lesson was about." (R teacher.)

Using the constant can help to establish the relationship with counting and "one more than"/"one less than". The teachers felt that they could work quickly on such ideas, as the children enjoyed themselves. The teachers all felt that they were making a much more conscious effort to work on language each time they decided to do mathematics using calculators with a class or group.

4. Creativity and exploration

Many of the activities already described demonstrate the opportunities for creativity and exploration stimulated by calculator use. The child who demonstrated "55555", like the rest of his class were working on *the answer is 25* (we had just done an assembly for the school's 25th anniversary), *what was the question.*

Some of the class produced lists such as 1+24=25, 2+23=25, 3+22=25, ... or 26-1=25, 27-2=25, ... others dotted around finding results such 6+6+6+6+1=25, or 12x2+1=25 or ...

Open tasks such as these and the CAN "numbers at the corners of a square", (Shuard et al., 1991, p22) became great favourites because everyone could begin and each child could work at his or her own level.

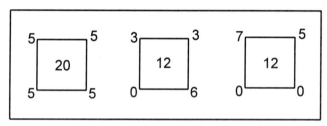

figure 2

Teachers began to ask children *how* they were working on a problem and found that there were many idiosyncratic methods. Even when working on apparently straightforward tasks, children interpreted the work to suit themselves. When I interviewed six children from one school (Perks 1990), they explained to me the rules of the games Bunny Hops and Teddy Bears Picnic (Mathematical Association, 1987). Two children play with a different number route each. I wrote on a piece of paper,

$$4 \rightarrow 6 \rightarrow 3$$

in an attempt to reflect the description of a number route given by one pair of children. The idea is to use the operations and numbers on the calculator to move from one number to the next without using the cancel button.

Working on the problem above, the children were all quite happy to enter four onto the calculator and then press + 2 =
The strategies for getting from the 6 to the 3, however, varied enormously from:

$2 + 1 =$ (using the fact that the calculator display can be overwritten after the use of the equals sign)

through $- 3 =$

to $- 6 = $ *nought!* Now $+ 3 =$

The teachers, from the pilot project, who had used this activity felt that this was very useful for the concept of difference (Coates 1989). For some children their working appeared to be closer to the empty box problem, eg

$$4 + \Box \ = \ 6$$

although the children had to identify the operation as well. For others finding a solution was closer to problems of the form "the answer is 3, what was the question?"

The games also became popular in the North area with children making up snakes of numbers for other pupils to play with. However they did not limit themselves to the one digit numbers of the original games:

> One snake had 37 -> 89 -> 100 -> 150 -> 180 and even more. I said it was too hard so two or three of them showed me what to do, managing to do each move in no more than two attempts. They set each other much harder problems than I would dare! (Y2 teacher.)

There were many examples of children working hard to check ideas; an odd number plus an odd number, adding two two-figure numbers will give you two-, three- and four-figure numbers, you can count to 36 in twos, threes, fours,.... The children shared their ideas, they talked about what they were doing, they experimented, studied their results and asked themselves lots of questions.

> The presence of the calculator not only provides children with the opportunity to engage in mathematical investigation, but also enables them to share their discoveries with teachers and other children by providing an object which can become the focus for genuine mathematical discussion. (Groves, 1994, p39)

Conclusion

Those teachers who worked with calculators regularly felt that a significant extra dimension to aid their pupils' learning had been made available. Children could use much larger numbers, they could approach activities which allowed them to predict and test. They could explore, create their own problems and make discoveries. Information about the operations no longer had to be limited to simplified incorrect ideas, but could be investigated using decimal and negative number. Pupils using calculators showed their mathematics more readily, they talked about their ideas and were prepared to experiment. The teachers felt that they had often underestimated what pupils could do.

> The calculator lifts teacher-imposed barriers. (Coates, 1989, p36).

The teachers described here shared a common view, they were prepared to commit their classes to using calculators. They demonstrated their enthusiasm by meeting regularly and sharing their ideas. Without the giving of their time to devise and discuss activities and their willingness to offer open activities to their pupils the richness of the children's work would not have been possible. The excitement which was generated in meetings was because of the mathematical discoveries their young pupils were making, that pupils were exploring and learning in a way not possible without the calculator.

The teachers tried hard to spread their enthusiasm amongst other teachers. Certain ideas transferred to other classes but generally the idea of working with calculators regularly did not. Those teachers, who met regularly, organised meetings and discussions in their schools, but their belief that open access to calculators was valuable was not an idea which gained much credence amongst others. There was a feeling that many staff believed that children had to work in a traditional way with number or they would not understand the work. Evidence from other classes was seen as accidental and a fluke, rather than perhaps being indicative of further possibilities. Those who worked on devising and using different activities became more and more convinced about the value of using calculators, but found themselves faced with school policies which imposed the regular use of a published scheme offering number work at a far less challenging level.

Despite major projects such as CAN and smaller projects which have taken place in various areas of the country, both in infant and junior schools, there is little evidence of calculators being used to explore mathematics and help pupils learn in a large number of classrooms. There is also a strong myth that calculators prevent children from learning "the basics" of arithmetic:

> There has been much debate and criticism of the use of calculators in primary mathematics and the impact this has had on pupils' skills and understanding. (Office of Standards for Education, 1994, p21)

The report offers little extra on the debate other than:

> It is the view of HMI ... that calculators are not over-used in schools but that the teaching of their proper use is not yet covered sufficiently. (ibid)

There is evidence that calculators can help as a bridge between teaching and learning mathematics with young children but unless teaching takes full advantage of their power they will remain peripheral to learning.

Notes

1   Compulsory schooling in England and Wales begins in the academic year in which a child is five-years-old, children are placed in a reception class (R), Y1 denotes the second year of schooling, the academic year when a child becomes six years old and Y2 the year in which the child will be seven.

2   The levels describe achievement at the ages of 7, 11 and 14. Level 2 is described as the level which will be achieved by the average seven-year old, level 3 by the above average seven-year-old.

References

Coates, D (ed) 1989, *Central Area Calculator Project, Report from Pilot Schools 1988-89.* City of Birmingham Education Authority.

County of Durham 1981, *A Report: Using Calculators with Upper Junior Children.* County of Durham.

Department of Education and Science, 1991, *Mathematics in the National Curriculum.* London: HMSO.

Fielker D, 1976, 'Using electronic calculators,' *Mathematics Teaching,* 76, 14-19.

Groves, S., Ferres, G., Bergfeld S. & Salter S., '1990 Calculators in the infant school: The Victoria College Calculator Project' in Clements, M.A. (Ken) (ed) *Whither Mathematics?* Mathematical Association of Victoria, Twenty-seventh Annual Conference, 244-250.

Groves, S. 1994, 'The effect of calculator use on third and fourth graders' computation and choice of calculating device' in da Ponte J.P. & Matos J.F. (eds), *Proceedings of the Eighteenth International Conference for the Psychlogy of Mathematics Education,*, 3, 33-40, Lisbon.

Johnson, D.C., 1978 'Calculators: abuses and uses', *Mathematics Teaching,* 85, 50-56.

Mathematical Association 1987, *Integrating Calculators into the Primary Classroom,* Leicester.

Office for Standards in Education 1994, *Science and Mathematics in Schools: A Review,* London: HMSO.

Perks, P., 1990, *Calculators and the Six-Year-Old.* unpublished M.Ed. dissertation, University of Birmingham.

Shuard, H., 1988, *The Calculator Aware Number Curriculum,* Lecture to infant teachers, City of Birmingham.

Shuard, H., Walsh, A., Goodwin, J. & Worcester, V. 1991, *Calculators, Children and Mathematics,* London: Simon & Schuster.

10 Visualisation, Confidence and Magic

The Role of Graphic Calculators

Teresa Smart

The availability of graphic calculators has widened access to technology in the mathematics classroom. Of particular interest is to question whether this has benefitted girls. This chapter reports on a case study of a class of 13-year old girls using graphic calculators. The study showed the calculators enabled the girls to develop strong visual representations of the functions given in symbolic form and allowed them to become unusually confident in talking about mathematics. The study also looked at ways of dealing with a major danger in the use of calculators, namely that pupils may accept without question results produced, as if by magic, by the machine.

Introduction

'What do you think the solution of the inequality $7 - 3x^2 < 0$ will look like?' I asked. Ama, who was stretched across the table sharing a graphic calculator with two other girls, looked up and after a while said: 'Well, the graph is upside down because it's minus and it's up there because it's +7 so it will be all that inside the curve.' 'Up there' and the shape of the curve were accompanied by vivid sketches in the air. Like the rest of the girls in her class, Ama employed an unusually high level of visual strategies to solve algebraic problems. This resulted from a term of work with graphic calculators.

Research on both younger, primary school, pupils, and advanced pupils at the end of secondary school shows the importance of wise use of calculators and computers. But little has been done on calculators with pupils of 12 or 13 years, which is an age crucial both for the formation of mathematical ideas, and for the formation of gender divisions. I studied the impact of graphic calculators on schoolgirls of this age, and show that, as expected – and as shown by the example, above – schoolgirls were able to use the calculator to markedly to increase both their visual skills and their confidence with mathematics.

The Study

A class of thirteen-year-old girls (year-9) in Sarah Bonnell School, London, was given free access to graphic calculators and I observed their mathematics lessons for a term. The work was exploratory; I had no single research question, but wanted to see what happens when a class has free access to the graphic calculator. Previous work with small groups had led me to believe that the presence of a visual image on the calculator screen pushes the learner into further investigation, enabling them to talk with confidence about mathematics. I wished to see a similar growth of confidence when working with a whole class.

Research has shown that girls, when working with technology, benefit from being able to share their work (Hoyles et al., 1991, Elkjaer, 1992), so we organised the class so that each pair of girls had to share a graphic calculator. In practice, we found they did more than simply share the calculators; they collaborated in their planning and were happy to share their findings with their small group and also, with a teacher's encouragement, with the whole class.

Graphic calculators as a catalyst for change

There has been no large scale survey of the use of calculators in the secondary (11-16 years) school. However, there is substantial research for primary age pupils. The Calculator Aware Number (CAN) Curriculum Project found that:

> [T]he calculator operated as a catalyst for change. It provided the means to offer pupils the chance to explore mathematical ideas and play an active part in their own learning. The pupils developed confidence in tackling new ideas. They did not

appear to fear making mistakes and their endeavours as problem-solvers and constructors of meaning raised teachers' expectations of them. (Walsh, 1992, p. 56).

In the CAN project, 4000 pupils between the ages of 6 and 11 were given access to calculators whenever they wanted. What was innovative was the aim to develop a number curriculum around the use of the calculator. During the life of the project, the curriculum was to be developed without the use of textbook schemes and 'no teaching of the standard formal algorithms used for number computation would take place during the three years.' (Walsh, 1992, p. 53).

Pupils made clear mathematical gains, particularly as they were able to develop and use their own methods. 'They are learning very much as one learns in life and are building up their knowledge in a way which allows for change and development.' (Walsh, 1988, p. 21) The changes that came about were not only due to the availability of calculators, but also came out of the teachers' active involvement in the development project.

In the U.K., there has not been a similar large scale project investigating the role of calculators in the secondary (11-16 years) classroom. In fact, primary teachers involved in the CAN project feared their good work would be undone in the secondary school and felt the need to teach their children the standard algorithms and methods when they reached the top of primary school, because this was 'part of the expectations of the receiving secondary schools'. (Duffin, 1991, p. 42).

There has been a major study of older students and calculators. In a project funded by The National Council for Education Technology (NCET) to investigate the use of the more complex graphic calculator in the Advanced–level (post-16 years, A-level) mathematics classroom, it was found that 'the impact on the project students of unrestricted access to graphic calculators, either as a personal or classroom resource, was impressive.' (Ruthven, 1992, p. 5). A research study undertaken as part of the project showed that students who had had access to graphic calculators as part of their mathematics lessons did significantly better than students who had had no access to the calculator. The graphic calculator provided the student with experience moving back and forth between the symbolic/algebraic and the graphical/visual form of a graph, encouraging 'both students and teachers to make more use of graphic approaches in solving problems and developing new mathematical ideas.' (Ruthven, 1990, p. 447). Although these are the findings of a project in A-level mathematics, the effect is general: 'More than perhaps any other early mathematics topic, technology dramatically affects the teaching and learning of functions and graphs.' (Leinhardt et al., 1990, p. 7). However, the arrival of the graphic calculator has provided the opportunity to develop a secondary school curriculum that is calculator-aware. The graphic calculator is a powerful tool for learning mathematics; it has many of the facilities of a computer but is small and cheap enough to become personal to the user in the classroom. It can be picked up and used whenever required. And it has been noted that 'the availability and access to graphics software for the secondary mathematics curriculum has provided two powerful learning modalities for students: visualisations and investigations.' (Duren, 1990-91, p. 23). The role of calculators in the 11–16 mathematics classroom was the focus of this study.

Ways of working with the class

The pupils I observed belonged to a popular girls' school in the east end of London. The school had high expectations of the pupils; however in mathematics the examination results at 16 were below average. A new head of department was appointed in 1993 with the task of raising pupils' achievement in mathematics. One area that the department explored was the integration of technology into the teaching and learning of mathematics. A successful bid was made to British Telecom and 35 graphic calculators were bought with the funds. It was hoped that every pupil in the school would have an opportunity to use the calculator and this use would be monitored.

A decision was made to undertake a detailed study of one class, and I observed four out of their five maths lessons every week for a thirteen-week term. Using this class I wished to explore how the graphic calculator could be integrated into the mathematics national curriculum and what effect it would have on the curriculum and the pupils' attitudes to mathematics. Through this process of investigation and visualisation the class teacher and I hoped the pupils would gain confidence with a wider range of mathematics content not normally considered part of their curriculum.

The class had access to 15 calculators (one between two pupils) for every mathematics lesson. The girls in the case study class were a high achieving group. They all liked mathematics, although none said mathematics was their favourite subject. Their teacher was an experienced mathematics teacher who had given up her post as a deputy head of a school and returned to the classroom wanting to put her time and energy into teaching. Her vision of the mathematics classroom was one where students were all involved, self-directed explorers who were enjoying their study of mathematics. To achieve this type of classroom both she and the students would need to be involved in change. She hoped that the graphic calculator would act as the catalyst for change.

The teacher wanted the girls to change the way they viewed mathematics and mathematics learning – to be more actively involved in setting their own goals, in discovering the rules and in talking and arguing about what they had discovered. She hoped the graphic calculators and my presence with the calculators would help promote this change.

Before the project started we discussed our respective roles. Chris (the teacher) was clear that she wished to maintain her role as the teacher who planned and directed each lesson. She welcomed me as an observer of the process but also as an 'experienced' graphic calculator user. The class followed a text book scheme with other resources brought in to provide the pupils with a greater range of practical and investigational work. The teacher recognised that during the time of the project a certain number of chapters from the text would need to be completed in order for the class to keep up with the parallel year-9 group.

At first, the class followed through the book, using the calculator when needed or useful. As time went on, their familiarity with graphs and the ease of drawing and checking them led the students to mathematical discovery aside from what was expected by the text. The girls became more confident and they could talk about their discoveries to the whole class. I give two examples of their work below.

Later it was decided to use the class time for discovery with the calculator and discussion about the findings. The students completed the chapters and exercises at home and in the odd catch-up lesson. This way of working meant that at times the teacher was not directing the development of the lesson and I was no longer the graphic calculator expert. We found that our students developed in the same way as those in the primary classroom described by Angela Walsh (1992), who said:

> The pupils received encouragement to explore and develop their own ideas. To search for pattern and generalisation, to develop their powers of mental abstractions and algebraic thinking and to build up mathematical structures from within their own learning and thus construct meaning from their experiments. (p. 55)

Special problems for girls?

Previous work with graphic calculators in the secondary classroom had shown that girls benefit, in part, because they value the opportunity to use a more personal form of technology (see Smart 1992). Having free access to the calculator allows the student to make mistakes in private 'away from the ridicule of the boy computer experts' (p. 42). Also, Ruthven (1990) says that having the facility to try out different ideas and methods leaves the student feeling much more confident with new, unknown problems: 'the use of feedback from a graphic calculator can reduce uncertainty and thus diminish anxiety.' (p. 448) Indeed, the NCET project showed that, contrary to expectations, girls in the graphic calculator group outperformed boys in the group. (see Ruthven, 1992)

Nevertheless, the class teacher and I were aware that there are also significant problems associated with the integration of technology into the curriculum, and that this can have a particularly detrimental effect on the progress of girls (see Culley, 1988; Hoyles, 1988). We needed to think carefully about how the girls met the calculator and got involved in the project.

Unlike the simple 4-operations calculator in use in primary schools, for example in the CAN project, it is not possible to pick up a graphic calculator and find out how it works just by pressing buttons. Alan Graham (1992) noted that when students were given more complex calculators with no help but only the manual, they 'were noticeably less enthusiastic about the calculator. They were critical of its idiosyncratic logic, less able to exploit its special features and tended to fall back on their trusty Casio FX machines.' (p. 26) We wanted the pupils to be enthusiastic over their access to technology. To ensure this, we provided them with 'friendly' materials written specially for the graphic calculator and encouraged the girls to familiarise themselves with the calculators at their own pace, picking up the calculators when they wanted to.

We urged them to work collaboratively and they were expected to discuss any interesting results and patterns with the rest of the class. We ensured that the girls had to share one calculator between two. This would be beneficial to their mathematics learning because discussion can lead pupils towards 'generalisation and abstractions – processes at the heart of mathematics.' (Hoyles et al., 1991, p. 163). Also the mounting evidence shows that girls benefit from group work around a computer (see Johnson et al, 1985; Underwood and McCaffrey, 1990; Underwood, Underwood et al., 1993; Underwood, 1994; Underwood and Jindal, 1994). As Geoffrey Underwood (1994) notes:

> . . .regardless of instructions, the girls tended to work well together and to discuss the task. They clearly enjoyed collaborating, and the performance of any two girls was generally superior to that of any other pair. When given non-collaboration instructions they still co-operated. (p. 12)

For many of the girls in this class, English is not their first language so the encouragement of discussion and collaboration is a central aspect of education in this school. The graphic calculators were seen as providing an extra stimulus for collaboration and discussion.

> A central way of achieving understanding of mathematics is by talking, reading and writing about it. In order to do this we must provide students with the appropriate mathematics vocabulary and the appropriate stimuli for the use of language to take place. (Cox et al., 1993, p.9)

Findings

As well as observing the class during their mathematics lessons I collected other data: I interviewed some of the students; they all handed in a written report of a project on work with parabolas; they filled in short questionnaires; and at the end of the term I made a video of the girls in a series of small-group discussions talking about their work and their feelings on using the new technology. Below I report on the findings of the study using the words of the girls.

Visualisation

The motivation behind the use of graphic calculators in the classroom often is to help students develop a link between the symbolic/algebraic form and the visual/graphical form. But Cyril Julie (1993), who has worked a great deal with graphic calculators, warns that these two sides of the link are not seen as equal. The graphic/visual is viewed 'as a sort of bolstering mechanism for the symbolic/algebraic and so devalues the graphic/visual'. (p. 347) He proposes that 'graphical work and the graphical/visual strategies developed to solve problems graphically in the graphic calculator environment to do mathematics must be accepted as legitimate mathematical activity and should not be placed in positions subservient to other approaches and strategies, especially the symbolic/algebraic.' (p. 345)

I agree. Two examples below show how pupils in this class used the graphic calculators to build a more fundamental visual image of equations given in symbolic form. Then, in preference to manipulation of symbols, they employed their visual knowledge to help make generalisations and solve any new problems. In doing so, they extended their mathematics beyond what was expected by the teacher and the textbook.

Straight or curved lines?

Early in the study the class had become familiar with the graph plotting facility by exploring any graph. They worked in groups and each made a display in the classroom of what they had drawn. At the end of this time they were at ease with pressing the right buttons in order to obtain graphs. But in doing so they had come across functions they had never met before. Instead of continuing, the teacher spent part of the lesson getting the class to see if they could classify the graphs into families. She asked: What was the same and what was different about the graphs? A first answer was that some are straight line graphs and others curved line graphs.

As part of my study, I taped some of the discussions. Yetunde and Michelle were working together, sharing a calculator, and trying to describe the graphs:

> Michelle: $y=2x$ is straight because it is a simple formula like $3x$ and $8x$,
>
> Yetunde: Umm, yes but when the number is square then it will be a curve.

Yetunde and Michelle then tried on the calculator the graphs of $y=x^2$, $2x^2$, $3x^2$, x^2+4, x^2+8, x^2-4, etc. and noticed that all the graphs were curved.

> Yetunde: what shall we do next? x^3?

They drew the graph of $y = x^3$ on the calculator.

> Yetunde: So x^3 has got 2 curves, lets try x^4, I think a simple x will be a straight line and x^2 will be one curve and then x^3 will have 2 curves, it curves double and so x^4 will have 3 curves.

They tried.

> Michelle: It has one curve
>
> Yetunde: Maybe it is even numbers, shall we try x^5?

Yetunde: Yes it is - it is a double curve.

They continued to explore, and when asked to share what they had found with the rest of the class:

Michelle: When it [the power] gets larger then the curve will get smaller, that is, steeper with a larger number. .... and if the numbers are even then it is like a single curve and if the numbers are odd then it is like a double curve

They then went on to test this conjecture with higher numbers - x^{10}, x^{15}, x^{20}

At the end of the lesson, these two girls (and, as I found later, several others) had a good visual image of the graphs of $y=x^n$ developed out of the request to investigate which graphs are straight and which curved.

Curve fitting

Through use of the calculators the girls were gaining a good visual image of the effect of varying the values of *a* or *n* in the equation $y=ax^n$. This provided them with a useful tool in a later session when they were working on proportionality. The aim of a particular exercise was for the students to decide whether two quantities were in direct proportion by plotting their graph. Using the graphic calculator, it was straightforward to see if the relationship was linear by plotting the data and checking if a straight line could be drawn through it.

As a more challenging exercise, the girls were given data that did not lie on a straight line. By then they were so familiar with the calculator that this did not worry them. Lina was investigating data that was not linear ($y = mx$) but of the form $y = m \tan x$. The students had not yet met tangents, so would not be able to find a curve to fit the data. However, Lina used her graphic calculator to plot the data and after some time thought she had found a graph to fit the data. I had watched her exploration and asked her to explain to the class. Lina did not like talking to the class so her explanation was very brief, but I had seen her getting to an answer by a method of trial and improvement – often visualising in advance of drawing what the effect of a constant would have on the graph.

Lina told the class:

> Well I thought it was x^2. [$y=x^2$] but it wasn't and I knew it wasn't x^3 and x^4 because they are too steep. I knew it can't be plus or minus anything [$y=x^2+b$] because then it would move up or down the screen and I didn't want that. So I tried multiply [$y=ax^2$, $a>1$] but that got far away from what I wanted so it only left divide [$y=ax^2$, $a<1$]. So I tried to divide by 2 and 4 and 10 but that was not getting me anywhere so I gave up. But then I tried a big number and that was better. Finally, the graph of $y=x^2/43$ almost fits the data quite closely.

Lina already had a growing visual sense of graphs and the effect of changing the value of the constants a and b in $y=ax^2+b$. At times she changed a without much thought but normally she made use of the feedback from the graph and moderated her guess, becoming more refined, just as Kenneth Ruthven (1992) says about the process of trial and improvement: 'As the user becomes increasingly familiar with a particular type of mathematical situation, such judgements tend to become more sophisticated.' (p. 6)

Discussion and confidence

This process of investigation and discussion led the girls to an unusually high level of confidence about their work, which showed itself in an exceptional willingness to talk about it. During the term we asked pupils to develop an extended piece of project work on graphs. Their task was to explore parabolas, and they could use graph paper and/or a graphic calculator. They had to write and hand in a report of their explorations. Each girl produced a report of their investigation, giving sketches of the graphs, any rules they had found and how they had tested the rules. Two of the pupils tried to justify and prove their rules. All the pupils had been stretched and some – through using the calculator – were working confidently at mathematics normally only introduced in an Advanced-level mathematics class. The reports they handed in showed that not only had they seen patterns on their calculator screens, they were able to write coherently about the mathematics they had developed. Their ideas remained in hard copy after the screens had been cleared.

Finally, to end my participation in the project, I asked the girls to talk about the mathematics they had found, the predictions they had made that had or had not worked and also their feelings about using the calculator. I videoed the discussions of groups of six girls. On a table was a set of cards, each carrying a different question. A few examples of these questions were:

- What predictions did you make? Tell the rest of the group about two of them
- What is the same and what is different about the graphs of the following functions?
  $y = (x+2)^2 \quad y = (x-2)^2 \quad y = (x+6)^2 \quad y = (x-5)^2$
- Did you make a plan before starting your work? How did the plan help?
- Did you find the graphic calculator useful in your maths lessons? Can you explain why?

Each girl was asked to pick up a card and answer the question. They could refer to their written report on the parabolas. When they had finished, the question was put out for open debate. Ten weeks after meeting the graphic calculator for the first time, they were able to talk coherently and fluently about their mathematical development – particularly about the predictions they had made and the misconceptions they had held. They referred to their reports and traced diagrams on the desk or scraps of paper to illustrate their points.

Several girls explained how they 'saw' that 'a minus and a minus is a plus'. As Shibley explained: 'I did these two [graphs] and I discovered that if you take $y = x^2+b$ and $y=x^2$ minus a negative number, it is the same'. I asked her if she was expecting this – what did she predict the graph of $y=x^2-- 6$ to look like? 'I thought the whole [positive] numbers were supposed to be at the top of the screen. I thought the minus was going to be at the bottom. After I realised that a minus and a minus is a plus.'

As a final question for each group they were asked to explain how they would investigate the graph of $y=ax^3 + b$ and what they would expect to find. Here Yetunde, Michelle and Shibley talk about it.

> Shibley: Well, do the same as what we have already done. Make a prediction – what you think the graph will turn out to be – and then try to learn how to do cubed on the calculator.
>
> Yetunde: Well, like I know, with x^2 a curve just like one curve and with x^3 it is a double curve like an S so I know what it looks like.
>
> Michelle: Well we would know what to do because we have got a plan here – it's the same. Well with $ax^2 + b$, the a makes the graph steeper and the b moves the graph up and down so we would try that with the S curve.

The graphic calculator enabled these students to become confident that they could tackle a new problem. And they recognised the possibilities of the calculator as a means to make and test out predictions.

The 'magic' effect of calculators

Although the graphic calculators clearly increased the pupils' confidence, they also created a certain disquiet. In Lina's case, the finding of a curve to fit the data was only possible through the use of the calculator. Yet she felt that 'I probably shouldn't use the calculator because I am not sure if I believe the result is right. It's too easy, I didn't get the answer myself.'

Initially, the students accepted without questioning the picture on the screen of the calculator, as Leinhardt et al. (1990) warned:

> In computer-based instruction, ... the graph the machine produces is unquestionable. A teacher should be aware of the 'magic' effect this may have on students. Students may develop too strong a reliance on the machine. For example, they may not understand the underlying patterns and principles that drive the production of the graphs. (p. 7)

The 'magic' element of the calculator led to the students developing misconceptions about the graphs of certain functions. Goldenberg (1988) talks about the 'ambiguities of graphs' that 'appear to arise from processes akin to well known visual illusions' (p. 142). For example, many of the girls believed, as Lina said: 'the graph of $y = x^2 + 6$ is much smaller and a different shape from the graph of $y = x^2 - 6$'. This was because they could see more of one graph than the other on their calculator screen.

We found that this 'magic' production of graphs on the calculator led to many of the girls accepting what they saw on the screen as the whole graph rather than a window onto part of the graph. This caused problems in describing the graph of $y=(x-8)^2$ when the range of the graph screen is x taking values from -8 to +8. What appears on the calculator screen seems to be a different graph from that of $y=x^2$ because only part of it can be seen. When asked to describe the graphs of $y=(x+c)^2$ and $y=(x-c)^2$, Janet said 'well they weren't what I thought they were going to be, because like the graphs before when I had plotted $y=ax^2$ and $y=x^2+b$, they were all the same – they were all U shapes – but with these ones with say $y=(x+4)^2$ the curve was not the same so I couldn't make predictions.' After discussion with the teacher, Janet readjusted her pictures of the graphs of $(x+c)^2$, noting on the video some weeks later: 'Well the graphs are all the same shape but they looked different shapes because we did the axes different and the graphs went off the screen so you know you have to add it on yourself in your mind - you have to imagine moving the y-axis over to get the extra bit of graph.'

Janet succeeded in questioning the picture given by the graphic calculator and hence was able to use it as a useful tool in her mathematics development. Other girls started to fear that working with the calculator, especially in a mathematics examination, was cheating and this led to a lively debate. 'When you use a graphics calculator in a mathematics exam it is not really cheating because you have to understand the method' was one girl's view of the validity of using a calculator. Another was not so sure saying 'yes it is cheating because I think people should be forced to use their brains because the calculator is doing all the thinking for you'.

Many diverse views emerged when they filled in a questionnaire at the end of the project. They were asked if they enjoyed the work and whether they found the graphic calculator useful in their mathematics. The open-ended nature of the work proved uncomfortable for some. Maria wrote on her questionnaire: 'I did enjoy maths in primary school and in years 7 and 8 but now we do open-ended work, I don't.' Yetunde said 'I do like maths now because it is different'. She found exploring, finding and testing out rules for herself easier than doing routine exercises: 'I do get stressed up when I see a page full of complicated sums, it sometimes makes me feel stupid.'

The graphic calculator challenged the pupils and pushed them to reach more advanced levels, often without pupils realising it. Davina thought 'it's too modernised. I would prefer to use an ordinary calculator.' Nadia disagreed: 'it helped me because it makes things easier; I didn't have to draw up a table of values to plot the graphs. Once you knew the method it was really easy, you could zoom into a graph and you could draw lots of graphs.' Janet added that it was easier to see patterns: 'you can put them together and you could compare them and you could see they were the same or different.' Rahima added: 'you could draw the graphs more accurately and you could compare them and trace them around using the cursor.'

But all agreed the graphic calculator took time to get used to. As Maria wrote: 'a graphic calculator is very complicated. But once you get to know what to do, it can help you in statistics, Algebra, standard deviation and doing graphs.'

However, Ama showed the contradictory impact of the calculator most clearly. She said 'we should be using an ordinary calculator; this is all too confusing.' Yet it was Ama who could visualise the graph of $y = 7 - 3x^2$ and easily solve the inequality $7 - 3x^2 < 0$, work she could not have done without the calculator. When I returned to the class five weeks later, all the girls were still using the calculators whatever the views they had expressed at the end of the study.

In this chapter I looked at three issues – visualisation, confidence, and magic – and the important ways in which they are linked. The data collected showed that access to the graphic calculators helped the girls to develop a strong visual representation of functions given in their algebraic form. They were able to picture and describe the graph when given its equation. Rather than attempt to solve a problem by manipulation of the algebra, they employed their growing powers of visualisation. This meant that they were happy solving problems that cause difficulties to older, more mathematically-advanced students who employ only algebraic methods. After just one term, these girls in a deprived area of London showed a remarkable sophistication in their approach to investigating new areas of mathematics.

References

Cox, L. et al. Ed. 1993, *Talking Maths, Talking Languages*, Derby: ATM.

Culley, L. 1988, 'Girls, Boys and Computers.' *Educational Studies* 14(1):

Duffin, J. 1991, 'Written Algorithm.' *Mathematics in Schools*, 20(4), 41-43.

Duren, P. 1990-91, 'Enhancing Inquiry Skills Using Graphics Software', *The Computing Teacher* 18(4), 23-25.

Elkjaer, B. 1992, 'Girls and Information Technology in Denmark - an account of a socially constructed problem', *Gender and Education* , 4(1), 25-40

Goldenberg, E.P. 1988, 'Mathematics, Metaphors, and Human Factors: Mathematical, Technical, and Pedagogical

Representation of Functions', *Journal of Mathematical Behaviour*, 7, 135-173

Graham, A. 1992, 'Too many buttons', *Micromaths*, Spring 1992, 25-27.

Hoyles, C. (Ed.) 1988, *Girls and Computers,* London: Bedford Way Papers, Institute of Education.

Hoyles, C. et al. 1991, 'Children talking in computer environments: New insights into the role of discussion in mathematics learning' in K. Durkin and B. Shire (eds.) *Language in Mathematical Education,* Milton Keynes: Open University Press, 162-175

Johnson, R. T. et al. 1985, 'Effects of co-operative, competitive and individualistic goal structures on computer assisted instruction', *Computers in Education 5 - 13,* Oxford: OUP, 167-182

Julie, C. 1993, 'Graphic calculators: beyond pedagogical and socio-political issues' in C. Julie et al. (eds.) *Political Dimensions of Mathematics Education: Curriculum Reconstruction for Society in Transition,* Cape Town: Maskew Miller Longman. 342-347.

Leinhardt, G., O. et al. 1990, 'Functions, graphs, and graphing: Tasks, learning and teaching', *Review of Education Research,* 60(1), 1-64.

Ruthven, K. 1992, 'Personal Technology in the Classroom', *Graphic Calculators in Advanced Mathematics,* Coventry: NCET.

Ruthven, K. 1990, 'The influence of graphic calculator use on translation from graphic to symbolic forms', *Educational Studies in Mathematics,* 12, 431 - 450.

Smart, T. 1992, 'A Graphic Boost for Girls.' *Micromaths*, Autumn 1992, 41-42

Underwood, G. 1994, 'Collaboration and problem solving: Gender differences and the quality of discussion', *Computer Based Learning*, London: David Fulton Publishers, 9-19.

Underwood, G. and N. Jindal, 1994, 'Gender differences and effects of co-operation in a computer-based language task', *Educational Research*, 36(1), 63-74.

Underwood, G. and M. McCaffrey, 1990, 'Gender differences in a cooperative computer-based language task', *Educational Research*, 32(1), 44-49.

Underwood, G. et al. 1993, 'Children's Thinking During Collaborative Computer-Based Problem Solving.' *Educational Psychology*, 13(3 and 4), 345 -57.

Walsh, A. 1988, 'Letting Go and Hanging On', *Micromaths*, Spring 1988, 20-22.

Walsh, A. 1992, 'The Calculator as a Catalyst for Change', *Early Child Development and Care*, 82, 49-56.

I would like to thank Annie Gammon, Christine Atkinson and the year-9 pupils of Sarah Bonnell School, Newham, London UK.

11 Research Results on the Effect of the Graphic Calculator on Female Students' Cognitive Levels and Visual Thinking

Mary Margaret Shoaf-Grubbs

This paper reports on a study comparing graphic calculator use by an Experimental Group with a Control Group whose elementary algebra course was identical except for the absence of the graphic calculator. The statistics and scattergram analysis support the conclusion that the gain and 'positive momentum' in level of understanding and spatial skills exhibited by the Calculator/Experimental Group was not experienced by the Traditional/Control Group. All subjects were female.

Conceptual Framework

Numerous published articles acknowledge the fact that too many students enter college with weak skills in mathematics, and in particular, elementary algebra. Past investigators have studied the effects of various teaching strategies in the area of algebra, the nature of spatial ability and its relation to mathematical problem-solving ability, the relationship between cognitive level as defined by Piaget and success in mathematics, and the effect of a symbol manipulator in the teaching of mathematics. These studies have primarily investigated students in the middle and high school and not students at the college level nor, in particular, females.

Thurston (1938) has classified spatial-visual aptitude as a "primary mental ability". Researchers have tried to identify factors related to mathematics achievement and understanding in efforts to determine why some students learn and understand mathematics better than others. Because studies have reported positive correlations between spatial skills and mathematics performance (Tartre, 1990a and references therein), spatial skills are of special interest in mathematics education in particular when the development of mathematical concepts is the focus. Tartre (1990a) has further stated 'spatial skill may be a more general indicator of a particular way of organizing thought in which new information is linked to previous knowledge structures to help make sense of the new material.' (p. 227)

Tartre (1990b) further describes a three year study in which male and female subjects were classified as either high or low in both spatial visualisation and verbal skills. Females consistently obtained the lowest scores on the mathematics tests regardless of their classification in either spatial or verbal ability when compared with their male peers. When the groups were offered help in translating verbal information into a pictorial representation, females with low spatial scores required the most help, while females with high spatial scores required the least help. Furthermore, females with high verbal but low spatial skills fell further behind the other groups over the 3 year study.

Fennema (1975) has stated that spatial visualisation is logically related to mathematics. Sherman (1983) found spatial visualisation to be a discriminator more important for girls than boys in predicting the number of mathematics college preparation courses taken in high school. Fennema and Tartre (1985) suggest that low spatial visualisation skills may be more debilitating to girls' mathematical problem-solving skills than to boys'. By 1986 research results were reporting smaller but consistent gender differences favouring males in the cognitive domain of spatial visualisation (Tohidi, 1986). Although more recent meta-analyses no longer find significant gender differences in spatial visualisation (Tartre, 1990b and references therein), spatial visualisation skill remains an important predictive indicator for females within the realm of mathematics. Thus spatial visualisation skills could serve as a means, perhaps more often in the case of females, to identify those students weak in the skills acknowledged to predict success in mathematics (Tartre, 1990a).

Even after considering these meta-analyses one cannot ignore recent investigations supporting the hypothesis that gender differences in spatial visualisation exist and provide evidence that these skills can be learned, improved, and enhanced through training (Ben-Chaim, et al, 1988; Baenninger and Newcombe, 1989). Cartledge (1984) recommended improving females' spatial visualisation skill as one way to overcome the superiority (if, indeed, it exists) of males in mathematics achievement from grades eight through college. A correlation exists between students' participation in spatial activities and their measured spatial ability, with males participating in a greater number of spatial activities (Newcombe, 1983).

Studies in computer technology and mathematics education have confirmed the positive effect of calculators and computers in the classroom (Collis, 1989). Because gender differences in spatial visualisation are considered by some to be a reason for gender differences in mathematical achievement, research indicating improved spatial visualisation with calculator use is particularly encouraging. Vazquez (1991) reported gains in spatial visualisation skills for students using graphic calculators. Dunham (1991) found that pre-test differences favoring males on visual graphic items were not evident on the post-test after instruction with graphic calculators.

Mathematics **can** be learned by all types of students. **But**, a method of teaching or presenting a mathematical concept so that one person can successfully understand and use the mathematics may not work for another concept or another person. Attention has focused on how "mathematics teaching and learning can be improved by developing more powerful approaches to connecting thinking and mathematics" (Silver, et al, 1990, p. v.) New concepts should be taught as extensions of prior mathematics understanding, thereby enabling students to see connections from one concept to another and develop a stronger conceptual understanding during the learning process. Students receive higher scores on a mathematics achievement test when teaching methods include concrete, manipulative materials (Threadgill-Sowder and Juilfs, 1980). Horowitz (1981) supported this deduction and reported that the visualisability of a problem affects its solvability by lower performance subjects.

Technology might be a mechanism through which these goals could be realized. The Calculator and Computer Pre-Calculus Project (C^2PC) (Demana and Waits, 1990) indicates that mathematics instruction with graphic technology can have a positive impact on student achievement and mathematical understanding. Students using graphic utilities attain higher levels of understanding in graphic concepts than those in a traditionally taught mathematics classroom (Taylor, 1990; Flores and McLeod, 1990). Ruthven (1990) reported research conclusions in the area of gender illustrating that:

> ...access to information technology can have an important influence both on the mathematical approaches employed by students and on their mathematical attainment. On the symbolisation items, use of graphic calculators was associated not only with markedly superior attainment by all students, but with greatly enhanced relative attainment on the part of female students. (p. 431)

This study examined spatial visualisation and understanding in mathematics and the role technology played in their enhancement. The emphasis was to investigate the effect of the graphic calculator upon college level female students' cognitive development and spatial visualisation in an effort to devise a means by which more

216

support would be provided to females during their study of mathematics. It was hypothesised that if conceptual understanding could be increased in elementary algebra involving the concept of function, females would be better equipped to study higher mathematics. Harel and Dubinsky (1992) argue that the study of function "is the single most important concept from kindergarten to graduate school and is critical throughout the full range of education" (p. vii). The graphic calculator was seen as a means of enabling female students to develop a stronger understanding of function in elementary algebra by aiding in the development of stronger spatial visualisation skills and furthering abstraction of mathematical concepts. The goal was to create a multi-dimensional, interactive learning environment that would:

1. emphasise making mathematical "connections";
2. focus clearly on the mathematical concepts of the course and teach them more effectively;
3. provide multiple representation of the concepts being studied;
4. nurture and support students as they attempted to become better problem-solvers;
5. encourage students to develop a positive outlook on mathematics in the area of mathematics where too many students have experienced one or more failures during their educational experiences.

Procedures and Design of the Study

The subjects for the study were 37 students enrolled in two elementary college algebra classes at an all-women's, liberal arts college. Prior to registration the researcher randomly labelled the two course sections as Calculator and Traditional. The students randomly enrolled in one of the two class sections after scoring below the pre-calculus level on the college's in-house mathematics placement test. The section chosen by each student was determined by her other scheduled courses for the semester. Students were not aware at the time of registration that they would be participating in a research project. It was only after meeting the class that they were made aware of the research study or which section would be using the graphic calculator. Nineteen were in the Calculator/ Experimental Group using the graphic calculator, and 18 were members of the Traditional/Control Group in which the material was presented in as nearly identically a fashion as possible, but without the graphic calculator. No attrition occurred in either class section and all students took all tests.

The goal of the lesson presentation in both groups was that the topics should be introduced in a manner unfamiliar to the students through a highly visual exploration-of-concepts approach in which the student would be an active participant in the learning process and not simply a passive receiver of information. The only difference in the two research groups was the use of the graphic calculator by the Experimental Group whereas the Control Group used paper-and-pencil and overhead transparencies to draw and study graphs of identical functions. The instructor, lesson plans, and number of graphs examined during the class period were identical. Any discovery by a student in one group that had not been included in the lesson plans was prepared and presented to the other group. Sketches drawn by hand or those rendered by the graphic calculator enabled the instructor to create a highly visual exploration-of-concepts teaching and learning environment. All three methods of concept presentation and problem solution -- numerical, algebraic, and graphical -- were part of the lessons. Students actively engaged in class discussion by stating, re-defining, and/or modifying conjectures about the concepts being examined. This then aided them in drawing conclusions that would support a conjecture. Class dialogue helped the instructor to determine areas where some students encountered difficulty such that more time and explanation could be given to those topics. Because the class size was small and the students actively participated in the discussions, they did not work in groups during class time. However, they were encouraged to work in groups when doing homework.

Pre- and post-tests were administered to determine the amount of growth, if any, experienced by each student and, therefore, each group in the area of a) spatial visualisation, b) level-of-understanding in algebra and graphic concepts, and c) spatial visualisation and level-of-understanding in each of the three main topics taught during the semester course. To ensure further equality, both groups were permitted to use only paper-and-pencil during the pre- and post-testing sessions.

The testing instruments were selected to ensure that they measured the areas in mathematics learning and understanding being investigated in this study. The Paper Folding Test and Card Rotations Test (Ekstrom R.B. et al, 1976) measured general spatial visualisation. McGee (1979) has stated that spatial visualisation involves "the ability to mentally manipulate, rotate, twist, or invert a pictorially presented stimulus object" (p.893). In other words, the subject must mentally move or alter all or part of a representation. This is exactly what is required during the abstraction process when studying elementary functions.

The Chelsea Diagnostic Mathematics Tests for Algebra and Graphs written by the Social Science Research Council Programme 'Concepts in Secondary Mathematics and Science' (CSMS) based at the Centre for Science Education, Chelsea College (now King's College, University of London) measured the level-of-understanding in algebra and graphic concepts. The tests measure and identify a hierarchy of understanding, referred to as level-of-understanding in this paper, connecting many key concepts in secondary school (Brown, Hart, and Kuchemann, 1985). The tests were considered appropriate for this study since many of the topics in algebra and graphs taught in secondary schools are also included in an elementary college algebra course. Because the emphasis of the research during the development of the tests was on the understanding of concepts, the tests contain very few items requiring routine mechanical skills (Hart, Brown, Kerslake, and Kuchemann, 1985). The tests minimized the need for memorised and recalled solution techniques.

While these tests served as the instruments to measure general spatial skills and level-of-understanding, there was also a desire to measure spatial skills and level-of-understanding specific to topics presented during the course. For this purpose, three topics traditionally taught in an elementary college algebra course were identified for further detailed investigation. These topics were:

1. Linear Equations
2. Systems of Equations and Inequalities Involving Functions With Two Unknowns/ Absolute Value Functions
3. Parabolas

These three topics were selected because of the high amount of visualisation required in the presentation, learning, and exploration of the concepts. Modelled after the Chelsea Tests, the aim of the six 'mini' tests was to examine a student's higher-order understanding in each of the specific topics and the accompanying spatial visualisation skills relevant to the topic.

Analysis

All together nineteen sets of data (of which four were aggregates of the others) were examined to determine the effect of the graphic calculator on spatial visualization skills and conceptual level-of-understanding. The data were examined in two fashions. The first was the analysis to obtain the statistical mean for each test gain, i.e., post-test minus pre-test, for each student and class. The t-test statistic was calculated to determine the significance of the differences in the gains between the two groups.

For many types of research, it would seem quite natural to proceed immediately to a statistical analysis of the data, particularly in cases involving pre- and post-tests. In this study statistics alone would not reveal the entire story of the graphic calculator's effect, particularly for each individual student and her 'path' of progress from pre- to post-test. When dealing with small numbers, it is simple to use scattergrams. This second manner of analysis involved what is believed to be a new method of data representation through scattergrams. It was important to visualise and to understand both the **actual** gain made by each individual student on each test, and also the portion of **possible** gain which each student realised. A data analytic method of presentation which allows the simultaneous perceptions of **both** quantities was developed. This new form of scattergram is "packed" with data and information about each student and class while at the same time remaining easy to read and interpret by the reader. A discussion of the scattergram found on the following page for the Card Rotation Test will help clarify this new type of data presentation. In each scattergram the abscissa represented the pre-test and the ordinate the gain score. Each line of slope -1 passing through the point (K, 0) represents a set of (pre-test, gain) scores corresponding to the fixed post-test score K. Thus it is easy to read off the following two pieces of information from each scattergram:

(a)   each of the five students who gained more than 40 points from pre-test to post-test was a student in the Calculator Group;

(b)   of the twelve students who scored between 110 and 160 on the post-test, nine were students in the Calculator Group.

The visual details in the scattergrams add much to the understanding of the statistics for the study.

Findings and Conclusions

All together nineteen sets of data were examined for each of the 37 subjects to determine the effect of the graphic calculator on spatial visualisation skills and level-of-understanding. As listed below, fifteen sets of data were from the individual tests:

Card Rotation Test
Paper Folding Test

Chelsea Algebra Levels #1, #2, #3, and #4
Chelsea Graphs Levels #1, #2, and #3

CARD ROTATION

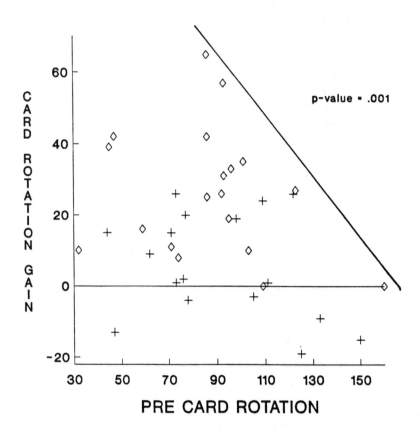

◇ CALCULATOR   + TRADITIONAL

CG CARD ROTATION MEAN GAIN ▪ 26.11
TG CARD ROTATION MEAN GAIN ▪ 5.59

Figure 1.

Spatial Visualisation (SV) for the 3 course topics
Level-of-Understanding (LOU) for the 3 course
topics

In addition four aggregated test combinations were examined as follows:

Chelsea Algebra Levels #1 + #2 + #3 + #4 =
Chelsea Algebra Total
Chelsea Graphs Levels #1 + #2 + #3 =
Chelsea Graphs Total
Spatial Visualisation for the 3 course topics =
SV Total
Level-of-Understanding for the 3 course topics =
LOU Total

The gains for each of the individual and aggregated tests were examined for significance using the t-test. These results are given in Tables 1 and 2 along with pre- and post-test means and standard deviations. Ten of the nineteen tests showed significance at the .05 or lower level in favor of gains made by the Calculator Group. There were three cases in which the Traditional Group had higher gains than did the Calculator Group. None of these gains was significant.

Table 2 gives a summary of the statistics for the Chelsea Algebra Total, Chelsea Graphs Total, Level-of-Understanding Total, and Spatial Visualisation Total. Because the number of test items was small for the individual tests, their aggregates were examined for a broader indication of patterns within each group with regard to pre-test, post-test, and gains scores. The statistical analysis supports the conclusion that the graphic calculator does have a positive and significant influence on the performance of female students. These results imply that spatial visualisation and mathematical understanding are strengthened through the use of the graphic calculator. The research indicates that the graphic calculator's visual representation of the mathematical concepts is important as both a heuristic and pedagogic tool---particularly for those weaker student in mathematics.

As with any new mode of instruction and teaching, time and patience is needed on the part of the instructor in integrating technology into the mathematics classroom. Teachers can no longer have complete control of what happens during class time, because the students **and** the teacher are examining and exploring mathematics in a completely new manner. Teachers themselves are learning and thinking about mathematics differently due to the

Table 1

Descriptive Statistics the Fifteen Individual Tests

Test	pretest	posttest	p-value gain	t-test
Card Rotation (max=160)				.001
Calculator (n=19)	86.9(29.25)	113 (31.08)	26.10	
Traditional(n=18)	91.4(30.57)	97 (30.04)	5.60	
Paper Folding (max=20)				.035
Calculator	7.40(2.93)	10.40(3.96)	3.00	
Traditional	9.60(3.38)	10.80(4.47)	1.20	
Chelsea Algebra Level 1 (max=6)				N.S.
Calculator	5.68(0.48)	5.74(0.65)	.06	
Traditional	5.89(0.32)	5.83(0.38)	-.06	
Chelsea Algebra Level 2 (max=7)				N.S.
Calculator	5.68(1.08)	6.05(1.31)	.37	
Traditional	6.00(1.14)	6.28(1.23)	.28	
Chelsea Algebra Level 3 (max=8)				N.S.
Calculator	4.21(1.93)	4.79(2.21)	.58	
Traditional	4.44(2.36)	5.78(2.37)	1.34	
Chelsea Algebra Level 4 (max=9)				N.S.
Calculator	3.26(2.33)	4.00(2.19)	.74	
Traditional	3.61(2.48)	3.78(2.34)	.17	
Chelsea Graphs Level 1 (max=7)				N.S.
Calculator	6.47(0.84)	6.58(0.61)	.11	
Traditional	6.00(1.28)	6.28(1.18)	.28	
Chelsea Graphs Level 2 (max=6)				001

Test	pretest	posttest	gain	t-test
Calculator	3.84(1.61)	5.53(0.84)	1.69	
Traditional	4.39(1.88)	4.50(2.07)	.11	
Chelsea Graphs Level 3 (max=11)				.015
Calculator	4.74(4.05)	8.68(3.16)	3.94	
Traditional	6.78(3.56)	7.89(3.50)	1.11	
LOU Linear Equations (max=13)				.050
Calculator	5.26(2.38)	9.53(3.36)	4.27	
Traditional	7.18(3.45)	9.94(3.69)	2.76	

			p-value	
Test	pretest	posttest	gain	t-test
LOU for Systems (max=11)				N.S.
Calculator	3.47(2.86)	5.90(2.96)	2.43	
Traditional	4.56(3.31)	5.33(3.60)	.77	
LOU for Parabolas (max=9)				N.S.
Calculator	1.84(1.84)	6.68(1.83)	4.84	
Traditional	2.17(2.28)	6.44(3.00)	4.27	
SV for Linear Equations (max=13)				.010
Calculator	6.53(2.32)	9.95(2.59)	3.42	
Traditional	8.22(3.04)	9.61(3.52)	1.39	
SV for Systems (max=8)				N.S.
Calculator	2.84(2.65)	3.53(2.01)	.69	
Traditional	3.06(2.16)	4.78(3.21)	1.72	
SV for Parabolas (max=10)				.005
Calculator	3.11(3.04)	8.47(1.51)	5.36	
Traditional	2.50(2.12)	5.17(2.38)	2.67	

Table 2
Descriptive Statistics For Four Aggregated Test Combinations

			p-value	
Test	pretest	posttest	gain	t-test
Chelsea Algebra Total (max=30)				N.S.
Calculator	18.84(4.54)	20.58(5.12)	1.74	
Traditional	19.94(5.56)	21.67(5.17)	1.73	
Chelsea Graphs Total (max=24)				.002
Calculator	15.05(5.66)	20.79(3.74)	5.74	
Traditional	17.17(5.52)	18.67(5.43)	1.50	
Level-of-Understanding Total (max=33)				.030

Calculator	10.58(6.28)	22.11(6.61)	11.53
Traditional	13.89(7.80)	21.72(8.90)	7.83
Spatial Visualization Total (max=31)			.023
Calculator	12.47(6.23)	21.95(4.70)	9.48
Traditional	13.78(6.13)	19.56(8.30)	5.78

graphic calculator, and, therefore, cannot always anticipate questions, insights, or misconceptions that students may have. Teachers who do not want to relinquish 'control' in the classroom or have difficulty with the fact that they 'may not know it all' could find themselves hesitant to bring technology into their classrooms. One might think of this new classroom environment as one in which the teachers takes on the role of 'coach' and the students are the 'investigators/explorers'. It also takes time and planning to develop effective lessons with technology, especially if the textbook being used does not incorporate technology. The responsibility for integration then falls on the shoulders of the teacher and is time consuming. Furthermore, lessons need to be revised as students and teachers learn from what happens in the classroom.

Computer Algebra Systems (CAS) have been used in the mathematics classrooms also and have been found to be effective. However, there are drawbacks to CAS. Hardware is needed and, often, no one software package does all the mathematics that the teacher desires in the manner that the teacher considers best for her students. In addition, computers may not be available to students after school and computer lab hours. The graphic calculator, however, is readily available at all times and is small enough to be conveniently carried anywhere with the student. Spreadsheets would offer strong mathematical abilities, but not the ability to graph functions.

Spatial visualization, or, if preferred, visual thinking should be nurtured and developed in students. It plays a significantly important role in the development of mathematical reasoning. Furthermore, this skill promotes the creative and insightful use of mathematical concepts that can transfer to other areas of learning. Through this visual, self-paced exploration process created through the use of the graphic calculator the student is able to use concrete imagery to move toward a higher level of understanding or abstraction. Through the use of the graphic calculator students are more likely to construct their own mathematical understanding through conscious reflection. The normally passive student becomes actively involved in the discovery and understanding process, no longer viewing mathematics as simply the receiving and remembering of algorithms and methods of solution.

References

Baenninger, M. & Newcombe, N. 1989. The role of experience in spatial test performance: A meta-analysis. *Sex-Roles*, 20, 327-344.

Ben-Chaim, D., Lappan, G., & Houang, R. 1988. The effect of instruction on spatial visualization skills of middle school boys and girls. *American Educational Research Journal*, 25, 51-71.

Brown, M., Hart, K., & Kuchemann, D. 1985. *Chelsea diagnostic mathematics tests*. Windsor, Berkshire, England: NFER-NELSON, Publishing Company.

Cartledge, C.M. 1984. *Improving Female Mathematics Achievement*. (ERIC Document Reproduction Services No. ED 250 198).

Collis, B. (1989. Research retrospective. *The Computing Teacher*, 16(9), 5-7.

Demana, F., & Waits, B.K. 1990. The Ohio State University Calculator and Computer Precalculus Project: The mathematics of tomorrow today! *The AMATYC Review*, 10(1), 46-55.

Dunham, P.H. 1991. Mathematical confidence and performance in technology enhanced precalculus: Gender-related differences.

(Doctoral dissertation, The Ohio State University, 1990). *DissertationAbstracts International*, 51, 3353A.

Ekstrom, R.B., French, J.W., & Harmon, H.H. 1976. *Manual for kit of factor-referenced cognitive tests*. Princeton, NJ: Educational Testing Service.

Fennema, E. 1975. Spatial ability, mathematics, and the sexes. In E. Fennema (Ed.), *Mathematics learning: What research says about sex differences* (pp. 33-44). Columbus, OH: ERIC Clearinghouse for Science, Mathematics, and Environmental Education.

Fennema, E. & Tartre, L. 1985. The use of spatial visualization in girls and boys. *Journal for Research in Mathematics Education*, 16, 184-206.

Flores, A., & McLeod, D. 1990. *Calculus for middle school teachers using computers and graphic calculators*. Paper presented at the Third Annual Conference on Technology in Collegiate Mathematics, Columbus, OH, November 1990.

Harel, G., & Dubinsky, E. 1992. In G. Harel & E. Dubinsky (Ed.) *The concept of function--Aspects of epistemology and pedagogy*. Mathematical Association of America Notes, Volume 25, Washington, D.C.

Hart, K., Brown, M., Kerslake, D., Kuchemann, D., & Ruddock, J. 1985. *Children's understanding of mathematical concepts*. London: John Murray.

Horowitz, L. 1981. Visualization and arithmetic problem-solving. Paper presented at the American Educational Research Association Annual Meeting.

Hyde, J.S., E. Fennema, & S. Lamon 1990. Gender differences in mathematics performance: A meta-analysis. *Psychological Bulletin*, 107, 139-155.

McGee, M. G. 1979. *Human spatial abilities: Sources of sex differences*. New York: Praeger Publishers.

Newcombe, N. 1983. Sex differences in spatial ability and spatial activities. *Sex-Roles: A Journal of Research*. 9, 377-386.
Ruthven, K. 1990. The influence of graphic calculator use on translation from graphic to symbolic forms. *Educational Studies in Mathematics*, 21, 431-450.

Sherman, J. 1983. Factors predicting girls' and boys'   enrollment in college preparatory mathematics. *Psychology of Women Quarterly*, 7(3), 272-281.

Silver, E.A., Kilpatrick, J., & Schlesinger, B. 1990. *Thinking through mathematics*. New York: College Entrance Examination Board.

Tartre, L. A. 1990a. Spatial orientation skill and mathematical problem solving. *Journal for Research in Mathematics Education*, 21, 216-229.

Tartre, L.A. 1990b. Spatial skills, gender, and mathematics In G. Leder and E. Fennema (Eds.), *Mathematics and gender* (pp. 27-59). New York and London: Teachers College Press, Teachers College, Columbia University.

Taylor, L.J.C. 1990. Assessing the graphic levels of understanding and quadratic knowledge of C^2PC students. In F. Demana, B. Waits, & J. Harvey (Eds.), *Proceedings of the Second Annual Conference on Technology on Technology in Collegiate Mathematics*(pp. 324-327).

Threadgill-Sowder, J. A., & Juilfs, P. A. 1980. Manipulative versus symbolic approaches to teaching logical connectives in junior high school: An aptitude X treatment interaction study. *Journal for Research in Mathematics Education*, 5, 367-367.

Thurston, L. L. 1938. *Primary mental abilities.* Chicago: University of Chicago Press.

Tohidi, N.E. 1986, Gender differences in performance on tests of cognitive functioning: A meta-analysis ofresearch finding. Paper based on doctoral dissertation, University of Illinois.

Trotter, A. 1991. Graphic Calculators are coming to class. *Executive Educator*, 13, 20-31.

Vazquez, J.L. 1991. The effect of the calculator on student achievement in graphic linear functions. (Doctoral dissertation, University of Florida, 1990). *Dissertation Abstracts International*, 50, 3508A.

Waits, B.K., & Demana, F. 1992. A case against computersymbolic manipulation in school mathematics today. *Mathematics Teacher*, 85, 180-183.

12 Pressing On

Towards Considered Calculator Use

Kenneth Ruthven

Calculators offer a cheap, portable and robust form of mathematical technology, often taken for granted and casually used. This chapter examines the current state of calculator usage in schools, and critically reviews evidence from research concerned with the effects of calculator use. As a basis for a more considered conceptualisation, five important facets of calculator use within mathematics education are analysed in more detail: procedure, strategy, disposition, curriculum and pedagogy.

The professional context of calculator use

Relative to other forms of computational technology, calculators are unusual in their public familiarity, ready availability, and widespread use. They seem to have many of the characteristics of an 'appropriate' technology, well attuned to its human and physical environment (Dunn, 1978), 'intermediate' in cost and sophistication between traditional artisanal technology and modern advanced technology (Schumacher, 1973), and 'convivial' in being readily available to ordinary people, and easy and effective for them to use in conducting their everyday business (Illich, 1973). Indeed, the arithmetic calculator has achieved an unusual degree of social diffusion. A study of mathematics in employment (Fitzgerald, 1985) reported that where calculation is required of employees, it is almost always carried out by calculator; and that most employees have little difficulty in learning to use the machine.

This cheap, robust and portable technology would seem particularly well suited to the school situation. Indeed, as the nineteen seventies

231

progressed, the arithmetic calculator became the focus of considerable professional discussion and innovative effort in mathematics education. By the early nineteen eighties, however, it had been upstaged by the microcomputer. Nonetheless, this early period of intensive interest left a threefold legacy: the findings of research into the effects of calculator use on student achievement and attitude (summarised in Roberts, 1980; Hembree and Dessart, 1986, 1992); a corpus of classroom calculator activities (exemplified in Eastwood et al., undated; Graham et al., 1986; Fielker, 1987); and some speculative writing about the reshaping of the mathematics curriculum (notably Girling, 1977; Plunkett, 1979).

More recently, a new generation of increasingly powerful and versatile graphic calculators has attracted interest in upper- and post-secondary mathematics education, offering a range of generic computational tools previously available only on computers: notably for analysing data sets and symbolic expressions, and for representing them in tabular and graphic form. As with its arithmetic predecessor, the advent of the graphic calculator has provoked research scrutiny, teaching innovation and curriculum speculation, if on a more modest scale. Reviews of research have started to appear (Hooper, 1993; Dunham and Dick, 1994); new teaching approaches exploiting the particular capabilities of the graphic calculator have been developed (exemplified in Demana and Waits, 1990; Ruthven, 1992a, Green and Pope, 1993); and speculative writing has suggested that, in its emerging form, the graphic calculator subverts much of the traditional secondary mathematics curriculum (Shumway, 1990; Dick, 1992; Ruthven, 1994a).

In the longer term, the convergence between advanced calculators and palmtop computers can be expected to produce machines of still greater power and versatility. For the immediate future, however, the contrasts in cost, robustness and portability between calculators and computers remain striking. While placing a single computer system in each classroom remains the height of most schools' aspirations, the graphic calculator seems set to displace the scientific calculator as a resource acquired personally by many secondary pupils, leaving schools to shoulder the relatively modest costs of complementing this personal ownership with small stocks of machines available in each classroom and in the library or resource centre. Moreover, whereas pupils' access to the single classroom computer or the school computer room is necessarily restricted, and predominantly on the teacher's initiative, the calculator can realistically become a genuinely personal technology, available for spontaneous use by the pupil within and beyond the classroom. Indeed, knowing that such a resource is readily available provides a powerful incentive to pupils both to make use of it and to learn how to use it effectively; and this learning can take place more privately and informally than on a classroom computer (Ruthven, 1992b, 1994c).

This, then, is an appropriate moment to take stock of calculators in mathematics education, drawing on the evidence and experience which have accumulated over twenty years of arithmetic calculator use, and which are starting to emerge from use of the newer graphic calculator. Although the following discussion will refer particularly to the British context, the issues that it addresses are not parochial. The evidence of comparative surveys is that pupil use of calculators has been more extensive in Britain than in many other countries (Lapointe et al., 1992; OFSTED, 1994); and that Britain has been an important site of innovation in this area (Gaulin et al., 1994; OFSTED, 1994). Hence, although the British experience has not been a typical one, it may prove instructive for a wider audience.

In Britain, calculators are widely available to pupils. Amongst those interviewed in the course of a national monitoring exercise carried out in 1987 (Foxman et al., 1991), 73% of 15-year-olds reported personal ownership of a calculator, and 51% of 11-year-olds. In addition, 98% of the older pupils reported themselves as making use of a calculator in school, and 69% of the younger; with 62% of the older reporting use as occurring several times a week or more, but only 27% of the younger claiming once a week or more. Evidence from an evaluation of information technology in British primary and secondary schools conducted between 1989 and 1991 (Watson, 1993) suggests that computer use by pupils is rather less frequent, even in mathematics classes where teachers plan regular use of computers and there are sufficient machines available to make this a realistic expectation. Around 60% of secondary pupils and 30% of primary pupils in such classes reported computer use as occurring once a week or more for at least one term over a two-year period; but only around 10% appeared to maintain this level of use throughout the period. In a comparison group of mathematics classes chosen to exemplify 'good curriculum practice' and innovative teaching without a particular emphasis on computer use, the proportion of pupils reporting computer use on this scale for at least one term was negligible at secondary level, and comparable at primary; although for none of these primary pupils was this scale of use maintained throughout the period.

Such evidence suggests that the limited use of both calculators and computers in the mathematics classroom cannot simply be attributed to difficulties of access to the technology. As significant are the conceptions of calculator and computer use within mathematics embodied in frameworks of policy and practice. According to school inspectors (HMI, 1991, 1992), schools rarely have considered policies on calculator and computer use in mathematics. On the one hand, the calculator is largely used, almost casually, as a convenient means of executing or checking computations, with little recognition of its potential in promoting knowledge and understanding of number. On the other, use of the computer tends to be more deliberate, predominantly employing teaching packages intended to develop or reinforce knowledge and understanding of specific topics, rather than making use of more generic tools to support problem-solving. In effect, there seems to be only limited recognition of the potential of both devices to fulfil the different functions of: *working implement* to carry out already standard mathematical processes; *teaching aid* to promote mathematical development; and *thinking support* to underpin novel mathematical strategies (Ruthven, 1994b).

Equally, however, it could be argued that national guidelines have encouraged inertia and caution. In the most prominent example, continuing public reservations about the use of calculators in the

234

classroom surfaced during the formulation of a national curriculum, with the character of the number curriculum, and the place of calculators within it, becoming a focus of controversy. For the Secretary of State, tempering the enthusiasm of the working group charged with devising the original mathematics programme:

> On the one hand, [calculators] can be a learning aid and pupils need instruction in their correct use to prepare them for new technology in the world of work. On the other hand, it must be important that pupils themselves understand and are proficient in the various mathematical operations that can now be done electronically. Your final report will need to recognise the risks as well as the opportunities which calculators in the classroom offer. (DES, 1988).

Recent technological developments seem set to broaden this debate. In Britain, although no systematic information is currently available on the spread of graphic calculators, ownership and use appear to have grown rapidly amongst pupils in advanced mathematics classes at upper-secondary level, with signs that interest is now emerging at lower-secondary level.

Research into the effects of calculator use

Although Britain has been the site of much interesting speculation and imaginative innovation on the educational use of calculators, there has been little systematic research into the issue. It is telling that in seeking authoritative evidence on the influence of calculator use on pupil achievement and attitude, the Cockcroft Report (Cockcroft, 1982) felt it necessary to turn to the major body of research into the use of calculators in education conducted in the United States over the period from 1976 to 1980. Here, the issue was conceptualised primarily as one of evaluating effects on achievement and attitude, through the method of comparing the prior and posterior attainment and attitude profiles of two groups of pupils: an experimental group 'using calculators', and a control group not doing so. In recent years, it is summaries of this early body of work,

235

notably Hembree and Dessart's (1986) meta-analysis, which have been widely cited in support of the claim that calculator use enhances mathematical learning and performance. In updating their work, Hembree and Dessart (1992) located only 9 further studies to add to the 79 in their original analysis, and reported that the new data either supported or enhanced the established findings. In short, their findings were that calculator use produced no effect on conceptual knowledge; consistently positive effects on computational and problem-solving skills when tested with calculators available, and broadly neutral to positive effects when tested without; no effect on anxiety towards mathematics; and positive effects on attitude towards mathematics and self-concept in mathematics (Hembree and Dessart, 1992).

Nonetheless, these conclusions deserve careful consideration. It is not only that 'there are numerous internal and external validity problems associated with many of the studies' (Roberts, 1980, p. 92) and 'most of the studies..suffered from serious design and sampling problems and few valid conclusions can be drawn' (Shumway et al., 1981, p. 119). More fundamentally, the summative construct of 'calculator use' is itself problematic. In an individual study, the term usually provides a convenient shorthand for a well defined pattern of calculator use which is more fully described elsewhere. When individual studies are synthesised, however, whether by review or meta-analysis, the term shifts from a specific reference of this type first to acquire a more generic connotation, and then, all too easily, to become a unitary abstraction. In the face of the accusation that meta-analytic studies are theoretically naive (Goldin, 1992), the defence has been that they inherit the orientations of the research surveyed (Hembree, 1992). But this is not an entirely satisfactory response.

In practice, the studies surveyed seem predominantly to have focused on the use of the calculator for functional computation within an otherwise unchanged classroom experience. Roberts (1980, p. 95) considered that 'most studies have not adequately integrated calculator use into the instructional process'; and Hembree and Dessart (1986, p. 97) concurred that 'special instruction [oriented towards the calculator] has been relatively unexamined by research'. Consequently, the main findings relate to the 400 achievement and 71 attitude effects in studies 'where calculator and non-calculator groups had received equivalent instruction except for calculators' (Hembree and Dessart, 1986, p. 89). In some respects this resembles the situation most commonly found in schools. In other respects, however, current pupil experience differs considerably from that analysed. In the studies surveyed, the duration of calculator use ranged from less than one class period to a full school year, with a median length of 30 school days. Furthermore, calculator use was, at least to some extent, shaped by a deliberate and explicit plan. These studies, then, do not replicate the conditions of extended and casual calculator use which now appear to be widespread within the school setting. For evidence about the longer-term effects of calculator use under typical school conditions, we must turn elsewhere.

The Office for Standards in Education (OFSTED) (1994, p. 21) notes that national monitoring of mathematical performance at ages 11 and 15 'did not find any evidence that the widespread use of calculators was associated with lower performance'. But again, these findings, from the Assessment of Performance Unit study (Foxman et al., 1991) deserve careful consideration. Because the great majority of secondary pupils reported calculator use in school, it was only possible to make comparisons at the primary age level. Here, the study found a significant positive association between school calculator use and pupil attainment. Of course, this evidence is not interpretable in causal terms; and caution is encouraged by the finding of a stronger positive association between calculator availability in the home and pupil attainment. In effect, these findings may reflect more complex relationships involving the social background of pupils and patterns of schooling, rather than any beneficial influence of calculator use.

However, it is the effects of innovative calculator-based approaches to school mathematics which are probably of most potential professional interest and importance. Here, Hembree and Dessart were able to draw on 34 achievement and 19 attitude effects produced by studies 'in which special calculator instruction was compared with traditional instruction without calculators' (1986, p. 89). Although the conflation of differing treatments to produce mean scores for a construct of 'special calculator instruction' is again problematic, where comparisons with the main study are possible, most of the trends appear to be more positive. In particular, effect sizes for computational and problem-solving skills assessed without access to calculators were larger, as was the effect size for attitude to mathematics.

Evidence of the longer-term effects of innovative approaches comes from more recent projects which have explored the development of a 'calculator-aware' number curriculum for the primary school. The *Calculator Aware Number* (CAN) project (Shuard et al., 1991) worked towards implementing such a curriculum through collaborative work with teachers based on the central principles of: devolving decisions about whether or not to make use of a calculator to pupils; not teaching standard written algorithms; emphasising mental calculation; investigating how numbers work; encouraging practical problem-solving and investigative work. The evaluation of the CAN project did not involve any controlled quantitative comparisons, on the grounds that 'misleading impressions might be gained from measurements of the comparative performance of children who worked in CAN against other children, as the two groups would have other differences in addition to the fact that they were following different curricula' (Shuard et al., 1991, p. 55). Nonetheless, these reservations did not inhibit the team from reporting the generally positive findings of an uncontrolled comparison, albeit with some cautions about the comparability of the pupils involved and the sizes of the groups.

For a more rigorous evaluation of a similar teaching approach, we must turn to findings from the Australian *Calculators in Primary Mathematics* project (Groves, 1993, 1994). Samples of grade 3 and 4 pupils were assessed through interview, after having followed the calculator-aware curriculum for three and a half years. Their performance was compared with that of a similar sample of pupils from the cohorts two years senior in the same schools, who had followed a conventional curriculum and had been assessed in the same way two years previously. The findings were that the project children were better able to tackle real-world problems and computational tasks; in particular, that while they did not make more use of calculators, they made more appropriate choices of calculating device and were better able to interpret their answers.

Within the smaller and more recent body of research into the educational use of graphic calculators, there has been a similar emphasis on developing innovative teaching approaches to exploit the capabilities of the calculator. Reviewing research in which such approaches have been compared with more conventional forms of instruction, Dunham and Dick (1994) cite three studies yielding positive effects, three showing neutral effects, and one producing a negative effect. It is notable that the positive findings came in those studies (reported in Ruthven, 1990; Harvey et al., 1992; Quesada and Maxwell, 1994) where students had access to calculators during testing, although in all three studies care does seem to have been taken to avoid setting items which would unduly favour calculator users.

As Dunham and Dick acknowledge, the effectiveness paradigm is problematic as a means of evaluating innovations which bring both new teaching methods and curriculum goals. More fundamentally, however, it marginalises questions about the actual processes of teaching, learning and thinking, at best accepting syncretic descriptions, at worst imposing superficial labels. The consequence of this lack of curiosity is a theoretical impoverishment which makes it difficult to interpret and generalise the substantive findings of research with any degree of confidence or subtlety. In effect, one conclusion emerges from the common weaknesses of research, policy and practice: the need for a more considered framework within which to analyse, plan and implement the use of calculators in mathematics education.

Towards an analytic frame for calculator use

Developing an adequate framework for analysing calculator use calls for further exploration of the roles played by the calculator within different facets of mathematical thinking, teaching and learning. Here it will only be possible to illustrate the kinds of issues which are likely to emerge as important. The discussion will necessarily reflect the fragmentary and tentative nature of much of the evidence currently available: it will give priority to speculative lines of analysis and potential dilemmas worthy of further investigation, rather than seeking to draw firm conclusions.

Procedure

To the expert user, calculators may appear to automate standard computational processes: typically, addition, subtraction, multiplication and division in the case of the arithmetic calculator; and the graphing, evaluation and analysis of symbolic expressions in the case of the graphic calculator. Certainly, calculator-based procedures can facilitate and expedite such standard processes, and increase their reliability. Nonetheless, adoption of a calculator for computational purposes continues to call for mathematical thinking on the part of the user; albeit not exactly the same thinking as that required for alternative mental or written procedures.

Amongst 11-year-olds tested in the course of national monitoring (Foxman et al., 1991), the success rate on the question $61 \div 4 =$ was 78% when a calculator was available (but not necessarily used); only 54% when a calculator was not available. Not all error in the calculator-available group can be attributed to pupils choosing not to use the calculator, or to their carelessness in using it. Classroom experience suggests that the response 1525, given by over 20% of the unsuccessful pupils in the calculator-available group, more often reflects a lack of appreciation of the significance of the decimal point displayed on the calculator screen, than careless transcription. Indeed, experience further suggests that some pupils, meeting what to them is a confusing calculator answer of 15.25, would switch to a mental or written procedure for the computation; typically leading to an answer in a different form, as evidence from the same study shows. 83% of successful pupils in the non-calculator group gave the response 15 remainder 1, only 19% in the calculator-available group (predominantly, one suspects, those who used a non-calculator method); whereas for the response 15.25, the corresponding figure was only 6% in the non-calculator group, compared to 78% in the calculator-available (predominantly, one suspects, those who used the calculator).

The role of the user is even more central when a calculator is used to graph symbolic expressions. Take, for example, $y = {}^{10x}/x{-}1$ (and imagine it written in fraction format with numerator over denominator). Here, more is require than simply transcribing the expression: it has to be translated from fraction format into the line format $y=10x\div(x-1)$ acceptable to the machine. Equally, the resulting graphic display needs to be interpreted. Here, whichever default setting is used for the range of the axes, the user needs to recognise that only a restricted portion of the graph is shown, and to find an appropriate rescaling of the axes. Far from being automatic, then, calculator graphing often takes on the character of an exploratory procedure in which the user modifies the range in response to the emergent characteristics of the graph.

These examples illustrate how, even when applied to computational processes corresponding to standard machine operations, calculator use is not wholly routine. The user has to formulate the computation for input to the machine, and interpret the output. Moreover, this may involve repeated computation during which the user makes important tactical decisions in order to arrive at an acceptable answer. Often, too, calculators are used in this way to carry out standard processes for which the machine does not offer an automated procedure; or to carry out, with more confidence and insight, processes for which it offers an automated, but opaque procedure. One common example is the use of a visually moderated procedure for solving equations on a graphic calculator, which involves a repeated cycle, first of roughly locating the graphic intersection corresponding to a solution of the equation, then of regraphing an appropriate neighbourhood of the point in more detail.

Strategy

Typically, the need for computation arises within some broader problem-solving task. It has often been suggested that the assistance that the calculator offers with computation should help pupils to tackle problems involving more complex and realistic data. Reports from three development projects, each emphasising mental calculation as well as calculator use, draw differing conclusions on this matter. At lower-primary level, Shuard et al. (1991, p. 13) observe: 'Because children who use calculators are able to handle large numbers, they can work in real-life situations in which the numbers have not been simplified'. At the same level, Groves (1993, p. 15) reports that 'children with long-term experience of calculators are better able to tackle 'real-world' problems which would normally be beyond their paper and pencil skills'.

Working, however, with upper-primary pupils, Hedren (1985, p. 177) reports:

> We were rather optimistic when we started the project and thought that with the help of hand-held calculators we could now use more realistic numbers in our word problems and were no longer restricted to numbers that were 'doctored' to give manageable calculations. But it soon turned out that the pupils got stuck, because they just could not handle the complicated [decimal] numbers despite their hand-held calculators.

Several factors might account for these contrasting conclusions: differences in the conceptual reorganisation required in moving from working with small numbers to large, in contrast to the shift from whole to decimal and composite numbers; differences between formulating a problem situation involving more complicated numbers, as against interpreting an answer involving such numbers; differing degrees to which pupils had been encouraged and accustomed to tackle problems conjecturally, in particular through exploratory calculation.

A further suggestion about the effects of using the calculator for computation in problem-solving is that it frees pupils to focus on strategical issues. Wheatley (1980) reports the outcomes of classroom work with upper-primary pupils, where there was explicit teaching of problem-solving techniques. Using clinical interviews to examine the solution processes used by pupils, she compared the performance of a class in which pupils did not have access to calculators, to one in which pupils were expected to make use of the machines for computation. Of those processes which occurred with reasonably frequency, the calculator group displayed a notably higher incidence of two: using unexpressed equations; and retracing the steps of a solution. On three more, there was a weaker trend in favour of the calculator group: making an estimate of the answer; checking that an answer was reasonable; and checking that the conditions of the problem had been met. On a final two processes, the groups showed very similar patterns of use: checking computations; and making successive approximations. The research team also judged that pupils using a calculator exhibited more exploratory behaviours in problem solving, and spent more time attacking problems and less time computing. But, as Fielker (1987, p. 421) has observed in analysing a classroom session with upper-primary pupils under more typical conditions, the calculator may simply enable pupils to pursue superficial strategies more efficiently: 'The trouble about the calculator is that its speed, which is its main advantage, drastically lessens the need for a systematic approach, [encouraging a] concentration on trying out a large number of guesses.'

Nonetheless, there are different degrees of guessing; indeed, guessing can take on its own systematic form. Strategies of successive approximation or trial-and-improvement with the calculator have been observed in use from the lower-primary school onwards. Smith (1991), for example, describes how a pupil presented with the problem of finding how much would be left from 84p after spending 27p, constructed this as '84 take something leaves 27'. He proceeded to trial a succession of guesses using the calculator: 84-44 giving 40; 84-36 giving 48; 84-67 giving 17. Then, in the light of his success in producing the desired 7 in the units digit, he tried 84-7 giving 77, confirming the pattern, but moving further away from his goal; followed, finally, by 84-57 giving 27. Ruthven (1990) found similar informal strategies amongst calculator users at upper-secondary level, when pupils were asked to find a symbolic expression to describe a given graph. Amongst graphic calculator users, this took the form of progressively refining their symbolic conjecture through feedback from graphing; amongst the comparison group of scientific calculator users, symbolic expressions were revised (with little success) in the light of feedback from calculation of particular values.

Although now recognised in the National Curriculum, trial-and-improvement strategies have not traditionally been encouraged in school, although they seem to be prevalent amongst pupils, even without the computational support provided by calculators. In the Wheatley (1980) study, for example, while successive approximation appears not to have been explicitly taught, its incidence was high. It is interesting that the calculator group did not make relatively more use of it; although this is perhaps explicable in terms of the influence of the teaching that they received. Trial-and-improvement tends to be adopted where the user is unable or reluctant to reformulate the problem situation so as to construct a more direct derivation of the answer from the information available. But, as well as leading, albeit rather indirectly, to a solution, use of trial-and-improvement can sometimes trigger the critical structural insight which then enables the user to devise a more direct path (Ruthven, 1990).

Arithmetic and graphic calculators have also been observed to support and extend the use of iterative strategies in which an answer is built up by repeated application of a simple process, often coordinated with counting (Ruthven, 1991; Shuard et al., 1991). These successive-trial and iterative-construction strategies resonate with Levi Strauss's concept of bricolage (Berry and Irvine, 1986) in which everyday thinking is characterised in terms of the search for solutions to problems by improvisation from the material most readily to hand. In functional terms, there can be little doubt of the value of such strategies in offering an accessible, if relatively lengthy, route to a solution; and of the power of the calculator in facilitating their use and enhancing their reach. In developmental terms, however, the question arises as to whether, by strengthening such approaches, calculator use may inhibit pupils' acquisition and objectification of the more curtailed strategies which provide the conceptual elements for later mathematics. In classrooms where calculator use is accompanied by an emphasis on pupils developing their own computational procedures and problem-solving strategies, this question assumes particular importance.

Disposition

Pupils' perceptions of calculators and their dispositions towards them, are likely to exert an important influence on their patterns of calculator use. Amongst the pupil population as a whole, evidence from national monitoring indicates that while only 3% of 15-year-olds expressed disapproval of the use of calculators for school work, a much higher proportion of 11-year-olds did so: 7% on the grounds that it was a form of cheating, and a further 22% in less censorious terms (Foxman et al., 1991). Not surprisingly, pupils who had not used calculators at school were much more likely to disapprove of their use. Amongst such pupils, views were commonly expressed that calculators 'stop you using your brains' or 'prevent you learning all sorts of sums'.

Where primary pupils have taken part in projects emphasising the development of mental methods of calculation alongside use of the calculator, it seems that such pupils do not, in general, make more use of calculators, but that they do make more appropriate choices of methods of calculation (Groves, 1993, 1994). In particular, 'most children..decide[ ] for themselves that they do not need, or want, to be dependent on their calculators for all calculation' (Shuard et al., 1991, p. 12). Indeed there are even reports of 'a kind of boycott against the.. calculator [breaking] out' (Hedren, 1985, p. 176). Similarly, some older pupils have been reported as reluctant to use the graphic calculator because they see doing so as involving a loss of their intellectual autonomy, in particular as surrendering control of

245

the development of a mathematical argument (Ruthven, 1992b).

These observations are consistent with the model which emerged from an analysis of pupils' views of number work and calculators as they transferred from primary to secondary school (Ruthven, 1995). Preference for not using a calculator was related to confidence in, and enjoyment of, number. In particular, for pupils with less confidence or enjoyment, the calculator seemed to provide a means of matching the demands of school work to their capabilities and interests. It is for this group that the role of the calculator is likely to be particularly influential, offering them new power over their experience of mathematics, albeit a power which is double edged. On one side, use of the calculator can make tasks more accessible, and enable these pupils to tackle them with greater success; on the other, it makes it possible for them to avoid important challenges which might help to develop their mathematical expertise.

Curriculum

The immediate, and apparently consensual, curricular shift associated with calculator use has been a greater emphasis on the checking of calculations. Often born of concerns about the atrophying of pupils' number skills, and sustained by the twin spectres of the malfunctioning machine and the careless user, checking has tended to take the form of independent estimation; although less commonly taught strategies, such as working back to an original item of data using the calculator itself, would seem at least as functionally effective. National monitoring of mathematical attainment has produced evidence (which must be interpreted cautiously in view of the possible intervening factors discussed earlier) that school calculator experience at primary level is associated with better performance on two types of number item corresponding closely to these checking strategies: estimating the missing value in a multiplicative problem; and identifying the operation needed to invert a calculation so as to get back from the answer to one of the original values (Foxman et al., 1991).

The major curricular controversy surrounding calculator use concerns the degree to which emphasis on traditional mental and written calculation procedures should be reduced. In immediately functional terms, the case for emphasis on these alternative modes is not strong. On developmental grounds, however, the importance of mental calculation is now widely accepted: current debate focuses on the contribution of written algorithms, particularly in standardised form. Some claim that:

> the exercise of algorithms..may have many positive side effects..pupils practice estimations..and get an idea of the divisibility properties of number (Hedren, 1985, p. 165).

Others are highly sceptical:

> Arguments that learning the school arithmetic algorithms contributes to mathematical knowledge ring hollow.. Just remembering a set of steps in computing is counter to constructing number relationships (Wheatley and Shumway, 1992, p. 2).

Much depends, of course, on the nature of the learning which is actually taking place: some consider that:

> the process [of learning the short division algorithm] is not at all the mechanical transmission of a mechanical procedure that it is reputed to be. In fact the process appears to be quite creative, perhaps one of the few almost universally experienced creative experiences in elementary school (Newman et al., 1989, p. 113).

A more refined position can be found in the principle of progressive schematisation which emphasises the formative value to pupils, first of developing, and then shortening, some generalised written schema (ter Heege, 1983).

With the coming of the graphic (and increasingly symbolic) calculator, a similar debate is erupting over the traditional routines of secondary mathematics. This raises many interesting questions which parallel those provoked by the arithmetic calculator. For example, what might constitute appropriate strategies for checking a graph computed by the calculator from a symbolic expression: predicting in advance, or scrutinising after the event, the overall shape of the graph, using knowledge of graph types and transformation effects; or checking critical values by comparing readings from the graph with results from substituting in the original expression? What types of mental manipulation of algebraic symbolism might it be important for pupils to develop to complement use of a symbolic calculator: and what are the relative merits of conceiving manipulation in terms of informal images of spatial movement and combination of terms, rather than in terms of formal principles of algebraic structure?

We should bear in mind that these debates are being conducted against a more fundamental shift within the mathematics curriculum, predating the calculator in origin, but deriving strength from its arrival. At the heart of this shift is a changed conception of mathematical thinking which places more emphasis on flexibility of analysis and less on mechanical skill, gaining particular influence over the course of the nineteen-eighties. Consequently, many recent calculator projects have incorporated calculator use within a curriculum philosophy which emphasises the place of conjectural thinking in mathematics and a greater devolution of intellectual initiative to pupils (Shuard et al., 1991; Dick, 1992; Ruthven, 1992b; Groves and Cheeseman, 1993). Under such conditions, calculator use is particularly likely to disrupt the careful regulation of mathematical experience within the traditional curriculum, with, for example, primary pupils meeting large, negative and decimal numbers much earlier, and lower secondary pupils encountering diverse and resolutely non-linear patterns of covariation.

There is, then, an important incommensurability between traditional models of progression and those emerging in classrooms where the calculator plays a part in supporting a new curriculum paradigm. Shuard et al. (1991, p. 44) report that:

> teachers began to notice that children's mathematics learning did not seem to progress in the ordered linear way in which it was traditionally structured. Individual children seemed to be putting together the network of mathematical concepts in their own individual ways. The differences between different children's mathematics were

more fundamental than mere differences in speed of learning.

Pedagogy

A new curriculum celebrating mathematical exploration and intellectual autonomy calls for correspondingly novel pedagogical forms. Here, the calculator has been assigned a variety of roles. On some occasions it is intended simply to accelerate the execution of routine computational tasks, releasing more time for conjectural thinking; and to provide valuable support for pupils who could otherwise be defeated or overwhelmed by the computational demands of a task. On other occasions the calculator is seen as providing a neutral authority against which pupils can check their findings or test out mathematical conjectures. Finally, the calculator can be taken as embodying mathematical ideas, making them available for exploration and analysis through interaction with the machine.

A provocative corollary is the challenge to orthodox views about the key role played by visual-tactile mediation in developing mathematical ideas. In a very early article on the electronic calculator, Fielker (1973, p. 29) records what has become the received position: that it is:

> strictly a machine, and not an aid to understanding. Unlike its mechanical ancestor, all the tactile, visual and dynamic processes take place in the solid-state chips hidden inside the case, and all that appears to the operator is the display.

Yet, as teachers observed their pupils' learning with the arithmetic calculator within the CAN project they noted that:

> children seemed to be overleaping the need for apparatus, using it only to demonstrate or explain their thinking. They went determinedly in abstract directions, experimenting with all the buttons on all the numbers they could think of. We didn't have millions of Multilink cubes, and the children wouldn't have bothered with them if we had (Shuard et al., 1991, p. 11).

Whereas the arithmetic calculator is resolutely symbolic, the graphic calculator links representations based on numeric and algebraic symbols with those centred on graphic images. Indeed, its advocates have emphasised the relative accessibility of mathematical ideas when treated graphically, in place of, or at least prior or parallel to, a treatment through algebraic symbolism (Dick, 1992; Ruthven, 1992b); here the principle that mathematical meaning resides in the interplay of representational systems is often invoked (Dick, 1992; Vonder Embse, 1992); and with it, the suggestion that the semantics of algebra are more important than its syntax (Kaput, 1989). In an apparent echo of the CAN teachers observations, Kirshner (1989, p. 197) is sceptical of this apparent dismissal of symbolism, considering that 'the human mind is uniquely fashioned to learn syntax as syntax..[and]..that the natural predisposition of the mind is to approach new, structured domains syntactically'. The arrival of calculators on which algebraic symbolism becomes active, in much the same way as numeric symbolism does on the arithmetic calculator, offers the fascinating prospect of these issues being explored further.

In conclusion

In policy, practice and research, too much has been taken for granted about calculators; too often, they have become a casually used and poorly understood resource. Yet, more considered use of calculators is probably the most realistic medium-term strategy for bringing the distinctive opportunities of sustained use of computational technology to teachers and pupils across the educational system. This article has sketched out the issues at the heart of any considered calculator use, and with them an agenda for further exploration and analysis.

References

Berry, J. W. and S. H. Irvine 1986. 'Bricolage: savages do it daily', in Sternberg, R. J. and R. K. Wagner (eds.) *Practical intelligence: Nature and origins of competence in the everyday world*. Cambridge: Cambridge University Press.

Cockcroft, W. H. (ch.) 1982. *Mathematics Counts*. HMSO: London.

Demana, F. and B. Waits 1990. *Precalculus*. New York: Addison Wesley.

Department of Education and Science 1988. *Mathematics for ages 5 to 16*. London: DES. Appendix 1.

Dick, T. 1992. 'Super Calculators: Implications for Calculus Curriculum, Instruction, and Assessment', in Fey, J. and C. Hirsch (eds.) *Calculators in Mathematics Education*, Reston VA: National Council of Teachers of Mathematics.

Dunham, P. H. and T. P. Dick 1994. 'Research on Graphing Calculators', *The Mathematics Teacher,* 87, 6, 440-445.

Dunn, P. D. 1978. *Appropriate Technology: Technology with a Human Face*. London: Macmillan.

Eastwood, M., B. Bagnall, M. Blows, P. Dann, J. Duffin, J. Holmes, P. MacConnacher undated. *Calculators in the Primary School*. Derby: Association of Teachers of Mathematics/Leicester: Mathematical Association.

Fielker, D. 1973. 'Electronic Calculators', *Mathematics Teaching,* 64, 28-32.

Fielker, D. 1987. 'A Calculator, A Tape Recorder, and Thou', *Educational Studies in Mathematics*, 18, 417-437.

Fielker, D. (ed.) 1987. *Calculators*. Derby: Association of Teachers of Mathematics.

Fitzgerald, A. 1985. *New Technology and Mathematics in Employment*. Birmingham: Faculty of Education, University of Birmingham.

Foxman, D., G. Ruddock, I. McCallum and I. Schagen 1991. *APU Mathematics Monitoring (Phase 2)*. Slough: National Foundation for Educational Research.

Gaulin, C., B. Hodgson, D. Wheeler and J. Egsgard 1994. Sections on 'The Impact of Calculator on the Elementary School Curriculum' and 'Miniconference on Calculators and Computers', in *Proceedings of the Seventh International Congress on Mathematical Education*. Quebec: Les Presses de l'Université Laval.

Girling, M. 1977. 'Towards a Definition of Basic Numeracy', *Mathematics Teaching*, 81, 4-5.

Goldin, G. 1992. 'Meta-analysis of Problem-Solving Studies: A Critical Response', *Journal for Research in Mathematics Education*, 23, 3, 274-283.

Graham, A., J. Baker, J. Daniels and K. Tyler 1986. *Calculators in the Secondary School*. Cambridge: Cambridge University Press.

Green, D. and S. Pope (eds.) 1993. *Graphic Calculators in the Mathematics Classroom*. Leicester: Mathematical Association.

Groves, S. 1993. 'The Effect of Calculator Use on Third Graders' Solutions of Real World Division and Multiplication Problems', *Proceedings of the Seventeenth International Conference for the Psychology of Mathematics Education*, University of Tsukuba.

Groves, S. 1994. 'The Effect of Calculator Use on Third and Fourth Graders' Computation and Choice of Calculating Device', *Proceedings of the Eighteenth International Conference for the Psychology of Mathematics Education*, University of Lisbon.

Groves, S. and J. Cheeseman 1993. 'Young children's number concepts - The effect of calculator use on teacher expectations',

Proceedings of the Sixteenth Annual Conference of the Mathematics Education Research Group of Australasia, MERGA.

Harvey, J. G., B. K. Waits and F. D. Demana 1992. 'The Influence of Technology on the Teaching and Learning of Algebra', paper presented at the Seventh International Congress on Mathematical Education, Université Laval.

Hedren, R. 1985. 'The Hand-Held calculator at the Intermediate Level', *Educational Studies in Mathematics*, 16, 2, 163-179.

ter Heege, H. 1983. 'The Multiplication Algorithm: an Integrated Approach', *For the Learning of Mathematics*, 3, 3, 29-34.

Hembree, R. 1992. 'Response to Critique of Meta-Analysis', *Journal for Research in Mathematics Education*, 23, 3, 284-289.

Hembree, R. and D. J. Dessart 1986. 'Effects of Hand-Held Calculators in Precollege Mathematics: A Meta-Analysis', *Journal for Research in Mathematics Education*, 17, 2, 83-99.

Hembree, R. and D. J. Dessart 1992. 'Research on Calculators in Mathematics Education', in Fey, J. and C. Hirsch (eds.) *Calculators in Mathematics Education*. Reston VA: National Council of Teachers of Mathematics.

Her Majesty's Inspectorate 1991. *Mathematics: Key Stages 1 and 3*. London: HMSO.

Her Majesty's Inspectorate 1991. *Mathematics: Key Stages 1, 2 and 3*. London: HMSO.

Hooper, J. 1993. 'Issues of Mathematics Classroom Use of Graphing Calculators', *The Mathematics Educator,* 4, 2, 45-50.

Illich, I. 1973. *Tools For Conviviality*. London: Calder and Boyars.

Kaput, J. 1989. 'Linking Representations in the Symbol Systems of Algebra', in Wagner, S. and C. Kieran (eds.) *Research Issues in the Teaching and Learning of Algebra*. Reston VA: National Council of Teachers of Mathematics.

Kirshner, D. 1989. 'Critical Issues in Current Representation System Theory', in Wagner, S. and C. Kieran (eds.) *Research Issues in the Teaching and Learning of Algebra*. Reston VA: National Council of Teachers of Mathematics.

Lapointe, A. E., N. A. Mead and J. M. Askew 1992. *Learning Mathematics*. Princeton NJ: Educational Testing Service.

Newman, D., P. Griffin and M. Cole 1989. *The construction zone: Working for cognitive change in school*. Cambridge: Cambridge University Press.

Office for Standards in Education 1994. *Science and Mathematics in Schools: A review*. London: HMSO.

Plunkett, S. 1979. 'Decomposition and all that Rot', *Mathematics in School*, 8, 3, 2-5.

Quesada, A. R. , and M. E. Maxwell 1994. 'The Effects of Using Graphic Calculators to Enhance College Students Performance in Precalculus', *Educational Studies in Mathematics*, 27, 2, 205-215.

Roberts, D. M. 1980. 'The Impact of Electronic Calculators on Educational Performance', *Review of Educational Research,* 50, 1, 71-98.

Ruthven, K. 1990. 'The Influence of Graphic Calculator Use on Translation from Graphic to Symbolic Forms', *Educational Studies in Mathematics*, 21, 5, 431-450.

Ruthven, K. 1991. 'Calculator Strategies', *Micromath*, 8, 1, 31-32.

Ruthven, K. 1992a. *Graphic Calculators in Advanced Mathematics*. Coventry: National Council for Educational Technology.

Ruthven, K. 1992b. 'Personal Technology and Classroom Change: A British Perspective', in Fey, J. and C. Hirsch (eds.) *Calculators in Mathematics Education*. Reston VA: National Council of Teachers of Mathematics.

Ruthven, K. 1994a. 'Supercalculators and the secondary mathematics curriculum', in Selinger, M. (ed.) *Teaching Mathematics*. London: Routledge.

Ruthven, K. 1994b. 'Computational tools and school mathematics', *Micromath*, 10, 3, 28-32.

Ruthven, K. 1994c. 'The graphic calculator as a personal resource: a study of lower secondary pupils in two schools', *British Journal of Educational Technology*, 25, 2, 147-148.

Ruthven, K. 1995. 'Pupils' views of number work and calculators', *Educational Research*, forthcoming.

Shuard, H., A. Walsh, J. Goodwin and V. Worcester 1991. *Calculators, Children and Mathematics*. London, Simon and Schuster, London.

Schumacher, E. F. 1973. *Small is Beautiful: A Study of Economics as if People Mattered*. London: Blond and Briggs.

Shumway, R. 1990. 'Supercalculators and the Curriculum', *For the Learning of Mathematics*, 10, 2, 2-9.

Shumway, R. J., G. H. Wheatley, T. G. Coburn, A. L. White, R. E. Reys and H. L. Schoen 1981. 'Initial Effects of Calculators in Elementary School Mathematics', *Journal for Research in*

Mathematics Education, 12, 2, 119-141.

Smith, R. 1991. 'Trial and Improve: Slowly', *Mathematics Teaching*, 136, 36.

Vonder Embse, C. 1992. 'Concept Development and Problem Solving Using Graphing Calculators in the Middle School', in Fey, J. and C. Hirsch (eds.) *Calculators in Mathematics Education*. Reston VA: National Council of Teachers of Mathematics.

Watson, D. (ed.) 1993. *The Impact Report*. London: Centre for Educational Studies, King's College.

Wheatley, C. L. 1980. 'Calculator Use and Problem-Solving Performance', *Journal for Research in Mathematics Education*, 11, 5, 323-334.

Wheatley, G. H. and R. Shumway 1992. 'The Potential for Calculators to Transform Elementary School Mathematics', in Fey, J. and C. Hirsch (eds.) *Calculators in Mathematics Education*. Reston VA: National Council of Teachers of Mathematics.

Section Four: How May Technology Support School Algebra?

The papers in this section address issues in the learning and teaching of school algebra raised by the influence of various kinds of computer software. These include spreadsheets, Logo, programming in PASCAL or BASIC, and Computer Algebra Systems such as DERIVE. In each case, the research provokes important questions for teachers about the algebraic concepts they expect students to conceptualise, and the relevance or compatibility of computer algebraic forms.

In Chapter 13, *An Analysis of the Relationship Between Spreadsheet and Algebra*, Giuliana Dettori and colleagues debate issues between traditional conceptions of algebra and those operating in a spreadsheet environment, arguing that spreadsheets do not isomorphically model some of the fundamental algebraic transformations which students need to know. For example, '=' in a spreadsheet represents an operation of *assignment* rather than *equality* or *relation* as in traditional algebra. Is spreadsheet algebra an extension to formal algebra, or is it simply lacking in its ability to model the traditional forms? The authors, nevertheless, acknowledge the value of spreadsheets in providing an environment for starting to address algebraic ideas, and emphasise the important roles the teacher needs to play in linking spreadsheet and formal algebraic transformations.

In Chapter 14, *Algebraic Thinking: the role of the computer*, Ros Sutherland examines the development of students' algebraic thinking as a result of generalising in computer environments. Her research shows that students who are able to write computer code to represent

Logo variables or spreadsheet locations seem to progress readily to expressing unknowns in traditional algebra. This is potentially at odds with the suggestions of Dettori et al above. Do spreadsheets (for example) model the traditional algebraic transformations which we expect students to know, or are the spreadsheet forms alternatives or reductions of these? If so what are the implications for students' learning? As Sutherland makes clear, she does not claim that 'these environments provide "the answer" to teaching algebra'. She acknowledges the importance of the teaching role in relation to computer work, but suggests that the relation between a spreadsheet expression (eg. 3A5+7) and an algebraic one (eg. 3x+7) is not as problematic as might have been supposed.

In Chapter 15, *Bridging a Gap From Computer Science to Algebra*, Jean-Baptiste Lagrange discusses the use of symbolic forms in computer programming, using strings and boolean expressions, to study students' difficulties and associated teaching approaches. The paper supports a movement to extend traditional notions of algebra to include the alternative forms common to computer programming. What is the special nature of computer programming: are its requirements in terms of precise symbolic structures either indicative of students' difficulties with more traditional algebra, or of use in enabling students to become more generally aware of algebraic processes and symbol use?

In the final Chapter, 16, *Using a computer algebra system with 14-15 year old students*, Mark Hunter and colleagues report on quantitative aspects of a research study designed to explore ways in which a Computer Algebra System (CAS) can improve the learning of selected algebraic and graphical concepts and skills. The chapter suggests that a CAS may support learning where students are 'mathematically ready to use it', while more 'traditional' methods may be more appropriate up to this stage. Moreover, the study highlights advantages of the use of the CAS as a 'tool' rather than as a teaching agent. It points out that the technology acts as an intermediary between the student and the mathematics, and that knowledge (or lack of it) of the intricacies of the technology affects study such as this are related to the particularities of the school, teaching approaches and students involved. The authors recognise that the ability of the students involved may have influenced particular responses.

Together, these papers alert us to the potential importance of technology in developing students' conceptions of algebra. They also make it clear that much more research is needed.

13 An Analysis of the Relationship between Spreadsheet and Algebra

G.Dettori, R.Garuti, E.Lemut, L.Netchitailova

In recent times, the spreadsheet has been suggested as a tool to teach algebra in intermediate high school. Our a-priori analysis of the relationship between spreadsheet and algebra shows the inadequacy of this tool to express the fundamental characteristics of algebra, that is, the manipulation of algebraic variables and relations, which make algebra suitable as a formalism for describing models. However, with the attentive guidance of a teacher, the spreadsheet can become a useful tool for motivating the introduction of some concepts of algebra and for reflecting on different resolution models.

Introduction

In the last years, several studies pointed out both potentialities and problems related to using a spreadsheet to teach some typical topics of mathematical curricula in intermediate/high school, such as algebra, approximate calculus, statistics (see Malara et al. (1992) for a review). The decision to use the spreadsheet to teach a discipline (like, for instance, algebra) sometimes seems implicitly determined by the assumption that an item of software can be successfully used to teach/learn any topic which is involved in the use of that software. We consider this point of view arguable, for two reasons; first, this position does not specify how deep it is necessary to know that topic to start using the software; second, it does not distinguish between "learning something about" and "learning the most important aspects of" a topic. Moreover, as concerns algebra in particular, some researchers (Capponi et al. 1989, Capponi 1992) point out that the spreadsheet is not really based on algebraic calculus, since the expressions it uses as formulae, though containing literals which recall

261

algebraic expressions, do not have algebraic character. The design of spreadsheet solutions contrasts with the approach of algebraic ones, in that algebra essentially gives an operative language to analyse or to manipulate relationships; its aim is not to perform computations (Booth 1984, Chevallard 1989). Moreover, the non-algebraic nature of the spreadsheet is the origin of some difficulties in transferring algebraic competencies when solving problems of different kinds.

In this paper we are concerned with the relationship between spreadsheet and algebra. We analyse in which measure the spreadsheet can really help students to learn algebra, considering not only a first approach to it, but also, and in particular, its most characterising aspects. In fact, we think that experimental researchers that propose the spreadsheet as a positive tool to learn algebra limit their observations to a first approach, that is, to solving simple problems which are typical of the elementary school algebra (Sutherland 1993, Sutherland et al. 1993). Moreover, the aim of these researches seems more to see how to use the spreadsheet to compute the solution of a problem rather than to check if the applied solving process is of algebraic nature.

Based on the above considerations, we analyse the spreadsheet from the point of view of learning the main concepts of algebra. This entails determining what it means to learn algebra, in particular for students under 16. Our work differs from other research in this field in that we emphasise which parts of algebra can be, or can not be, tackled with the spreadsheet, instead of pointing out how much algebra is involved in the spreadsheet. Our analysis emphasises that the spreadsheet can be useful to introduce some elements of algebra, but that its results are inadequate, if not misleading, for a deep learning of the fundamental aspects of algebra. In our opinion, some limits of the spreadsheet can be overcome if we do not use it as a tool for solving problems, but rather consider its underlying resolution model as a tool for reflecting on higher levels of abstraction, synthesis and generalisation, under a teacher's guidance.

Our work has the characteristics of an a-priori analysis, but takes also into consideration both our own classroom experiences and those mentioned in the referred papers (Chiappini et al. 1991, Dettori et al. 1993).

School algebra at age 11-16

Before discussing how suitable is the spreadsheet to teach/learn algebra, we need to define what we mean for algebra and what is relevant to teach/learn at age 11-16. We are aware that there are various conceptions of what is algebra (Usiskin 1988) and that different countries have different school algebra traditions. These discrepancies can induce different evaluations of the spreadsheet's influence in teaching/learning algebra. In order to remain as general as possible, we took into consideration both a classical conception (like that proposed by O. Terquem in the last century (Chevallard 1989, pg. 36) and a modern view of algebra which emphasises, as a crucial starting point, the conceptual break with arithmetic (Chevallard 1984, Chevallard 1989, Cortes et al. 1990). This break is characterised by entering into a modelling process that changes the nature of a problem's resolution (from problem to equation, from equation to equation's solution by producing a formula, from formula to calculation). In this view, the algebraic resolution of problems implies the construction and resolution of algebraic equations, and this requires the ability to perform algebraic transformations.

As concerns school algebra at age 11-16, we consider meaningful:

- to understand what are variables and unknowns;
- to understand the meaning of formulae (e.g. $A=LW$), equations ($4X=48$), identities ($sinX=cosX*tanX$), properties ($n*(1/n)=1$), functions ($Y=kX$);
- to learn to manipulate algebraic equations and inequalities according to the rules of literal calculus;
- to learn to apply algebra, that is, to formulate equations which model problems or classes of problems;
- to learn to apply the algebraic calculus to the demonstration of simple theorems.

Using a spreadsheet for solving some algebraic problems

We want first to analyse if the use of the spreadsheet induces that break between arithmetic and algebra that is at the base of high level functions characteristic of algebra (such as synthesis, generalisation, transformation) in relation to its objectives (that is abstraction, formalisation, modelling). We address this issue by discussing the resolution of some meaningful school problems. All examples have been implemented in Excel 2.2a on a Macintosh (A1-style relative reference system and $A$1 absolute reference system).

Using relations

Let us consider the following problem: *We want to distribute 100 books among three persons so that the second one receives four times the first one, and the third one as the second one plus 10.*

Though problems of this kind are often assigned in the introductory phase of algebra, they are certainly not the most meaningful, since they can be directly solved using arithmetic, hence missing the point of the conceptual break between arithmetic and algebra. However, this problem can be solved algebraically by solving the equation X+(4X)+(4X+10)=100, where X is the number of books given to the first person. This equation describes the problem, and, at the same time, the solution of the problem in implicit form. Transforming it by means of the rules of algebra we obtain the equation X=10, which represents the solution explicitly. Using a spreadsheet, we can only seek a solution of the problem by trial-and-error, that is, by repeating several times some numerical computations (more precisely, by computing the value of some formulas), until the required value is found. Fig. 1 shows the beginning of a schema of solution often used by students.

	A	B	C	D
				books.1
1	first	second	third	sum
2	1	=4*A2	=B2+10	=A2+B2+C2
3				

	A	B	C	D
		books		
1	first	second	third	sum
2	1	4	14	19
3				

Fig. 1 - a) Data and formulas; b) Data and computed values

A numerical result can be found by inserting different values in column A, either by copying the line or by successively substituting different values in the cell A2, until the value in the corresponding cell in column D is 100. The equation X+Y+Z=100 is not explicitly expressed. The student solves it implicitly by checking, by himself, the value of the sum in the last column, but he may never become aware of the equation. This depends on the fact that spreadsheets do not allow one to express equations, which, on the other hand, are basic tools of algebra.

This observation leads us to point out a first fundamental discrepancy between spreadsheet and algebra: the sign of equality used in spreadsheets is actually the assignment of a computed value to a cell, while the equal sign in algebra represents a relation. The inability to write relations in a spreadsheet implies that it is not possible to use it to completely represent algebraic models. Hence, the resolution approaches of algebra and spreadsheets are strongly different: in algebra the solution of a problem is found by formal manipulation of equations describing it, while with a spreadsheet successive numerical approximations must be performed until a numerical solution is reached. This basic discrepancy can even lead students to misunderstand what is algebra if they are told that, using a spreadsheet, they are learning algebra.

Synthesising equations

Using a spreadsheet leads students to recognise which elements are involved in a problem, and to express part of the relationships among them, hence getting used to the kind of problem analysis which is necessary for algebra applications. Passing from the spreadsheet style of resolution to the algebraic one requires learning to synthesise the partial relationships presented in the spreadsheet into one or more equations describing the problem. The plain use of a spreadsheet, without teacher's help, will hardly lead the student to acquire this synthesis capability. At the same time, the possibility to find a solution simply by numerical attempts (which are easy to perform by computer) can discourage the student from making an intellectual effort toward synthesis, hence missing a fundamental tool for the development of mature cognitive capabilities.

	A	B	C	D
				books.2
1	first	second	third	sum
2	1	=4*A2	=B2+10	=A2+B2+C2
3	2	=4*A3	=B3+10	=A3+B3+C3
4	3	=4*A4	=B4+10	=A4+B4+C4
5	=A4+1	=4*A5	=B5+10	=A5+B5+C5
6	=A5+1	=4*A6	=B6+10	=A6+B6+C6
7	=A6+1	=4*A7	=B7+10	=A7+B7+C7
8	=A7+1	=4*A8	=B8+10	=A8+B8+C8
9	=A8+1	=4*A9	=B9+10	=A9+B9+C9
10	=A9+1	=4*A10	=B10+10	=A10+B10+C10
11	=A10+1	=4*A11	=B11+10	=A11+B11+C11

Fig. 2. Formulas for problem 1

Variables and Unknowns

Another component which is absent from the spreadsheet solution is the unknown X. Since formulae computed by spreadsheets are not relations but functions, the involved cell names at most play the role of functional variables rather than algebraic unknowns. However, the functional variable is the result of an abstraction process, which is the capability to figure out an object beyond the possible values that can be substituted for it. As in the case of synthesis, abstraction capability will hardly be achieved by students with the only help of a spreadsheet, without support of a teacher who can suggest it by pointing out the analogies and differences between rows obtained from one another with a copy instruction (see fig. 2).

Proving results

Let us now consider another problem, slightly more complex than the previous one: *The theatre of a country town has 100 seats, divided into front section and rear section. The price of front seats is 8$, that of rear seats is 6$. When all seats are sold, the total income is 650$. How many front seats and rear seats are there in the theatre?*
The algebraic resolution is made up of two equations in two variables, $X+Y=100$ and $8X+6Y=650$, which give in a simple way the solution of the problem as $X=25$, $Y=75$.

If the spreadsheet solution is designed without implicitly solving the first of the above equations with respect to one variable, we obtain a table which is clearly too cumbersome, but bright students will of course almost immediately notice that for each X value it is necessary to try only one Y value, otherwise their sum may not be 100 (see fig. 3). However, being unable to make formal proofs, it will be difficult to argue that the solution found is the only one, while the algebraic solution gives this certitude. On the other hand, tackling situations of this kind with a spreadsheet, the teacher can stimulate students to reason about the range and relationships of possible solutions to a problem.

Introductory study of functions

An interesting aspect of spreadsheet resolutions, as appears from the above example and considerations, is that these tables are potentially dynamic, that is, columns and rows can be added (e.g. row 12 in fig. 3) or deleted in a much easier way than with paper-and-pencil tables, hence making it possible to solve more problems with a same table, or to seek a more accurate solution of a problem. Moreover, the facility of building and modifying tables leads more easily to their use as a tool to make and to test conjectures about a function's trends. For these reasons, the spreadsheet can be a good introductory tool for the study of functions.

	A	B	C	D
1	1 sector	2 sector	seats	income
2	10	90	100	620
3	20	80	100	640
4	30	70	100	660
5	40	60	100	680
6	50	50	100	700
7	60	40	100	720
8	70	30	100	740
9	80	20	100	760
10	90	10	100	780
11	100	0	100	800
12	25	75	100	650
13				

theatre.1

	A	B	C	D
1	1 section	2 section	seats	income
2	10	90	=A2+B2	=8*A2+6*B2
3	20	80	=A3+B3	=8*A3+6*B3
4	30	70	=A4+B4	=8*A4+6*B4
5	=A4+10	=B4-10	=A5+B5	=8*A5+6*B5
6	=A5+10	=B5-10	=A6+B6	=8*A6+6*B6
7	=A6+10	=B6-10	=A7+B7	=8*A7+6*B7
8	=A7+10	=B7-10	=A8+B8	=8*A8+6*B8
9	=A8+10	=B8-10	=A9+B9	=8*A9+6*B9
10	=A9+10	=B9-10	=A10+B10	=8*A10+6*B10
11	=A10+10	=B10-10	=A11+B11	=8*A11+6*B11
12	25	75	=A12+B12	=8*A12+6*B12
13				

Fig. 3. Numeric resolution and formulas for problem 2

Generalisation of problems by means of parameters

The spreadsheet can be useful to introduce the concept of generalisation of a problem and to learn to distinguish between variables and parameters. For instance, the problem considered above can be used, with the same structure but with different data, to describe the case of different theatres, and it is important that students learn to recognise its unchanged structure. The presence of values that can be considered as parameters can be emphasised in the spreadsheet by writing them separated, before the description of the problem (see cells A2, B2, C2 in Fig. 4). These cells are then referred by using the "absolute notation" (e;g. $A$2), which shows that the cell name is not to be changed during the copy-and-paste operations. The use of relative and absolute references emphasises the difference between parameters and variables.

	A	B	C
			theatre.1.1
1	1 price	2 price	seats
2	8	6	100
3			
4	1 section	2 section	income
5	10	=$C$2-A5	=$A$2*A5+$B$2*B5
6	=A5+10	=$C$2-A6	=$A$2*A6+$B$2*B6
7	=A6+10	=$C$2-A7	=$A$2*A7+$B$2*B7
8	=A7+10	=$C$2-A8	=$A$2*A8+$B$2*B8
9	=A8+10	=$C$2-A9	=$A$2*A9+$B$2*B9
10	=A9+10	=$C$2-A10	=$A$2*A10+$B$2*B10
11	=A10+10	=$C$2-A11	=$A$2*A11+$B$2*B11
12	=A11+10	=$C$2-A12	=$A$2*A12+$B$2*B12
13	=A12+10	=$C$2-A13	=$A$2*A13+$B$2*B13
14	=A13+10	=$C$2-A14	=$A$2*A14+$B$2*B14
15	25	=$C$2-A15	=$A$2*A15+$B$2*B15
16			

Fig. 4. Emphasising parameters in problem 2

The relationship between parameters and variables can be stressed by considering the dual of a problem. For example, in relation with the previous problem, let us consider the following one: *On the occasion of a special show, more expensive than the usual ones, the management of the previously considered theatre needs to increase the price of all seats. The prices ratio must remain 4 to 3, and the total income must be $975. At what price must front and rear seats be sold?* In order to resolve this problem using a spreadsheet, it is sufficient to exchange the roles of variables and parameters in the previous table, as shown in fig. 5. Though the formulas in column B are slightly changed, those in column C are exactly the same.

	A	B	C
1	1 section	2 section	price ratio
2	25	75	0,75
3			
4	1 price	2 price	income
5	8	6	650
6	9	6,75	731,25
7	10	7,5	812,5
8	11	8,25	893,75
9	12	9	975
10			

theatre.1.

	A	B	C
1	1 section	2 section	price ratio
2	25	75	0,75
3			
4	1 price	2 price	income
5	8	=$C$2*A5	=$A$2*A5+$B$2*B5
6	=A5+1	=$C$2*A6	=$A$2*A6+$B$2*B6
7	=A6+1	=$C$2*A7	=$A$2*A7+$B$2*B7
8	=A7+1	=$C$2*A8	=$A$2*A8+$B$2*B8
9	=A8+1	=$C$2*A9	=$A$2*A9+$B$2*B9
10			

Fig. 5. Numeric resolution and formulas for problem 3.

Problems with more than one solution
A different case is that of false generalisations. Let us consider for instance this problem:

The theatre of a country town has 100 seats, divided into first, second and third class seats. First class seats cost 9$, seconds class cost 7$ and third class cost 5$. When all seats are sold, the income is 700$. How many first, second, and third class seats are there in the theatre?

Seeking a solution by using the spreadsheet, it is evident that it is more difficult to chose suitable values by trials, but it is not evident where the difficulty comes from. It is clear that one of the three values is determined by the other two in order to make their sum 100, but this does not help to find a solution, and certainly does not make clear that in this case, unlike in the previous one, there are many possible solutions (see fig. 6). An algebraic treatment of this problem, on the contrary, shows that the two problems are structurally different, since in this case we have three variables and only two equations, hence the problem has an infinite number of solutions in R, all characterised by the relations $X=Z$, and $Y=100-2X$, which allows us to chose a finite number of reasonable solutions in N. This is certainly not easy to see if the problem is tackled only by spreadsheet, without any algebraic consideration.

	theatre.2			
A	**B**	**C**	**D**	**E**
1 section	2 section	3 section	seats	income
10	20	70	100	580
20	30	50	100	640
30	40	30	100	700
40	50	10	100	760
50	50	0	100	800
20	60	20	100	700
30	40	30	100	700
40	20	40	100	700

Fig. 6. Numeric resolution for problem 4

Conclusions

This a-priori analysis pointed out that the spreadsheet can be useful to introduce algebra, since it leads to recognising the elements involved in a problem and to expressing part of the relationships among them. On this basis, numerical solutions of simple problems can be easily found. The limitations of this environment as concerns teaching/learning algebra are due to several factors:
- spreadsheets deal essentially with numbers, or addresses of numbers, and functions;
- algebraic variables and relations can not be directly handled in a spreadsheet; only assignments are made;

- spreadsheets operate from "knowns" to "unknowns", which is the opposite of what characterises the algebraic thinking.

Moreover, spreadsheets are unable to formally manipulate relations, hence they are useless for learning to construct formal demonstrations, which is one of the main applications of algebra.

However, using a spreadsheet, which by itself would lead students to solve problems only by trials, under the wise guidance of a teacher can lead:

- to activate a modelling process for problem resolution;
- to understand what it means to solve an equation, even before knowing what an equation is (that is, to find a value such that an expression is valid);
- to reason on the constraints of a problem in order to decrease the number of trials necessary to find a solution (from casual to focused trials);
- to introduce the concept of approximate calculus.

Further steps toward a real learning of basic algebra can be made through a reflection, strongly guided by the teacher, on the resolution model implemented by means of the spreadsheet. In fact, the teacher's role appears essential:

- to guide her students to abstract the concept of algebraic variable, not present in the spreadsheet, by remarking the analogies of formulas repeated in different rows;
- to show to her students how to synthesise equations describing the problem, based on the formulas used in the spreadsheet and on direct numerical checks on them;
- to make her students aware of the fundamental diversity of the operator "=" in the spreadsheet and in algebra, that is, assignment vs. relation (without this distinction, using a spreadsheet can even be misleading);
- to introduce problem generalisation by differentiating parameters and variables;
- to reinforce the student's algebraic competence by comparing, as a metacognitive activity, different problem solving methodologies, such as arithmetic, algebra and spreadsheet.

References

Booth, L., 1984, *Algebra: Children's Strategies and Errors*, Windsor: Nelson Pub.

Capponi, B., Balacheff, N., 1989, "Tableur et calcul algébrique," *Educational Studies in Mathematics*, 20, 179-210.

Capponi, B., 1992, "Désignations dans un tableur et interaction avec les connaissances algébriques", *Petit X*, Grenoble: IREM.

Chiappini, G., Lemut, E.,1991, "Construction and Interpretation of Algebraic Models", *Proc. XV PME Conference*, Assisi, Italy.

Chevallard, Y., 1984, "Le passage de l'arithmétique à l'algèbre dans l'enseignement des mathématiques au collège," *Petit X 5*, 51-94, Grenoble: IREM.

Chevallard, Y., 1989, Arithmétique Algèbre Modelisation, Étape d'une Recherche, Report n.16, Aix-Marseille: IREM.

Cortes, A., Vergnaud, G., Kavafian, N., 1990, "From Arithmetic to Algebra: Negotiating a Jump in the Learning Process," *Proc. XIV PME Conference*, vol. 2, 27-34, Mexico.

Dettori, G., Lemut, E., Netchitailova, L., 1993, "Spreadsheet: a tool toward Algebra?," Rendiconti del Seminario Matematico e Politecnico di Torino, vol. 51, n.3.

Malara, N., Pellegrino, C., Tazzioli, R., 1992, "I fogli elettronici in attività di matematica per gli allievi dagli 11 ai 16 anni", Pubbl. Comune di Modena, Italy.

Sutherland, R., 1993, "Thinking Algebraically: Pupils Models Developed in Logo and a Spreadsheet Environment," in Lemut, du Boulay, Dettori eds., *Cognitive Models and Intelligent Environments for Learning Programming*, NATO ASI series F, vol. 111, 270-283, Berlin: Springer-Verlag.

Sutherland, R., Rojano, T., 1993, "A Spreadsheet Approach to Solving Algebra Problems," *Journal of Mathematical Behaviour*.

Usiskin, Z., 1988, "Conceptions of School Algebra, and Uses of Variables", in Coxford A.F. (ed.), *The Ideas of Algebra, K-12, 1988 Yearbook*, Reston, Va: National Council of Teachers of Mathematics.

14 Algebraic Thinking: the role of the computer

Rosamund Sutherland

This paper presents an overview of results from Project AnA, a longitudinal study of pupils' developing algebraic approaches to problem solving in mathematics. The study was carried out in a primary and a secondary school. Within the study pupils worked on Logo and spreadsheet activities. Results from the study highlight the importance of the computer-based language in supporting pupils expression of generality and development of an algebraic method.

Introduction and Background

Within this paper I discuss some of the results from a recent project (Project AnA, Sutherland, 1993) which investigated the ways in which pupils' experiences with Logo and a spreadsheet environment influenced their learning of algebraic ideas. Longitudinal case studies were carried out with three groups of pupils, aged 10-11, 11-13 and 14-15. The first two groups were chosen from pupils with a range of mathematical attainment and the third group was chosen from pupils with low mathematical attainment. Parallel studies were also carried out with groups of pupils in Mexico (Sutherland & Rojano, 1993). Pupils worked on a range of spreadsheet activities (see for example Fig. 4 and Healy & Sutherland, 1991) and Logo activities (for example Fig. 2) all aimed at provoking the development of an algebraic approach to problem solving.

Algebraic thinking focuses on the idea of the general — general mathematical objects, general relationships and general methods. This generalising is constrained by certain rules, rules which derive from the problem situation or from algebra itself. Whereas pupils readily generalise in the mathematics classroom, they very rarely take into account the necessary constraints on their generalising. Central to algebraic thinking is the general and algebraic method which involves expressing relationships between general mathematical objects and manipulating these relationships. This emphasis on general mathematical objects, general relationships and general methods can be contrasted with the characteristics of arithmetic thinking which focuses on situation specific methods and specific values. The emphasis on structure in algebraic thinking can be contrasted with an emphasis on process in arithmetical thinking. Algebraic thinking does not replace arithmetic thinking — it supersedes it, becoming a new vantage point from which to view arithmetic. The movement backwards and forwards between arithmetic and algebraic thinking is an important aspect of problem solving in mathematics.

The theoretical rationale for the study was influenced by previous research on pupils' understanding of algebra (for example Booth, 1984; Filloy & Rojano, 1989; Küchemann, 1981) and a consideration of the potential of computer-based environments for learning algebra. Results from work with computers often conflict with established results on pupils' difficulties with traditional algebra, particularly with respect to pupils' understanding of literal symbols. For example in computer environments pupils accept the idea that a literal symbol represents a general number and readily use unclosed algebraic expressions such as 'x+3' (Sutherland, 1992, Tall, 1989).

The Role of Interacting with a Symbolic Computer Language

Analysis of text books currently used in UK classrooms suggests that school mathematics tends not to take advantage of the mediating role of algebraic symbols. Natural language is assumed to be the mediator of mathematical thinking. So for example pupils are usually asked to express a mathematical relationship in natural language *before* they are asked to express it in algebraic language. Work with computers is provoking a re-questioning of this dominant anti-symbol ideology. Algebra-like computer languages support pupils in their problem solving constructions. When working on new and challenging problems pupils formulate general relationships by interacting with the computer language and in this way use the computer language to express their mathematical ideas. Pupils incorporate Logo or spreadsheet code into their talk. This is particularly interesting in the case of spreadsheet code because within Project AnA pupils were taught to use the mouse to enter spreadsheet formulae and were never explicitly taught the spreadsheet-algebraic code (although we referred to the code when talking to the pupils). The following is a typical example of two pupils' discussion whilst working on a spreadsheet problem.

Jo   *Equals now what do you click on Sam....so what*
            will it be.... B2 minus 4...
Sam   *Yes B2 minus 4.....no wait a minute ....*

We have previously reported many other similar examples of pupils incorporating Logo code into their talk (Hoyles & Sutherland, 1989). The Logo or spreadsheet code is used as a means of communicating with other pupils and with the computer. In this sense the computer-based language is supporting inter-mental functioning, which from a Vygoskian point of view is a precursor of intra-mental functioning.

Expressing Mathematical Generality

Expressing mathematical generality involves both thinking with a general object and thinking about general relationships between objects. The fact that most pupils do not interpret traditional algebraic symbols as representing general objects (Küchemann, 1981) is a barrier to their use of these symbols to express general mathematical relationships. When working in a Logo or a spreadsheet environment pupils learn to view a symbol as representing a general number and this may be the most important aspect of work with these computer environments. It seems to be the interaction with the computer-based algebraic code which supports pupils' development from thinking in terms of a specific object to thinking in terms of a general object. In Logo it is the naming of the variable in the title line of a procedure which is important — pupils learn from experience that changing the value of this variable changes the size of the geometric image being generated. In the spreadsheet it is the cell itself which represents a general number.

The following example (Fig. 1) illustrates the way in which pupils focus on a specific number when first faced with an activity which involves constructing equivalent expressions in a spreadsheet. Andrew and Graeme (aged 10) were finding expressions equivalent to "multiply an unknown number by 4". They entered the number 3 in cell A2 and then in another cell entered the rule A2*0.5*8 which produced a value of 12. In a new cell they then entered a new expression A2*3 + 3 which also produced a value of 12. The second expression suggests that they were thinking with the specific number 3 in order to construct the equivalent expressions. When they changed the number in A2 they realised for themselves that the second rule was not equivalent and changed it without any need for intervention from the teacher. In this sense their work at the computer had pushed them from thinking of a cell as representing a specific number to thinking of a cell as representing a general number.

	A	B	C	D
1				
2	3	=A2*0.5*8		=A2*3+3
3				
4				
5				
6				

Fig. 1 Andrew and Graeme's equivalent expressions for "Multiply an unknown number by 4".

As part of Project AnA pupils were interviewed individually before, during and after they carried out the computer activities. They were asked to solve some paper-based algebra and some computer-based questions. At the beginning of the study most of the pupils said that they could not answer the algebra questions because they had never seen anything like them before. Many of them interpreted a letter as representing the position in the alphabet. In subsequent interviews some pupils began to refer spontaneously to their Logo or their spreadsheet work when presented with the algebra questions *"because we've done it on the computer.....because we've been using like ...ah ...variables and such and working out with the variables...... it just gave me a picture of how that might be done.* When probed more about what this picture was this pupil said *"a formula with the letter and then it looked like that but with the little colons"* (referring to the Logo notation for signifying a variable). This pupil had been working on problems like the one presented in Fig. 2. Another pupil Rachel wrote down the following solution to an algebra question :m x 3 + 1, again using the Logo notation for a variable. When Rachel was asked (in the beginning interview) the question *"If John had J marbles and Peter had P marbles what could you write down for the number of marbles they have altogether?"* she gave an answer of 25 (saying that the letters represented their position in the alphabet). In the final interview she responded *"Well J marbles could be anything and P marbles could be anything ....say J could be 10 and P could be 12....so the answer could be any number".* I suggest that this represents an important move in the development of algberaic thinking and that this development has been mediated by the algebra-like symbols of the computer environment.

Developing an Algebraic Method

Booth suggests that "many of the difficulties which children have in algebra are not difficulties in algebra as such, but are rather difficulties in arithmetic. Many children i) do not explicitly consider method in arithmetic ii) may in fact use procedures which are not formalised, iii) tend to interpret expressions in terms of context so that the need for precision and rigour in the form of mathematical statements is not appreciated, and iv) even in the case of those mathematical procedures of which they are explicitly aware are not always proficient in their symbolisation" (Booth, 1984). Despite work on situation-specific mathematical strategies in the workplace (for example Lave, 1988) there has been very little attempt to classify these informal strategies in the mathematics classroom. In this paper I shall use the word situation-specific to mean a method which involves thinking with the objects of the situation. Lins (1992) in an experimental study investigating the ways in which Brazilian and British pupils solved algebra word problems, found that non-algebraic strategies almost always involved an underlying whole-part model. This is illustrated by the way some pupils solved the measurement of a field problem (Fig. 3).

Measurement of a Field

The perimeter of a field measures 102 metres. The length of the field is twice as much as the width of the field. How much does the length of the field measure? How much does the width of the field measure?

Figure 3: Algebra Word Problem used in pre- and post-interview

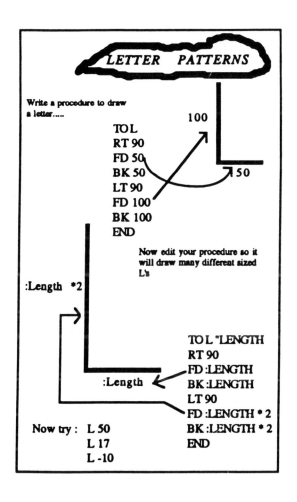

Fig. 2 The "L" Task.

Graeme used a situation-specific method to solve this problem *"I did 102 divided by 6 ... I just did two of the lengths to make it sensible .... I just thought there must be two of those in one length.........".* This solution involves working with a known whole (the perimeter) and dividing this into parts to find the unknown lengths of the side of the field. Another method used by the pupils can be described as a systematic "trial and error" approach *"well when I tried 40 it was 120 ..so I knew it must be smaller than that ....in the 30s ....and when I tried 36 and it was 108... I knew it couldn't be 35 ....so it must be 34"*.

A possible formal and algebraic method involves working with the unknown lengths (called for example X and L) and expressing these unknowns in terms of the known perimeter as illustrated by the following example:

$$
\begin{aligned}
\textit{Let the width of the field} \quad &= \quad X \textit{ metres.} \\
\textit{Let the length of the field} \quad &= \quad L \textit{ metres.} \\
\textit{Then } L \quad &= \quad 2X - (1) \\
\textit{and } 2L + 2X \quad &= \quad 102 - (2) \\
\textit{So by substituting (1) in (2)} \quad & \\
4X + 2X \quad &= \quad 102 \\
6X \quad &= \quad 102 \\
X \quad &= \quad 17 \textit{ metres.}
\end{aligned}
$$

The teaching approach which we used for the algebra story problems also involved working from the unknown to the known but within a spreadsheet (see for example Fig. 4).

The unknown is represented by a spreadsheet cell (this could be called x in the algebraic solution). Other mathematical relationships are then expressed in terms of this unknown. When the problem has been expressed in the spreadsheet symbolic language pupils can vary the unknown either by copying down the rules or by changing the number in the cell representing the unknown (in a paper and pencil algebra approach this would be the equation solving part). This approach is similar to the formal algebraic method outlined above and only differs in the final solution stage (the final stage of a paper-based solution involves solving an equation, the final stage of this spreadsheet algebraic approach involves using trial and error).

Rectangular Field

The perimeter of a rectangular field measures 102 metres.

The length of the field is twice as much as the width of the field.

Use a spreadsheet to work out the width and the length of the field.

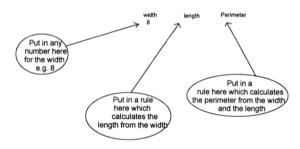

Then change the number for the width until you can answer the question.

Figure 4 : A spreadsheet-approach to the rectangular field problem

At the beginning of the study none of the 14-15 year old group of pupils were able to solve problems like the Measurement of a Field Problem. However by the end of the study all of these pupils could solve this problem using a spreadsheet and many could solve it on paper without a computer present (similar results were found for the Mexican group of pupils, see Rojano & Sutherland, 1993). The methods pupils used on paper were influenced by the spreadsheet activities as illustrated by Jo's paper-based solution to the following problem: "100 chocolates were distributed between three groups of children. The second group received 4 times the chocolates given to the first group. The third group received 10 chocolates more than the second group. How many chocolates did the first, the second and the third group receive?".

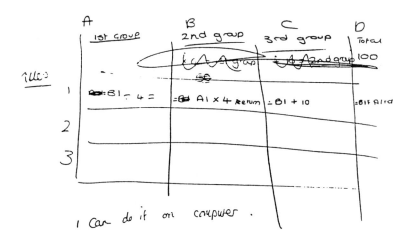

Figure 5: Jo's solution to the Chocolates problem (on paper) in post-interview

She had drawn a spreadsheet on paper to support her solution processes and had correctly written down all the rules represented in the problem. She had not specified the unknown and if she had been working at the computer the "circular reference" error message would have provided feedback on this error. When interviewing Jo I asked her *"If we call this cell X what could you write down for the number of chocolates in the other groups"* and she wrote down:

284

$$= X \qquad\qquad = X \times 4 \qquad\qquad = X \times 4 + 10$$

Jo, who had always been unsuccessful with school mathematics, had successfully carried out what is considered to be the most difficult part of solving an algebra story problem, that is representing the problem in algebraic code.

Some Concluding Remarks

In conclusion there are a number of possible reasons why computer environments support pupils to develop an algebraic approach to problem solving in mathematics. The most important is that pupils use the computer-based symbolic language to construct their own mathematical generalisation, which derives from their previous experience of arithmetic. In addition the computer frees pupils from the process activity of evaluating an expression, thus enabling them to focus more on the structural aspects of a situation. Within this chapter I have suggested a number of reasons why work with Logo or spreadsheets can help pupils develop algebraically. I am not claiming that these environments provide "the answer" to the teaching of algebra and clearly work with other environments can also be effective. The issue of how pupils make sense of their computer work in non-computer settings depends very much on the ways in which the teacher supports them to make links between the two settings. Results so far suggest that it is not as problematic as we might have supposed to transfer from, for example, a spreadsheet expression $(3A5 + 7)$ to an algebraic one $(3x + 7)$. This seems to be because both the spreadsheet symbol and the algebra symbol come to represent "any number" for the pupils.

References

Bednarz N, Radford L, Janvier B, Leparge A (1992) Arithmetical and Algebraic Thinking in Problem-Solving in the *Proceedings of the Sixteenth International Conference for the Psychology of Mathematics Education.*

Booth L (1984) *Algebra: Children's Strategies and Errors*, Windsor: NFER-Nelson.

Filloy E & Rojano T (1989) Solving Equations: The Transition from Arithmetic to Algebra, *For the Learning of Mathematics,* 9(2):19-25.

Hart K (1981), *Children's Understanding of Mathematics 11-16,* London: John Murray.

Healy, L. & Sutherland, R (1991) *Exploring Mathematics with Spreadsheets,* Oxford: Blackwell.

Hoyles C & Sutherland R (1989)*Logo Mathematics in the Classroom*. London and New York: Routledge.

Küchemann, D (1981) 'Algebra' in HART, K., (Ed), *Children's understanding of Mathematics: 11-16,* John Murray: London.

Lave J (1988) *Cognition in Practice*, Cambridge: CUP.

Lins, R. (1992) *A Framework for Understanding what Algebra Thinking is,* Unpublished PhD Thesis, Shell Centre for Mathematics Education, Nottingham University.

Rojano T & Sutherland R (1992) A New Approach to Algebra: Results from a study with 15 year old algebra-resistant pupils, paper presented at *Cuatro Simposio International Sobre Investigacion en Educacion Matematica,* Caudal Juarez Mexico.

Sutherland, R (1992) Some Unanswered Research Questions on the Teaching and Learning of Algebra, *For the Learning of Mathematics*, 11, No 3, pp 40-46.

Sutherland R & Rojano T (1993) A Spreadsheet Approach to Solving Algebra Problems, *Journal of Mathematical Behaviour*

Sutherland R (1993) *Project AnA — The Gap Between Arithmetical and Algebraic Thinking,* Final Report to the ESRC, Institute of Education University of London.

Tall D (1989) Different Cognitive Obstacles in a Technological Paradigm, in Wagner, S & Kieran C, *Research Issues in the Learning and Teaching of Algebra*, Hillsdale, USA

Acknowledgements — The author would like to thank the Economic and Social Science Research Council for their support for this project (Grant number: R000232132).

15 Bridging a Gap from Computer Science to Algebra

Programming non-arithmetical objects with 10th and 11th grades pupils

Jean-Baptiste Lagrange

In order to observe how pupils understand symbolic forms on non-arithmetical objects, we designed programming tasks. Many pupils failed at first in those tasks, because they did not conceive the expressions as actual computable objects. They understood string expressions as means to express actions and Boolean expressions as conditions in an alternative statement. Suitable understanding of expressions appears therefore to be deeply related to the ability to think of expressions at a "structural" algebraic level (Kieran, 1991). Our experiment indicates that teaching may help pupils to get a better algebraic understanding through a process of interaction between the use of the symbolism by pupils, and their conceptions of the objects.

Introduction

Computer technology does not only introduce new methods of teaching mathematics. It also induces new conceptual developments, promoting computer science as a new field for teaching secondary level pupils. Knowledge acquired by pupils in this field could be of some interest to mathematics learning insofar as it is related to algebra.

This view originates from a study of pupils learning to program non arithmetical objects like strings and Booleans to solve "real world" programming problems. The goal of that study was to bring new directions in early computer science teaching, and results were

289

obtained about pupils' conceptual difficulties, and teaching strategies toward the acquisition of programming knowledge (Lagrange 1991, 1992). The results also appeared to have many connections with the conceptualisation of algebraic ideas like variables and expressions. Furthermore, there seemed to be a consistency between those results and issues of a research on mathematics teaching in a computer environment (Artigue 1991).

In most studies in algebra, pupils try to use the symbolism to express relations on numerical objects of which they have arithmetical conceptions. The question that we want to clarify in this paper is how pupils may consider an algebraic symbolism on objects of which there is no arithmetic conception. The next paragraph specifies this question and presents its usefulness, from an analysis of the algebraic abilities required in a task of programming problem solving.

Analysing a programming task from an enlarged conception of algebra

We designed tasks to face pupils with the use of variables and expressions through modelling relations on various objects. Figure 1 is an example. A constant string is assigned to the string variable SCALE.

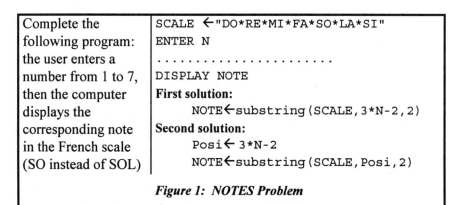

| Complete the following program: the user enters a number from 1 to 7, then the computer displays the corresponding note in the French scale (SO instead of SOL) | SCALE ←"DO*RE*MI*FA*SO*LA*SI"
ENTER N
. . . .
DISPLAY NOTE
First solution:
        NOTE←substring(SCALE,3*N-2,2)
Second solution:
        Posi← 3*N-2
        NOTE←substring(SCALE,Posi,2) |

Figure 1: NOTES Problem

The value of the numeric variable N is entered from the keyboard. The pupil has to write one or more program lines to calculate the string NOTE using the substring function. In the first solution, the constant string SCALE is the first argument, the position of the first character of NOTE is the second argument, and its constant length is the third. The position is computed from the variable N using a nested sub-expression. In the second solution, the position is first computed, then the NOTE string is again calculated using the substring function, but without nested sub-expressions.

What abilities are required in this task ?

- First, pupils have to name varied general objects: strings and positions, constants and variables.

- Also pupils must handle expressions with varied significances: the variable N naming a position, the expression 3*N-2 is related to "counting in three" then "translate two positions to the left". So the basic operations have different significances in this context of the positions, as compared with the pupils' familiar algebraic context of "glorified arithmetic". We have also expressions on objects like strings, which have no computable statute in ordinary mathematics teaching.

- Furthermore, pupils are required to formalise relatively complex relations, and then to express the formalization by means of complex expressions (with nested sub-expressions: first solution) or alternatively, by splitting it, using successive assignments to intermediate variables (second solution).

The above abilities have little relation to algebra, if this is limited to using unknown variables and deriving expressions to solve abstract equations. Nevertheless a trend in research suggests thinking of algebra in a less limited way. From numerous papers, two enlarged conceptions of algebra emerge which seem much related with the abilities we try to develop through a computer science teaching :

- Chevallard (1989) considers that modelling general relations in a computable way is a fundamental task for pupils to learn algebra.

He argues that pupils' arithmetic conceptions of numerical operations may bring difficulties in accessing generality.

- Kieran (1989) stresses that, to access algebraic thinking, a pupil must be able not only to see the general in the particular, but also to express it by means of an algebraic symbolism, and Kieran (1991) argues that algebra requires a "structural" understanding of this symbolism: pupils must be able to refer at a symbolic expression as if it was a real thing, in order to manipulate it as a whole.

Following Chevallard (1989), we assume that programming problems asking pupils to model relations on objects may be valuable algebraic tasks and that the problems could be even more interesting if there are no arithmetic conceptions of those objects. We expect that, using data like strings or Booleans to program general relations, pupils will spontaneously consider generality, and face the task of expressing it by means of the symbolism of the programming language. Their challenge will be to build proper significance for the expressions within this symbolism. Following Kieran (1991), this challenge is deeply related to algebra, because building a significance for the expressions implies considering them as actual symbolic computable objects.

The goal of this paper is therefore to study the behaviour of pupils in a programming task where a "structural" understanding of expressions is needed

1. How do they spontaneously understand the expressions ?

2. What form of computer science teaching may help them to build a suitable understanding?

Answers to these questions may actually contribute to the perception of how algebraic understandings are built, illustrating current theory on learning algebra. Furthermore, mathematics teaching may find those answers immediately useful when introducing algebraic symbolism with non-arithmetical objects (like functions, sequences, logical values, probabilities...), and when using pieces of software requiring an algebraic language to operate non-arithmetical items.

In the next paragraphs, examples of difficulties encountered by pupils using strings or Boolean data types demonstrate how pupils may spontaneously (mis)understand algebraic symbolism. These proved consistent with difficulties which appeared when using a computer language on geometrical items. Afterward, a classroom experiment illustrates how computer science teaching may help pupils to build better understandings.

Study of pupils' spontaneous (mis)understanding of expressions

Pupil's difficulties in tasks involving strings
The string data type is a "basic" type, taught in most early computer science courses, and useful to organise non numerical items in a computable way. To pupils, expressing calculations on items that they consider to be intuitive objects, is a very new thing. In addition, expressing calculations on positions within strings requires giving a different significance to the well known basic operations. It is therefore interesting to observe pupils trying to use algebraic symbolism with both strings and positions.

Tenth grade pupils were observed in tasks involving strings, by means of paper tests and directed work on computer. Pupils were in their first year of computing science teaching and had very varied levels of abilities. Pupils programmed an imperative computer language, PASCAL or BASIC.

The HOELL problem (figure 2) is an example. It was set, in a paper test, to 54 pupils at the end of the year (pupils had 75 hours teaching in computer science including the string data type).

A program is requested: the user enters a string X and a number N lower than string's length; then the computer displays a string R built from X, with the last character being moved to position N.

Example: user enters the word HELLO and the position 2, computer displays HOELL.

The aim in the following program is separating X into three parts (A is the part of the word to the Nth character, then B is the part from the N'th character to the last but one, and C is the last character), and then building the result from A, B, C. Complete this program :

		Solution	
A ← substring(X,...,...)		1,	N-1
B ← substring(X,...,...)		N,	L-N
C ← substring(X,...,...)		L,	1
R ←		A + C + B	

Figure 2 HOELL Problem

The pupils' task was to complete the numerical arguments in the substring functions, and to build the concatenation of those substrings in the suitable ordering (C between A and B).

Among 54 pupils, only two were successful. Wrong answers are classified in two groups of difficulties.

- The first difficulty is writing correct arguments in the substring function, and in the operation of concatenation. 42% calculations of A, B or C and 66% calculations of R, are wrong. For instance, pupils give substring(X,L,N) for the first part of the word, to the Nth character. From recorded interviews with pupils trying such programs on a computer (see Lagrange, 1991), we know that when a variable like L represents the length of a string, pupils may not consider it as a number, but as a description of a counting process through the string. Therefore substring(X,L,N) stands for "count the characters of the string and stop before the Nth".

Another example is the calculation of the result R by concatenating the three parts in initial order: A + B + C. Many studies of programming (for instance Rogalski and Hé, 1989) stated that pupils tend to think that whenever a statement is not completely explicit, the computer will understand it in an intelligent way. Answering A + B + C, the pupils give the computer the information that the result is made from the substrings A, B and C, and think that it is smart enough to arrange them in the suitable ordering.

- In another class are the difficulties to express calculations on positions (83%). For instance, in a resistant error, frequently observed and analysed in interviews, the sub expression L-1 is given for the position of the last character (C) in a string of length L. To pupils, as reported above, L may stand for a counting process on the string. We observed also that the operations (+, -) may be understood as actions on the string when calculating positions of characters within a string. Therefore L-1 expresses an action : "Counting the characters in the string to the end, and cutting the last one".

The above difficulties are representatives of misunderstanding encountered in many other problems, even after formal teaching about string data types (see Lagrange, 1992). Spontaneous conceptions of strings and positions are not computable, and therefore algebraic symbolism is misunderstood:

- arguments in string expressions stand for information with a vague statute, that pupil gives to the computer to obtain a result; and not for components within a computable symbolisation.

- string and ordinal expressions, particularly when being nested, are thought to express actions on objects, instead of being actual objects. In consequence, pupil may give no significance to the assignment statement.

295

Pupil's difficulties in tasks involving Booleans

The purpose of the Boolean data type is to express a logical condition in a computable way. Being a "basic" type, it is a possible subject in an early computer science course. In contrast, pupils in France have no mathematics teaching about Boolean calculations. Aiming to answer questions about understanding of algebraic symbolism, when no arithmetical conception exists, it was interesting to observe how pupils would react to algebraic symbolism on logical values. Eleventh grade pupils were observed in tasks of modelling with Boolean data types. Pupils were in their second year of computer science, and mostly from scientific divisions.

The INVITE problem (figure 3) asks the pupils to model a familiar situation of social constraints by building Boolean expressions. Each "friend" is represented by a Boolean variable, with the value `True` if he or she is "invited" and `False` otherwise. The logical value of each clause has to be expressed by means of a Boolean expression, including one or more logical connectors. Assignment of Boolean sub-expressions is useful to split complex calculations (see the solution to Clause 3).

The problem has been set to 9 pupils in an 11th grade scientific division. Pupils had 100 hours computer science teaching including an initiation into the Boolean type. Analysing pupils' answers, it appears that they gave up using algebraic symbolism on Booleans, using instead alternative statements: `IF <condition> THEN <ACTION1> ELSE <ACTION2>`. However, to avoid complicated nested alternative statements, some pupils made use of Boolean connectors to build complex conditions. Some also used "short" alternative statements (without the `ELSE` part), `<ACTION1>` being an assignment of a logical value.

I have got 5 friends: Marie, Marc, Luc, Janine and Jean
I wish to have some of them to dinner, but there are preferences or incompatibilities, which I express in the following clauses :

- Clause 1: "Marie and Jean are not suited to each other. I cannot invite them together."
- Clause 2: "Marc and Marie will not come one without the other: If one is invited, then the other must be also invited".
- Clause 3 : "If Janine is invited, Luc or Jean must be also invited, but it is impossible to invite the three of them together".

I want you to write a program to know whether it is possible to invite a group of friends together: the program is to ask 5 questions ("Is Marie invited (Y/N)"...) then compute and display the logical value for each clause.

A Solution

```
Clause1 ← NOT (Marie AND Jean)
Clause2 ← (Marie = Marc)
Clause3a ← (NOT Janine) OR (Luc OR Jean)
Clause3b ← NOT (Janine AND Luc AND Jean)
Clause3 ← Clause3a AND Clause 3b
```

Figure 3: INVITE Problem

We have therefore four classes of solution (figure 4). The class "*Full nested alternative statements* " uses no algebraic knowledge. No answer exists in that class, because solutions are very complicated especially for clauses 2 and 3. At the other end of the table, the class "*Short alternative statements with complex condition*" is close to the expected solution which can be completed using Boolean symbolism.

	Action: display a message	Action: assignment of a Boolean value

Simple Condition	*Full nested alternative statements*
	`IF Marie`
	`  THEN IF Jean`
	`    THEN DISPLAY`
	`      'Clause1 not valid'`
	`    ELSE DISPLAY`
	`      'Clause 1 is valid'`
	`  ELSE DISPLAY`
	`    'Clause 1 is valid'`
	No pupil's solution
		Nested short alternative statements
		`Clause1 ← TRUE`
		`IF Marie THEN`
		`  IF Jean THEN`
		`    Clause1 ← FALSE`
		
		One pupil's solution

Complex Condition	*Conditional display*
	`IF Marie AND Jean`
	`  THEN DISPLAY`
	`    'Clause1 not valid'`
	`  ELSE DISPLAY`
	`    'Clause 1 is valid'`
	Three pupils' solutions
		Short alternative statements with complex condition
		`Clause1 ← TRUE`
		`IF Marie AND Jean`
		`  THEN Clause1← FALSE`
		
		Three pupils' solutions

Figure 4: pupils' solutions to INVITE Problem (Clause 1)		

However pupils encounter two difficulties in tackling that:

- first, to write the expected solution (`Clause1←NOT(Marie AND Jean)` for clause1), pupils would have to use the Boolean connector "`NOT`". Indeed, they do not like to use connectors to which they cannot assign a significance from familiar conditional propositions.

- second, pupils have difficulties in considering the direct assignment of a Boolean expression to a Boolean variable. Indeed most pupils already understand the assignment of a Boolean constant (`Clause1 ←True` or `Clause1 ←False`) and consider easily a Boolean expression when it is the "condition" part in an alternative statement `if <expression> then <action>`, because those forms are close to natural language statements. In contrast, assigning a Boolean expression like `NOT(Marie AND Jean)` to a Boolean variable means referring a general object by way of a variable. Therefore, to consider that assignment, pupils should think of an expression of this kind as an actual Boolean object. Interviews of pupils (Lagrange 1991) confirmed that pupils understood Boolean expressions preferentially as "conditions" issued from familiar conditional propositions.

Pupils think of non-arithmetical objects at a familiar non computable level

The experiment indicates that most pupils give to the string expressions as well as to the calculations on positions, significances issued from usual conceptions of those objects: information, action.... In the same way, most pupils think of Boolean expressions from everyday "conditions". Trying to use the algebraic symbolism to express relations on objects that they conceive at a familiar non computable level, they tend to give the symbolism significances pertaining to that level of conception.

Results are consistent with research like that of Samurcay (1989) about the way pupils give a meaning to variables when included in algorithmic structures, specially regarding difficulties with the assignment statement. However, the use of symbolism, to which our results are related, could be wider and closer to algebra, because within tasks in our study, difficulty comes from expressions and not from the algorithmic structure.

Studying pupils learning to program drawings with the aim of making a generalisation, Sutherland (1989) points out that difficulties to express generality arise when a gap exists between "pupil's informal methods and the formal representation of this method". This seems close to our study, however, in Sutherland's study the gap is between a method where the drawing is not thought of to be general, because ratio relations between its components are not identified, and a formal expression of the relation between proportional items. Therefore, the gap is not only of an algebraic kind, but also strongly related to the ratio concept.

Other researchers, like Capponi (1989), observed how pupils use a spreadsheet to express general relations on arithmetical objects. In that context, arithmetical conceptions arise, inhibiting the access to generality, and the teaching strategies are directed toward overcoming that obstacle. In contrast, in our study pupils have no difficulty to conceive generality, but they fail to express it by way of algebraic symbolism because of their familiar conceptions, and therefore teaching strategies may be turned toward the acquisition of computable conceptions of objects and structural understanding of the expressions.

Using a computer language on mathematical objects

A consistent analysis may be made from difficulties encountered when using a computer language in mathematics teaching, for instance, from observations of pupils using *EUCLIDE* (Artigue, 1991). *EUCLIDE* is a French product designed to be a "geometrical computer language". Pupils use it to build expressions on geometrical objects to define and draw variable geometrical figures, with the aim to perceive their general properties. For instance, given three variable points, a program (figure 5) defines and displays the fourth point in a parallelogram. *EUCLIDE* is written as a non ergonomic super set of *LOGO*, and therefore, its use by pupils is not very realistic. However, several mathematical packages, like *MATHEMATICA* or *DERIVE*, whose use is fully realistic, may be extended with a similar geometrical language, and above misunderstandings, that Artigue observed, are likely to appear with those extensions.

Difficulties related to the language are:

* Pupils using EUCLIDE tend to ignore formal rules governing the position of arguments in expressions: for instance, to

`SOIT "D1 DRPP :A :B`
`SOIT "D2 DRPP :A :C`
`SOIT "D3 DRPAR :D1 :C`
`SOIT "D4 DRPAR :D2 :B`
`SOIT "D INTDD :D3 :D4`
`DES :D`
	`SOIT` is the assignment statement: LET...BE..
	`DRPP :A :C` Straight Line containing points A and C
	`DRPAR :D :C` Straight line containing point C, and parallel to straight line D
	`DES` is the display statement.

Figure 5: An example of EUCLIDE program (Allard, Pascal 1986)

express the image of point A in the line of symmetry with respect to line D, pupils persist in writing the expression `SYM :A :D` instead of `SYM :D :A`, not understanding why the device does not distinguish itself which argument is the point, and which argument is the line.

• Some pupils use the EUCLIDE assignment statement `Soit <variable> <expression>` like the mathematical definition: *Let* <variable> *such that* <property>. For instance to define point C as symmetrical to point A with respect to point B, they write `Let "B be MIDDLE :A :C`, instead of `Let "C be SYMP :B :A`. Other pupils do not distinguish the assignment statement from the display statement: they cannot understand the difference between assigning and displaying an object.

The first difficulty is linked to the idea of expression on geometrical objects. Pupils do not conceive that computer expressions are governed by formal constraints which do not exist in the usual geometrical language. The second difficulty results from structural differences between the computer language and the usual geometrical language. In the former language, a value is named from a result in a calculation, obtained by means of an expression. In the latter, naming an object consists of choosing within a (non empty) set of objects owning a property. Furthermore, naming an object is little distinguished from drawing it on the figure.

As above, pupils try to use a symbolism to express relations on objects they do not think of in a computable way. However, in this situation, conceptions of objects are not issued from a familiar everyday context, but are built through mathematical teaching. Difficulties arise from this gap between existing conceptions of objects, and computable conceptions required to use the symbolism properly. As reported in Artigue's observations, difficulties may inhibit mathematical activity when pupils are not prepared to cross this gap.

A teaching experiment to bring better conceptions of objects with help of computers feedback

In a programming activity on "non-arithmetical objects", pupils have no difficulty to consider generality, but they fail to use properly the algebraic symbolism, because their conceptions of objects are not appropriate to a symbolism. Aiming to build a teaching experiment to bring better use of symbolism, we assumed that teaching may use both symbolism's constraints and computer output as a resource to help pupils adapting their conceptions when completing a programming task (figure 6). But to make it work, the task must be designed to get meaningful feedback from symbolism and computer running.

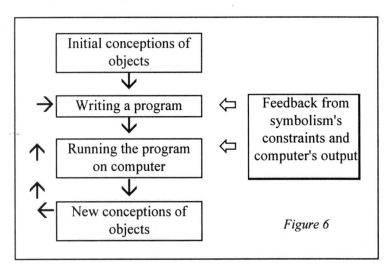

Figure 6

In an exploratory attempt, we designed a teaching experiment related to the strings and positions calculations. It consisted of 9 sessions from one to one and a half hours. The idea of 'type' was assumed to provide a meaningful symbolism constraint, and tasks were designed to require a special attention to it: to process large numbers, pupils had to make an explicit choice of a type (string or number) for the objects in the program, verifying the consistency between type and operations. The use of the assignment statement was presumed to be another meaningful feature. Within task, pupils had to split complex string processings into serial calculations, assigning intermediate results to variables. They were therefore faced the constraints of the processing of variables and with expressions in the computer. For instance, organising a serial calculation, pupils had to give a significance to the variable in the first part of the assignment statement.

We had special observation of 4 pupils, initially getting great difficulties in programming. At the end of the sessions, they passed a paper test to compare them with the pupils that we tested before. The test proved a good understanding of the type constraints, and a more computable conception of the function of each variable when building an expression. The pupils used properly more often the assignment statement, getting then a computable conception of the way expressions are processed in the device. Better use of the symbolism was consequently obtained.

Conclusion

From programming tasks concerning non-arithmetical items, we attempted to offer a bridge between early computer science teaching and algebra learning. In those tasks, pupils tried to express generality using the symbolism of the programming language, but they failed because they did not at first conceive the expressions as actual computable objects. This confirms that in those tasks, a "structural" understanding of expressions is needed, deeply related with the ability of thinking objects at an algebraic level (Kieran, 1991).

Pupils faced constraints resulting from the type of the objects, and from the use of the assignment statement, then they recognised that their spontaneous conceptions of the objects were not consistent with the symbolism. From appropriate programming tasks, and using computer feedback as a resource, teaching then helped pupils to build computable conceptions.

In this process of interaction, the computer provided feedback as well as representations. Understanding built through this process is nevertheless algebraic because it is at a "structural" level which is required in algebra. Consequently, this process of interaction could be of some interest when aiming algebraic skills more directly linked to mathematics.

References

Allard, J.C., Pascal C. 1988, *EUCLIDE Un langage pour la géométrie* Paris: CEDIC/NATHAN Ed..

Artigue, M. 1991, 'Analyse de processus d'enseignement en environnement informatique' *Petit x*, 26, 5-27.

Capponi B., Balacheff N. 1989, 'Tableur et calcul algèbrique', *Educational Studies in Mathematics* 20, 179-210.

Chevallard Y. 1984, 'Le passage de l'arithmétique à l'algèbre dans l'enseignement des mathématiques au collège' *Petit x* 19, 43-72.

Kieran C. 1989, 'A perspective on algebraic thinking', *Proceedings of the XIII P.M.E. Conference,* 163-171.

Kieran C. 1991, 'A Procedural-Structural Perspective on Algebra Research', *Proceedings of the XV P.M.E. Conference,* 245-253.

Lagrange J.B. 1991, 'Représentations mentales et processus d'acquisition dans les premiers apprentissages en informatique' *Thèse* Université Paris VII.

Lagrange J.B. 1992, 'Mental Representations of String Data Type: an experimental study on pupils learning to program' *Proceedings of the fifth workshop of the "Psychology of programming interest group"* F. Détienne Ed., INRIA, Rocquencourt France.

Rogalski J. , Hé, Y. (1989). Logic abilities and mental representations of the informatical device in acquisition of conditional structures by 15-16 years old students. *European Journal of Psychology of Education,* 4, 71-82.

Samurcay R. 1989, 'The concept of variable in programming: its meaning and use in problem solving by novice programmers', *In* Soloway and Spohrer (Eds) *Studying the Novice Programmer.* N.J: Lawrence Erlbaum.

Sutherland R. 1989, 'Providing a computer based framework for algebraic thinking', *Educational Studies in Mathematics* 20, 317-344.

16 Using a Computer Algebra System with 14–15 year old Students

Mark Hunter, Paul Marshall, John Monaghan and Tom Roper

This paper reports on an experiment in which 14-15 year old students used a Computer Algebra System to study quadratic functions. Two classes of middle and high ability students in two English schools used palmtop and laptop computers as the principal learning medium for this topic for three weeks. Data was collected via tests and was supported by video-tapes of student activity and interviews. Issues concerned with student learning, technical problems, public versus private technology and student motivation are addressed.

Introduction

This paper describes research carried out on students' understanding of linear and quadratic equations using the Computer Algebra System (CAS) *Derive*. Monaghan (1993) and Hunter & Monaghan (1993) report on a pilot study for this work. The sample consisted of middle and high ability year 10 students (14-15 years of age) in two state comprehensive schools. The research design was that of control and experimental groups.

The aims of the study included ways in which a CAS can improve the learning of selected algebraic and graphical concepts and skills, particularly with reference to the link between the two representations of the same concept. The research also focused on whether the use of a CAS enabled the development of concepts at an earlier stage of mathematical development. The main reasons for choosing the quadratic topic and the age and ability range of the students are that: i) the topic is suited to algebraic and graphical representations and *Derive* illustrates the interconnections particularly well, ii) very little research has been published on CAS use with students under 16 years of age, iii) the topic is usually reserved for high ability students. It was anticipated that the use of a CAS could open up routes to understanding higher mathematics to students of average ability.

The majority of the students in the experimental groups used palmtop computers, with *Derive* ROM cards, which they were able to take home. The remainder worked in pairs on laptop computers which they could not take home. One aspect of the study was whether personal or public technology had an effect on students' understanding, motivation and styles of use. The technical difficulties inherent in using a CAS in general and *Derive* in particular were also addressed.

This paper concentrates on the quantitative aspects of the research, although qualitative data from the video tapes and interview is used where it adds to the discussion. A fuller account of the qualitative interpretation of the technical difficulties can be found in Wain (1994)

Methodology

Two mixed ability state schools in the north of England were involved in the research project. School A was in Leeds. The top set formed the experimental group and the second set (out of six) formed the control group. School B was in Rochdale. The top set formed the control group and the second set (out of three) formed the experimental group. School A takes students from a range of backgrounds. School B is an 'inner city' school. The teaching groups were of approximately 30 in school A and 25 in school B.

In each school, the experimental group was taught by a member of the research team who was also on the staff of the school concerned.

The experimental group followed a programme of instruction based on a CAS that started with one week on linear equations. This was intended to introduce the pupils to the CAS whilst working on a familiar topic. Plotting graphs, finding the x- and y-axis intercepts and solving the equations were studied. Students were encouraged to look for connections between their results and the original linear equations. These patterns were then interpreted in algebraic and geometric terms by the students where possible. These and other results were then summarised and interpreted for the class by the teacher. For the next two weeks the students followed a similar programme for quadratic functions: plotting graphs, finding minimum points and x intercepts, solving, expanding and factorising were all studied in the same way. All work was done using the CAS, with results and connections being recorded in writing. Students were encouraged to try their own quadratics once they had completed those on the sheets. Homework was not set during the research period, but students were encouraged to use the CAS outside the classroom.

In order to try to make the control group similar in style to the experimental group, the students were expected to draw a number of quadratic curves to collect results and find patterns and underlying structure. To facilitate this and to introduce innovative materials, an OHP was used with quadratic curves drawn on overlays. The students used a standard quadratic curve on tracing paper to assist with the production of graphs. The teacher took the role of summarising results and explaining the mathematics that gave rise to the data.

The research tool used for this project was innovative in that it used three methods of data collection. Both experimental and control groups were given a pre- and a post-test (for each of algebra and graphs) based on the CSMS tests (Brown, 1984). In addition each group was given a post-test focusing on the linear and quadratic content. The experimental groups were given this test twice, once with only pencil and paper available and once with a CAS (only those using palmtops took the second administration). Each experimental lesson was videoed, with one of the research team asking questions of the students. One third of the control group lessons were observed. In most lessons, two or three individuals or pairs of students were videoed working. At the end of the research period, six students from each experimental group were interviewed using a hierarchical focusing technique (Tomlinson, 1989). The purpose of the interviews was to establish the extent of home use and the students' perception of the computer work, from both motivational and problem solving viewpoints.

Results

The three categories of tests (algebra, graphs and quadratic) contained many questions. Only results informing subsequent discussion are noted here. A full account is available in Wain et al (1994). For the algebra and graphs tests, only those students who took both the pre- and post-tests were included in the analysis. The tests were marked according to the CSMS criteria, using a coding of 0 for no attempt, 1 for correct and codes 2 to 9 to classify different errors and misconceptions. For the purpose of this data analysis, the answers were classified only as correct or incorrect, but the additional information is available for more detailed analysis. Each question was then considered to see if there had been an improvement (i.e the student had answered incorrectly in the pre-test but had then answered correctly in the post-test), a decline (i.e. correct in the pre-test and incorrect in the post-test) or no change (the category of error was not important). The total number of changes for each question was then considered, both up and down. The statistical test used was a sign test. The null hypothesis was that there was no change in performance, so that any changes occurred randomly and were binomially distributed with probability of increase being 0.5. A two tail test was used at the 5% significance level.

Significance calculations were performed to enlighten discussion. As prior hypotheses were not made concerning outcomes in every case, significant results should be considered as tentative and in need of further investigation.

Algebra Test

A1) 4 added to *n* can be written as *n* + 4. Add 4 onto *n* + 5.

A2) What are the areas of these shapes?

    a) 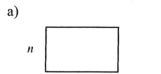   b)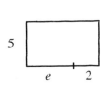

A3) What can we write for the perimeter of this shape?
    Part of the figure is not
    drawn. There are n sides
    altogether, all of length 2.

A4) What can you say about u if $u = v + 3$ and $v = 1$?

A5) Mary's basic wage if £20 per week. She is also paid another £2 for each hour of overtime that she works. If *h* stands for the number of hours of overtime that she works and *W* stands for the total wage (in £s), write down an equation connecting *W* and *h*.

Table 1 below shows the numbers who improved and declined on the questions above. It also shows the numbers who could have improved or declined (data not accounted for in the sign test).

Table 1: Student post-test performance in relation to pre-test performance on questions A1 to A5 above.

Entire refers to all students, *experimental* and *control* refer to CAS and traditional groups, *higher* and *middle* refer to sets, and *(A)* and *(B)* refer to schools.

Qn.	Group	Number who could improve	Number who did improve	Number who could decline	Number who did decline
A1	Entire	31	19	52	5
	Experimental	18	11	23	2
	Middle	21	12	19	2
A2a	Entire	16	10	67	2
A2b	Control	38	8	4	1
A3	Control	19	8	23	1
	Middle Control (A)	12	6	9	0
A4 Mid. Exp. Girls (B)		0	0	9	6
A5 Higher Exp. (A)		19	6	3	0

Graphs Test

There were no significant falls in the individual questions and any significant gender differences were restricted to the boys. The only significant changes concerned three questions.

G1) John leaves home to go to a disco in Cambridge 3 miles away. He walks one way and takes the bus the other way. This is a graph of his journey:

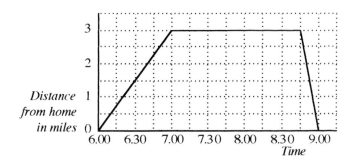

What do you think he was doing between 7.00 and 8.45?

G2) Two of the lines A, B and C represent the same information.

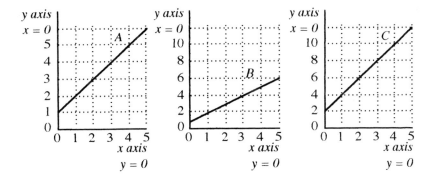

Put a ring round the letters of the two lines that represent the same information:

A   B   C

G3) Which of the graphs below represents the line $y = 2x$?

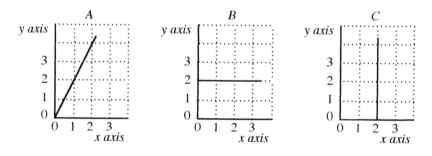

Table 2: Student post-test performance in relation to pre-test performance on questions G1 to G3 above.

Entire refers to all students, *experimental* and *control* refer to CAS and traditional groups, *higher* and *middle* refer to sets, and *(A)* and *(B)* refer to schools.

Qn.	Group	Number who could improve	Number who did improve	Number who could decline	Number who did decline
G1	Entire	28	12	54	0
	Middle Control (A)	7	6	12	0
G2	Entire	37	20	45	7
	Middle Control (A)	9	6	10	0
G3	School (B)	34	8	9	1

Quadratic Test

The test was administered twice after the period of instruction. The first time was to both control and experimental groups, using pencil and paper. The second administration, two weeks later, was to those who had used palmtops only and they had access to the CAS for the test. There were 20 questions in the test. Only results informing subsequent discussion are noted here and full details are available in Wain et al. (1994). The global test results are shown in table 3 below.

Table 3: Number of students taking the quadratic tests and mean number of correct answers (out of 20).

Experimental and *control* refer to CAS and traditional groups respectively and *(A)* and *(B)* refer to the schools.

School	Group	Number present			Mean number of correct responses	
		First	Second	Both	First	Second
A	Exp.	24	20	18	1.96	8.5
A	Control	20			3.25	
B	Exp.	17	13	12	0.35	2.4
B	Control	20			3.7	

The table shows that the control groups performed better than the experimental groups in the pencil and paper sitting. Using a binomial model there is a significant increase in School B ($p < 0.0005$) but not in School A ($p = 0.133$). Note, however, that the School A control group was set two out of six whereas the experimental group was the top set. The table also reveals a significant increase in the performance of the experimental groups in their second sitting ($p < 0.0001$ for School A, $p = 0.005$ for School B).

The number of correct responses for School A were ranked and comparisons were made. This was not possible for School B as the first administration had a very low number of correct responses.

The rank correlation coefficient for the first and second administration for the experimental group was 0.685 (n = 20) which is significant at the 5% level. Only three questions had large rank differences, they were:

Q1   Draw $y = 2x + 1$ (ranked 3 first time, 13 second time)

Q2   Solve $2x + 1 = 0$ (ranked 6 first time, 1 second time)

Q3   Factorise $x^2 + 3x - 4$ (ranked 12 first time, 5 second time)

The increased performance in Q2 and Q3 can be ascribed to 'button pushing' e.g. press 'L' for Q2, press 'F' for Q3.

The rank correlation coefficient for the first administration of the experimental group and the control group was 0.595 (n = 20), which is significant at the 5% level. Only two questions had large rank differences, they were:

Q3   Factorise $x^2 + 3x - 4$ (ranked 12 for the exp., 4 for the control)

Q4   Factorise $3x - 12$ (ranked 4 for the exp., 14 for the control)

It is possible to ascribe these results to the effect of instruction, as the control group practised factorising quadratic expressions by hand but did not do any linear work.

Qualitative Results
Students experienced problems using graphic commands appropriately (scale, centre, zoom)

- Students experienced problems interpreting graphic and algebraic output e.g. solution of $3x - 5 = 0$ is 5/3 but the root of the graph of $y = 3x - 5$ appears to be 1.64444.

- Students experienced problems with *Derive*'s language e.g. 'cannot do implicit plots'. However, there were several examples of students taking in *Derive* terminology e.g. 'I'm going to Delete this graph'.

- Opinions as to the usefulness of *Derive* were mixed. Some students felt that it had assisted their progress in the topic whilst others felt that they had spent their time simply pushing buttons. Several students expressed their frustration with the technicalities of operating the machine.

- Home use of the palmtops was almost exclusively reserved for playing a built in 'Pacman'-like game.

- Palmtop users tended to work in isolation.

Discussion

The results of A1 and A5 in the algebra test suggest that the higher attaining experimental group had an improved understanding of the use of a letter to stand for a variable compared to the control group. Ascribing to students an understanding of the concept of a variable is difficult for, as Küchemann (Hart, 1982, p110) states, 'many items that might be thought to involve variables can nonetheless be solved at a lower level of interpretation.'

There was also evidence of a decline in ability to interpret the substitution question A4 correctly by the girls in the middle set experimental group. The relationship $u = v + 3$ is a functional one, one which the CAS effectively masks. To respond correctly, an understanding of the role of substitution is required - a skill for which the CAS has removed the need. This interpretation can be cast directly in terms of procepts (Gray and Tall, 1993). Function is a procept, that is both a process and an object. Students need to move flexibly from one form to the other. Substitution reinforces the process aspect of the function procept and the CAS, by bypassing this, obscures the true nature from the student. In a similar vein Kaput (1992, p.535) grants that 'computer graphing shortcuts the numerically mediated graphing experience' but knows of no studies directly testing this.

Question A2b suggests an improvement in expanding expressions amongst the control groups. This was shown mainly by the set two pupils in School A and seems to represent the teaching style of that group rather than the fact that there was more room for improvement with the lower attaining pupils. This improvement in expanding expressions can, however, be seen in the light of improved performance in generalised arithmetic items.

Questions A2b and A3 can be viewed as generalised arithmetic items and the control groups exhibited an increased facility in both questions. A point to consider here is that an understanding of generalised arithmetic is a necessary prerequisite for meaningfully operating a CAS but the concept is not developed by the use of a CAS. A more radical way of expressing this is that an understanding of generalised arithmetic is concerned with ascribing meaning, that is a semantic feature, whereas a CAS works at a purely syntactic level. It appears unlikely that it will, on its own, assist students' semantic problems. Further evidence for this thesis comes from the responses to Q2 and Q3, indicating that the improved performance by the experimental groups in the second quadratic test could be ascribed to 'button pushing', that is to syntactic aspects.

It seems that the use of the CAS has both advantages and disadvantages particularly with the concepts that surround the use of variables. A CAS will work correctly whether or not the student has an understanding of the mechanism that it uses. Generalised arithmetic (the use of letters to stand for unknown numbers) is an intermediate step between arithmetic and the use of letters to stand for variables. The CAS uses letters to stand for variables, regardless of the students readiness for this approach. The weaker pupils seemed to benefit from the control approach to the use of letters within algebra and the ensuing exposition or discussion of their

318

meaning. However, the more able pupils seemed to benefit from the more abstract approach, probably because they understood the prerequisite steps. The implication is that there is a stage before which a pupil does not have the necessary algebraic concepts to understand the use of a CAS. Beyond this point, however, the CAS does confer advantages for learning.

This syntax/semantic divide, however, is not fully supported by the graphs test results. The results for question G1 show an overall improvement and no real differences between the groups on a question requiring the understanding of a distance-time graph. This is surely a question concerned with interpretation (semantic considerations) and stands in contrast to a thesis that CAS use emphasises only syntactic features. The results, however, are compatible with the results of Ruthven (1990) who, in a study concerned with graphic calculator use, found that graphic calculator users did not score better than non-graphic calculator users in similar interpretation questions.

The results of question G2 indicate little difference between groups for the scaling aspect of graph work though a significant improvement in the School A control group emerged. Students of this age commonly experience problems choosing suitable scales for graphs and video evidence revealed CAS users experiencing considerable problems with scaling - in one case two higher set students spent 20 minutes unable to view their graph because they could not obtain the correct scale. It is possible to partially ascribe problems here to the software, for *Derive* appears to encourage students to believe that 'it will do the thinking for them', which it does not. Choosing a suitable scale involves the transformation

The graph window environment interposes a further set of mental decisions and physical acts that interrupt this 'flow' i.e. **Plot, Scale, Centre,** ... and the transformation becomes

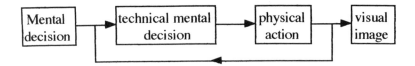

It is quite reasonable to assume that these further actions generate 'cognitive noise' that interferes with the thinking required in the basic transformation.

Question Q1 produced some interesting answers consistent with G3. It was expected that the responses would show an improvement when the students were using a CAS, although this was not the case (indeed, in the rank order this question dropped from third to thirteenth out of 20 questions). A detailed examination of the students' responses seems to suggest that the CAS allows the students to focus on the main detail (the shape of the line being parallel to $y = 2x + 1$) whilst not being forced into interpreting other graphical features such as, in this case, the y-axis scaling. This was also observed in other responses, for example there was no increase in students' performance in items that required them to sketch a quadratic with particular roots.

These cases are examples of a CAS masking some of the detail that would traditionally be needed, for example that of deciding on scales before plotting points. This is a similar result to the one that Hurd (1993) reports, where students studying the Newton-Raphson method via spreadsheets showed a better understanding than those using *Derive* for the same purposes, because the use of a spreadsheet make the mathematics more transparent. Any piece of technology places an intermediary between the student and mathematics. If the student does not master the intricacies of the technology, then the learning process is bound to be affected. This applies both to short term progress and longer term motivation. From qualitative evidence, instruction must be given both on the technical aspects of using the software and on the interpretation of the forms of output a CAS is likely to produce.

There was a lower than expected performance of the experimental groups. A possible reason for this is the teaching and learning styles employed in both groups and the relationship of these styles to performance on pencil and paper tests. The control groups were taught in a style in which methods were presented and exercises set to reinforce the skills inherent in these methods. This is supported by the results from school A on Q3 and Q4 where the control group practised factorising quadratic expressions by hand and overall performed much better at manual factorising questions. The control group did not study linear work and therefore did comparatively worse in a linear factorisation question. However, the experimental group did better when they were able to push buttons to reach their answers.

The approach to the CAS supported work of an investigative nature which encouraged students to explore rules generated from CAS patterns. Moreover, due to the nature of the different approaches, the control group were often addressed as a group so that work could be corrected and results related to overall patterns. The experimental group, however, mainly progressed at their own pace with relatively short periods of correcting and summarising results. The research team felt that the approach employed with the control group generally lead to repeating learnt procedures in a closed-test situation whereas this link is less obvious with students in the experimental group.

The investigative intention of the experimental groups' work was not fully realised in practice. This is in contrast to the results of the pilot studies reported in Hunter and Monaghan (1993), where students regularly created and tested their own rules. On one occasion in the pilot study in school A, when the class were brought together for an explanation of the rules for factorising a quadratic, the teacher acted as a mere agent for writing down rules discovered by the students. Nothing like this happened in the same school in this study. It appears that this is due to motivation (the pilot groups were, to all observers, better motivated) and to the smaller amount of student-student debate when they used palmtops as opposed to laptops. The use of personal technology in the form of palmtop computers did not assist student performance or generate independent work at home. Problems connected to the use of palmtop computers were; difficulties with teacher monitoring of student work and student complaints that the screen was uncomfortably small. The investigative approach was also hampered by difficulties in interpreting the data that the CAS produced.

Conclusion

From the graphical results, there are no significant differences between the performance of the control and experimental groups. This seems to indicate that, on the whole, experiencing graphical work through graph sketching and drawing or through computer generated graphs makes little difference to the student's progression. However, the algebraic results do show some significant differences that indicate that a CAS can be advantageous in some aspects of algebra, whereas a more traditional approach is advantageous in other aspects. A CAS can be of benefit for the students' learning of abstract elements of algebra as long as the students are mathematically ready to use it. However a more traditional approach with its attention to constructive detail is more appropriate up to that stage.

There is a body of opinion that the proximity of functional and graphical representation, the ease of movement between the two and the speed at which many graphs can be drawn have great potential for enhancing the graphical understanding of the symbolic. Here *Derive* is being used more as a tool, and as such it would appear to be more effective than when used as a teaching agent. Further research in this area could pay dividends especially as it would be building upon an already established body of research concerning graphical understanding. The necessity to understand scaling was highlighted in this research and further study comparing this aspect of graphic calculators to CAS would be beneficial.

The study generates a number of other questions for further research. From the algebraic and graphical results, there is scope for further research on the progression of understanding from arithmetic, through generalised arithmetic to the understanding of a variable. In particular, the role that substitution plays in this progression would merit further study.

Future studies can learn much from the outcomes of this study. We recommend that such work ensures that students are given a general introduction to the CAS they will be using and are prepared for problems with graphical and numeric output.

The project reported here was carried out in conjunction with Peter Johnson and Geoff Wain and was funded by the University of Leeds Academic Development Committee.

References

Brown, M., Hart, K. & Küchemann, D. 1984, *Chelsea Diagnostic Mathematics Tests*. Windsor: NFER-Nelson.

Foxman, D., Ruddock, G., McCallum, I. & Schagen, I. 1991, *APU Mathematics Monitoring (Phase 2)*. Slough: NFER

Gray, E. & Tall, D. 1993, 'Success and Failure in Mathematics: the flexible meaning of symbols as process and concept', *Mathematics Teaching*, 142, 6-10

Hunter, M. & Monaghan, J. 1993, 'Quadratics Made Easy?', *MicroMath*, 9(3). 20-22

Hurd, M. 1993, 'Student Learning with a Computer Algebra System and a Spreadsheet' in Monaghan J. & Etchells T. (eds.) *Computer Alg-ebra Systems in the Classroom*, 54-67. Leeds: University of Leeds

Kaput, J. J. 1992, 'Technology and Mathematics Education', pp. 515-556 in D. A. Grouws (ed.) *Handbook of Research on Mathematics Teaching and Learning*, New York: Macmillan.

Küchemann, D. 1981, 'Algebra' in K. M. Hart (ed.) *Children's Under-standing of Mathematics: 11-16*, 102-119, London: John Murray.

Monaghan, J. 1993, 'Using Computer Algebra Systems to Teach Quadratic Functions' in Bohm, J. (ed.) *Teaching Mathematics with Derive* (Proceedings of the Krems International School on the Didactics of Computer Algebra), 51-55. Bromley: Chartwell-Bratt.

Ruthven, K. 1990, 'The Influence of Graphic Calculators Use on Translation from Graphical to Symbolic Forms', *Educational Studies in Mathematics*, 21(5), 431-450.

Tomlinson, P. 1989, 'Having it Both Ways: hierarchical focusing as a research interview method', *British Educational Research Journal*, 15(2), 155-176.

Wain, G. T., Hunter, M., Johnson, P., Monaghan, J. & Roper, T. 1993, *Report of the Leeds Year 10 Computer Algebra Study*, CSSME, University of Leeds.

Wain, G. T. 1994, 'Some Technical Problems in the Use of *Derive* with School Pupils', *The International Derive Journal*, 1(1), 49-55.

Section Five:
Advanced Mathematics -
Some Perspectives

There were many papers submitted to this book which could have fallen within this section. Unfortunately most authors did not come close to fulfilling the criteria for seeing technology as a bridge between teaching and learning, rather discussing the technology with regard to the mathematics to be taught. There were also many conceptualisations of ways in which computer environments might address mathematical objectives. However, most of these failed to consider the position of the learner or the role of the teacher. A possible conclusion is that the state-of-the-art in teaching at this level is to focus on mathematical concepts without addressing pedagogical considerations.

The five papers which are included do address issues in learning and teaching, albeit somewhat diversely. Issues include the effect of algebra packages on the perceived curriculum from the point of view of the teacher and autonomy of the student; the excitement, for both teacher and students, which can emerge from exploring a mathematical concept using an algebra package; the use of supercalculators in the teaching of linear algebra ways in which spreadsheets can help technology students with conceptually difficult mathematics; and the use of MATLAB in the *assessment* of a mathematics course for electrical engineers.

In Chapter 17, *Computer Algebra and the Structure of the Mathematics Curriculum*, Andrew Rothery addresses the role of a computer-algebra package (in this case DERIVE) to support students in solving optimisation problems without direct use of calculus. Study of his own students' responses to questions in their introductory mathematics course provides evidence of successful problem-solving resulting from use of the computer package. He suggests that rather than *replacing* the need for techniques of calculus, the freedom gained allows students to take more responsibility for their own learning needs. Some might choose to learn calculus as a result of their experiences, deciding for themselves when they should learn a particular topic and to what depth. He argues, further, that computer use allows greater freedom for the teacher in devising curricula more suited to their students' needs. What are the implications of these claims for the curricula of the future?

In Chapter 18, *Prime (Iterating) Number Generators with Derive,* Patrick Wild describes his use of the *Derive* package to stimulate investigation of iterative sequences generating prime numbers. Many questions were raised providing exciting mathematical challenges for himself and his students. The investigation led to his own developing awareness of alternative learning and teaching approaches. The chapter highlights the role of *Derive* in stimulating the described mathematical activity. To what extent did the package contribute to the excitement and enthusiasm generated? Was this more that just one serendipitous experience?

The teaching of linear algebra is the focus of Donald LaTorre in Chapter 19, *Using Supercalculators to Affect Change in Linear Algebra.* Students use the calculators in 'an active constructive environment' with evidence of their enhanced conceptualisation. The calculators relieve the tedium of matrix manipulation. However, this raises issues about curriculum content. Students no longer need to practice their matrix manipulation skills, so time can be given to higher level thinking processes. What are the implications for the linear algebra curriculum?

Chris Bissell presents, in Chapter 20, *Models of Technology,* a perspective of the spreadsheet as a medium to enhance the understanding of technological models for which conceptually difficult mathematics proves problematic for students. He offers, most lucidly, a number of examples to illustrate the potential of spreadsheets, particularly their graphical capabilities, and suggests to teachers ways in which students might be supported and challenged. Does this medium offer an avoidance of mathematics rather than an increased conceptual understanding? This might be set against Andrew Rothery's suggestions that the technology can motivate students' own approach to conceptual exploration. What are the implications of mathematics being seen primarily as a technological tool?

In the final chapter of this section, 21, *Using Technology In Examinations*, Jim Tabor discusses the use of *Matlab* in developing the mathematics needed by electrical engineers, arguing that a good understanding of the relevant mathematics is required. He makes a strong case for the use of computing facilities in the *assessment* of a course for which a computer package is an integral part of its teaching. He addresses practical issues both in building a package into the teaching and into the assessment of the course, showing that standard examination procedures can be observed. An interesting question which the reader might consider is whether use of a computer environment could be a catalyst to consideration of *new* forms of assessment, perhaps valuing the experimentation by students which the package encourages, rather than being limited by traditional forms of assessment.

The messages which these papers offer to teachers of higher mathematics seem to encompass two important phenomena:

- the crucial nature of a pedagogic awareness in teaching which encourages students to interact with technology and take responsibility for their own learning;

- a shift from seeing teaching as a transmissive process to a broader view of students as (co)-constructors of their own mathematics.

17 Computer Algebra and the Structure of the Mathematics Curriculum

Andrew Rothery

Ways in which the use of computer algebra systems will change the sequencing of the mathematics curriculum are the focus of this paper. Observations of teaching and learning experiences, in which students use calculus, are presented to illustrate the manner in which students work with computer algebra and in particular to indicate how the structure of the curriculum may change. Such changes arise in a working environment in which the use of computer algebra bridges the gap between the teacher's choice of learning activities and the student's ability to cope with the work involved.

The curriculum debate

Curriculum developments which arise from the use of technology in general typically offer opportunities

- to design new ways of teaching existing topics;
- to enable more effective work in modelling and the applications of mathematics to be carried out;
- to make advanced mathematical topics more accessible;
- to provide learners with an alternative approach to thinking about a particular piece of mathematics - whether a problem solving strategy, a concept or a proof.

Computer algebra systems can offer all these opportunities. Reports on curriculum developments arising from the use of computer algebra systems indicate that most aspects of the teaching and learning process will be affected by this particular technology. (For example, Ainley and Goldstein, 1993; Bohm, 1992) However, in

329

addition to enhancing the quality of the delivery of the curriculum, the use of computer algebra has a direct and fundamental effect on the nature and sequence of the curriculum itself. Since computer algebra systems carry out skills and processes which students are traditionally taught to do themselves, it becomes a realistic possibility that whole areas of the curriculum might become redundant and major changes in the nature of the curriculum will take place. For example, "Do we need to teach integration by parts any more?" Curriculum change is the key issue addressed in this paper.

When evaluating the potential use of a technology which sets out to improve teaching and learning, teachers have to decide whether the proposed use is going to be effective or not. Improvement is always welcome. However, if use of a technology changes the curriculum then there is likely to be some considerable debate. When a particular mathematics topic is threatened with redundancy, someone will defend it. Reasons for inclusion of a topic are often highly personal, based on individual perceptions about which topics are essential to the nature of mathematics or typify its essence. Agreement is not achieved easily, if at all.

Some teachers and lecturers see changes to the curriculum as a threat - a result of understandable caution about failing to teach certain ideas which may be needed at some later stage in mathematics. The curriculum is a complex system and if you remove a critical principle at a lower level you might remove support for a higher level topic. For instance, differentiation of composite functions (the 'chain rule') is initially taught as a technique for differentiating complicated functions, and is a candidate for redundancy since computer algebra systems can perform such differentiation easily. Yet in later mathematical work it is sometimes useful to change variable in a particular equation and a rule such as $\dfrac{dy}{dx} = \dfrac{dy}{dt}\dfrac{dt}{dx}$ is needed, not as a calculating technique, but to alter the *form* of an expression. Decisions on when to teach mathematical principles like this become difficult ones which people are reluctant to tackle in a hurried way.

However, I will try to explain in this chapter why a conventional curriculum debate on which maths topics to 'leave in' or 'throw out' is in many ways irrelevant. Whilst lecturers and teachers are struggling to decide how best to change the curriculum, they will be overtaken by their students, who will be the ones to decide what to learn and when.

A bridge between teaching and learning

Computer algebra can help create a bridge between leaning and teaching through its impact on the student's progress through the curriculum. Traditionally it has always been the teacher's responsibility to organise the curriculum and to structure the teaching and learning experiences of the learner. This can of course lead to a mismatch between the learner's progress and the pace and structure of the teaching programme. A student who fails to properly understand a particular concept, principle or strategy will find it difficult to cope with the next set of activities in the course being followed. Teachers have always been familiar with this problem and have a number of strategies for coping with it. They can study research into student's learning difficulties to better understand their misconceptions and re-structure their teaching approach. Or they might offer remedial help. They might plan self-teaching individualised materials. But unless the teaching is carried out on a one to one interactive basis, most teachers are, quite reasonably, unable to plan a teaching programme to match the rate of progress of each individual learner for whom they are responsible.

Computer algebra systems add a new dimension to the process of matching the teacher's programme of work to the student's programme of learning. Even if a student fails to master a particular skill they can still go on to the next stage of learning by using computer algebra as support. This process is sometimes called 'scaffolding'. Kutzler (1994) describes the scaffolding metaphor by likening the learning of mathematics to building a house. In this metaphor each stage of mathematical development is seen as adding a new storey to "the house of mathematics" but if a student fails to build a particular storey the new ones cannot be built properly.

However, computer algebra acts as a scaffolding and helps prop up the storeys so that the later ones can be built despite the early ones being incomplete.

As a result of using computer algebra as 'scaffolding', the ownership of the responsibility for the sequence of learning topics can be shared more equally between student and teacher. Students themselves can choose when to 'go back' and learn the things they missed. The bridge is formed by the closer partnership this sort of approach can bring. The following sections of the paper illustrate this idea through accounts of students' work.

Examples of teaching and learning with the aid of computer algebra

In the following sections, I describe and discuss some examples of using computer algebra in my own teaching. These examples illustrate a number of ideas about the use of computer algebra but they have been chosen to show how the sequence of learning is affected.

The students are first year degree students at Worcester College of Higher Education following an introductory mathematics course. The specific examples given refer to work carried out in the 1993/94 academic year with a group of 20 students. Their backgrounds are varied. In particular, in terms of calculus knowledge, they fall into three (not very distinct) categories:
1. those who are reasonably able and confident in calculus techniques,
2. those who have followed an A-level course but lack facility in their use of techniques, and
3. those who have followed access courses other than A level and have had little or no experience of calculus.

Teaching such a mixed group provides interesting opportunities to compare the differing levels of knowledge with the progress made.

Computer algebra was introduced into this course in the previous academic year, 1992/93, to see if this would allow students the chance to participate in mathematical modelling activities regardless of their ability in calculus. The computer algebra was successful for this purpose (reported in Rothery, 1994) and its use has been retained and indeed encouraged further.

The particular program chosen was *Derive*, mainly because of the combination of a good level of performance with a low cost. Licensing arrangements enable students to own their own copies at a very low cost.

Maximising a cone

Suppose you wish to make a cone from a circular piece of paper by cutting out a segment from the circle. What angle should you choose in order to maximise the volume of the cone? An account of school pupils working on this problem is given by Murray (1993) but for students working at a more advanced mathematical level, the problem looks as though it ought to be a straightforward application of calculus.

When my students tackled this problem they found that creating the algebraic model, i.e. the formulas to describe features of the cone, was not as simple as it looked but needed some thought. The students who enjoyed most success were the ones who started by constructing a spreadsheet to calculate volumes for different angles; they went on to use their experience of creating spreadsheet formulas to help them build the algebraic expressions needed for using calculus.

The volume of a cone, V, in terms of the radius of its base, r and the radius of the circular piece of paper, R, is

$$V(r) = \frac{1}{3} \pi r^2 \sqrt{R^2 - r^2} .$$
(1)

FIGURE 1 Making a cone from a circular piece of paper

When this is rewritten in terms of the angle of the segment, x radians, it becomes

$$V(x) = \frac{1}{3}\pi \left(\frac{R(2\pi - x)}{2\pi} \right)^2 \sqrt{R^2 - \left(\frac{R(2\pi - x)}{2\pi} \right)^2} \qquad (2)$$

which simplifies to

$$V(x) = \frac{R^2(2\pi - x)^2 \sqrt{R^2 x(4\pi - x)}}{24\pi^2}. \qquad (3)$$

As you would expect, students followed different methods of deriving the expression for $V(x)$ in equation (3), some more complicated than others. Some used degrees rather than radians and therefore had a slightly different version of the expression.

Even those students who had studied A level mathematics and were reasonably proficient in calculus techniques found the task of differentiating the expression for *V(x)* in equation (3) rather daunting. All students used *Derive* for this process since even those who might have coped without doing so recognised the task as a long one in which there was a high probability of making a simple mistake. For those students who lacked expertise in differentiation techniques or had never learned them in the first place, then using *Derive* was the only way they could solve this problem.

Some students avoided the expression for *V(x)* in equation (3) . They chose to work with the expression for *V(r)* in equation (1) to find the optimum radius even though this is not the explicit objective of the problem. Once the optimum radius is found, $\sqrt{\frac{2}{3}}R$, the corresponding angle can be calculated from the relationship

$$r = \frac{R(2\pi - x)}{2\pi}. \tag{4}$$

The optimum angle turns out to be $2\pi\left(1 - \sqrt{\frac{2}{3}}\right)$ radians. This strategy is a lot simpler than tackling *V(x)* directly. Since students were working in a workshop atmosphere, news of this simpler alternative approach spread quickly.

Working on this problem helped students learn to seek out different strategies for solving a problem and it improved their skills in handling algebraic expressions efficiently. All students chose to use *Derive* for most of the work, though this was mixed in with their own algebra 'on paper'. Those who had studied calculus at A level found that the algebraic expressions were more complicated than those they had previously been accustomed to and so they felt they could not work on their own without making a mistake somewhere - thus spending a lot of time checking for errors. The few who had not studied calculus in depth did not even know the product and composite function rules needed to differentiate *V(r)* and *V(x)*. Yet all students were able to participate in the activity, to learn the modelling skills in building the equations and to discuss the relative merits of whether to adopt *x* or *r* as the decision variable. Lack of fluency in or knowledge of differentiation techniques did not exclude students from the work. Yet without computer algebra this particular activity would have been impossible for all but the most able students.

This example illustrates the general benefits of using technology - for instance, to enable more effective work in applications of mathematics, and to make difficult activities more accessible. However it very much illustrates the way in which students are left to make their own decision as to when they need to learn a particular skill. The students who recognised that they themselves were unable to differentiate the expression could treat this lack of ability as a separate issue; they could say 'There's something I would like to learn sometime'. Without computer algebra their lack of skill would have simply led to an inability to proceed or to participate.

Happy families

Students were working in groups with families of curves (described in ATM, 1988). They were invited to devise the parametric equations for families which fitted the pattern indicated in the graphs and then compare features such as gradients, maxima and areas.

In previous years, this activity had been considered for inclusion in the course but it was decided not to use it. It took a long while for students to achieve results of any significance owing to the time taken to complete the algebra and calculus. As some students were weak in calculus and others had not yet studied it in any depth it meant that their attention would be restricted to the algebra of the parameterisation. The activity was considered useful but not useful enough to be worth spending a substantial amount of time on.

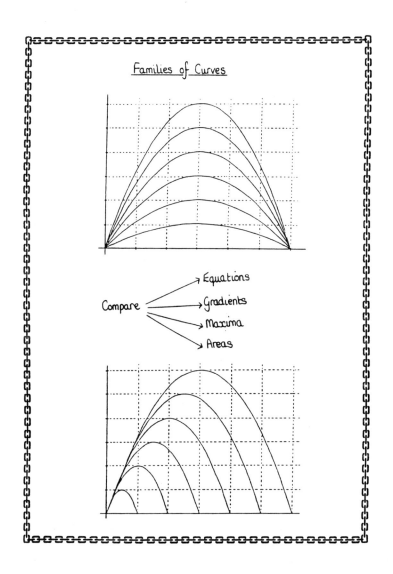

FIGURE 2 Families of curves worksheet

However, with the use of a computer algebra system students can complete the activity more quickly and complete the calculus tasks easily.

For the family of curves which cut the x axis at $x = 0$ and $x = 6$, students used a wide variety of ways (none using computer algebra) to work through the first phase of the investigation and obtain a parameterisation of the type $y = \dfrac{n}{9}(-x^2 + 6x)$. They could check for instance that when $x = 3$, $y = n$, confirming that the nth curve in the family has maximum value n.

Students who had not studied specific integration techniques did however know that integration was a process for finding area and so they immediately applied *Derive* to finding that the area under the nth curve,

$$A(n) = \int_0^6 \frac{n}{9}(-x^2 + 6x)\,dx, \tag{5}$$

was $4n$. Students more confident in calculus were able to find this result without using a computer. They preferred to work with a specific value of n since it was easier for them to apply their knowledge of integration techniques to an expression such as $-\dfrac{x^2}{3} + 2x$, which involves only x, rather than to one which includes both x and n. Students who did this then found that the areas were 4, 8, 16, 32, . . . and quickly arrived at a conjecture for A_n.

For the second family of curves, which intercept the x axis at $x = 0$ and $x = a$, the integral required is a little more difficult. Again, students who had not learned calculus techniques applied *Derive* immediately to the parametric expression for the curves to obtain an integral such as

$$A(a) = \int_0^a -\frac{4x(x-a)}{a}\,dx \tag{6}$$

and found the area to be $\dfrac{2}{3}a^2$. Most students who attempted to carry out the integration without using computer algebra again preferred to work with a specific value of the parameter, and in some cases without the factor $(4/a)$. Again, the reason for this was that it left a simpler expression, involving only the symbol x, to which it was easier to apply the integration rules.

Computer algebra was used eventually even by students who knew the techniques. In some cases this was to cope with a more difficult expression which included both the parameter a and the variable x; in other cases the computer was used because the student anticipated that they might make an error; in yet other cases, it was used to check their own results.

For instance, one error made in integrating the expression $-x(x-3)$ was to anti-differentiate the expression and obtain the incorrect expression: $-\dfrac{x^2}{2}\left(\dfrac{x^2}{2}-3x\right)$. Students with better facility in calculus knew to expand the initial expression to $-x^2+3x$ before going on to obtain the correct antiderivative, namely $-\dfrac{x^3}{3}+\dfrac{3x^2}{2}$. So for the weaker student, using *Derive* was an important check and alerted the student to mistakes.

This activity illustrates the style in which students work with computer algebra. Students adopt a mixture of their own 'pen and paper' work and *Derive* work, sometimes using *Derive* instead of doing it themselves, sometimes as well as doing it themselves. Students of different ability and background are able to participate in the same activity and are not denied the opportunity to build equations, organise their investigation, interpret results, make generalisations, and so on.

Students knowing calculus basic skills found that using computer algebra supported them in practising these techniques whereas those who had never acquired such skills coped quite happily without them. Naturally the latter were aware that they were unable to differentiate and integrate unaided by *Derive* but they could see that learning such skills was something they could tackle as a separate goal. Without computer algebra, lack of skill or even lack of facility in those skills becomes a stumbling block, conveying a sense of failure. With computer algebra, the identification of a lack of skill reveals an opportunity for further learning rather than a bar to progress.

Re-ordering introductory calculus

As an example of a more formal re-organisation of the sequence of the curriculum I shall outline the case of four particular students. The study of their work is currently in progress and a fuller account of this particular example is in press (Rothery, 1995).

Four students had not learnt calculus at all yet were required to carry out a modelling assignment which included optimisation of a function of a single variable. Would it be possible to offer them a brief amount of tuition which would enable them to cope with the task using *Derive*? How little do they need to know about calculus to be able to perform effectively?

They were shown a graphical interpretation of the notion of derivative which is quite commonly used in teaching to provide an introductory concept. They sketched graphs of derivatives given the graphs of particular functions by estimating steepnesses. They used this graphical experience to help them interpret the meaning of the functions which the computer algebra system produced in algebraic form by its process of differentiation. They did not themselves know the algebraic form for the derivative of any function; this was something they recognised that they had yet to learn.

The principle that maxima and minima of a function correspond to zeros in the derived function is one which proved easy to understand as it relates closely to work with graphs. It did not seem necessary to understand how to differentiate common functions algebraically in order to understand the strategy for finding maxima and minima. The whole method for using computer algebra to find the algebraic derivative, and then solving an equation to find the zeros of the derivative, and hence tracking down the positions of maxima and minima was easily learned. The amount of tutorial time for achieving this stage was three 30-minute sessions and initial indications are that students understood the topic. Further study of their progress is being carried out.

Though these students were soon able to find maxima and minima of a given function and had an appreciation of the nature of differentiation and its relation to optimisation, they themselves did not know how to find algebraic formulas for the derivatives of a given function. The students who adopted the computer algebra approach were very aware that they could not differentiate something as simple as $f(x) = x^2$ themselves yet did not regard this as a failure - it was simply something they could look into in due course. Indeed when they did look into it they used computer algebra to allow them to extend their own (non-computer) work on simple expressions like

$$\frac{(x+h)^2 - x^2}{h} = 2x + h \tag{7}$$

to more complicated expressions such as

$$\frac{(x+h)^5 - x^5}{h} = 5x^4 + 10hx^3 + 10h^2x^2 + 5h^3x + h^4 \tag{8}$$

and also much higher powers.

Traditionally, students are expected to learn the skill of differentiation before being able to go on to carry out optimisation, but the need for this sequence only arises because without computer algebra you have to be able to differentiate in order to be able to apply the principles of finding maxima and minima, not because the conceptual difficulty is any less. In fact, to learn to differentiate requires a formal approach to the definition of derivative, something which students find relatively difficult; it also requires proving and learning the principles of sum, product, quotient and composite function 'rules'. The ideas underlying the strategy for finding maxima and minima are not as demanding as the ideas underlying the formal process of differentiation. So teachers might find it useful to reverse the present order of topics more often.

Conclusion

In the examples, the curriculum debate did not centre on questions such as "should we no longer teach product and quotient rules for differentiation?", "do we need teach integration techniques". The fact that students could cope with optimisation without learning differentiation techniques did not mean they need never learn them. Instead, the use of computer algebra bridged the gap between the teacher's programme of work and the student's ability to cope with it; furthermore, the question of **when** the student should learn the techniques became more important than **whether**.

It will become less easy to define an 'official' or even a commonly agreed structure to the curriculum; decisions on the sequence of topics to be taught and mathematical problem-solving activities to be introduced can be taken in terms of the needs and backgrounds of particular groups of students, perhaps even sometimes negotiated with individuals. Conventional teaching sequences in terms of pre-requisite skills will often be completely re-organised. No-one need plan in advance whether or not certain mathematical topics should be discarded; however, some will inevitably become marginalised and eventually wither away. Since in many cases students will be able to carry out certain mathematical activities whether or not they have acquired a particular technique themselves, it will become even more difficult to describe the curriculum in terms of a syllabus of techniques. The challenge to curriculum planners will be to explore ways to define the curriculum so as to accommodate the more flexible approach which computer algebra will permit learners to take.

References

Ainley J., Goldstein R. (eds) 1993, 'Computer algebra systems special feature', *Micromath*, 9, 3.

ATM 1988, 'Families of Curves' in *Whatever Next?*, Derby: ATM.

Bohm J. (ed) 1992, *Teaching Mathematics with DERIVE*, Bromley: Chartwell-Bratt.

Kutzler B. 1994, 'DERIVE - the future of teaching mathematics', *International DERIVE Journal,* 1, 1, 37-48.

Murray J. 1993, 'Max cone', *Micromath*, 9, 3, 8-9.

Rothery A. 1994, 'Using computer algebra systems in teaching mathematical modelling' in Heugl H., Kutzler B. (eds) *DERIVE in education: opportunities and strategies*, Bromley: Chartwell-Bratt

Rothery A. 1995, 'Re-organising introductory calculus with computer algebra', *International DERIVE Journal,* 1 , 3.

18 Prime (Iterating) Number Generators with Derive

Patrick Wild

Prime Iterating Number Generators, or PINGs for short, are those mappings which successively generate a sequence of prime numbers where the initial value is itself prime. The mapping shown below produces a sequence of numbers where the first eleven terms are all prime. With a number of students aged 16-17 years I undertook, with the aid of DERIVE, to discover similar mappings. This paper is an account of our investigations. It attempts to describe how using a software package like DERIVE enabled those involved to explore uncharted areas of Mathematics

For example the mapping

$$x_{n+1} \rightarrow 3x_n + 16$$

$$\text{with} \quad x_1 = 587$$

generates the following sequence where the first eleven terms are all prime: [587, 1777, 5347, 48187, 144577, 433747, 1301257, 3903787, 11711377, 35134147]. The twelfth term is 105402457 which factorises to 67*137*11483.

It was under the umbrella of a General Studies sixth-form course, designed to demonstrate algebraic packages like DERIVE, that three Advanced Level Mathematics students began investigating the capabilities of DERIVE. During this familiarity process, a teaching colleague gave me a list of ten numbers to check to see if they were prime. Having checked them on DERIVE, by attempting to find factors and discovering that they were prime, there now came the first of a series of questions which was to lead the group along an exciting and rewarding route.

One of the students noted that all the primes ended in a seven. Almost as soon as one student had made this observation another remarked that the iteration always produced a number ending in seven as long as the first number ended in a seven. Multiplying by three would produce an integer ending in unity and finally adding an integer ending in six ensures a number ending in seven.

Well, now came a new flood of questions and discussion. Was such a sequence of primes due to random chance? Why concentrate on an iteration producing integers ending in a seven? Would an iterative scheme ending in unity, three or nine be better? Most of these questions came from the students and I found myself having to admit that I did not know the answers, but perhaps DERIVE could help us find the answers.

Until this stage I had only been demonstrating the built-in facilities of DERIVE. It was clear that in order to obtain the answers to these questions it would be necessary to write our own procedures. I gave a demonstration of instructions like ITERATES and VECTORS and within two lessons a lot of prime numbers were being generated.

One of the first questions that was investigated was which prime number ending was most frequent? Having decided in discussion that a thousand primes seemed to be a fair sized sample, the practicality of counting the frequency of each of the different endings seemed daunting. 'I thought that was what computers were for' challenged one of the students. I leapt to the defence of the new technology and I was determined somehow to get DERIVE to do the donkey work. The following procedures were developed to answer the above questions. One of the students wrote this procedure for producing a list of prime numbers.

PTAN G(n,p): = ITERATES(NEXT_PRIME(x),x,p,(n-1))

(where NEXT_PRIME (7) would output the next prime 11).

This procedure produces a list of n prime numbers beginning with the prime number p. It took 43.7 seconds to generate the first 1000 primes.

In order to analyse the unit column more easily the modulus to base ten was found of all primes in the previous list

(PTANG1(n,p):=MOD(ITERATES(NEXT_PRIME(x),x,p,(n-1)),10)

This now left a list of 1,3,7 and 9's to be counted. These values were now recorded in matrix form where [1,0,0,0], [0,1,0,0], [0,0,1,0] and [0,0,0,1] signify the values 1,3,7 and 9 respectively using the following procedure.

PSHAH(v,n):=
VECTOR (if(element(v,m)=1,[1,0,0,0], if(element(v,m)=3,[0,1,0,0],
 if(element(v,m)=7,[0,0,1,0], [0,0,0,1])))),m,1,n)

where v is the vector PTANG1 and m is a dummy variable within the routine.

Premultiplying with the unit row matrix

$$PRICE(n):=vector(x^0,x,n)$$

gives the total count of each of the values 1,3,7 and 9.

The totals seemed remarkably similar. One of the students wanted to know if there was a statistical test to confirm that each of the prime number endings was equally likely. I chose the χ^2 test as this was a topic which the students would eventually meet in their A-level course.

We had the following two hypotheses:

H_0 : distribution is rectangular.
H_1 : distribution is not rectangular.

Unit Value	Observed Frequency O_i	Expected Frequency E_i	Difference^2/E_i
1	245	250	25/250
2	254	250	16/250
7	255	250	25/250
9	246	250	16/250

$$\chi^2 \text{ calc} = 18/250 = 0.328$$

$\chi^2(0.95)=0.352$ 3 Degrees of freedom

We accept H_0 if χ^2 calc < 7.81 at 5% significance level.

χ^2 calc < 7.81 so we accept the null hypothesis H_0 that the distribution of prime number endings is rectangular.

Further questions arose from the students. Did we take too many primes so that the distribution evened itself out? Should we not look at the next thousand primes to see if the rectangular distribution is maintained? It was decided unanimously that those questions were best left for another day. For the moment it seemed we could search for a longer sequence of primes using iterative schemes involving any of the prime number endings. The burning question now was could we discover a simpler and longer PING than the original one?

Several iterative schemes were devised following the lead of the first example in that they produced the same value in the units column.

For example beginning with 11 the iterative scheme 2x-1 would always leave unity in the units column. At one stage the students were investigating an iterative scheme which would produce a sequence of numbers ending in an odd number but excluding five in the units column. At the time of writing no such scheme was found.

The previous investigations paved the way for the following procedure to be developed.

$$BECK(p,u,n) := VECTOR(ITERATE(u,x,p,m)m,1,n)$$

Where p is the initial prime, u is the iterative scheme, n is the number of iterations required and m is a dummy variable within the subroutine.

Hence BECK(41, 2X-11, 10) produced the vector

[71, 131, 251, 491, 971, 1931, 3851, 7691, 15371, 30731]

which when factorised by DERIVE gives

[71, 131, 251, 491, 971, 1931, 3851, 7691, 19 809, 79 389]

Notice how 15371 factorises to 19 x 809

This PING is simpler but only has nine successive prime numbers. With further investigation it became clear that a run of twelve or more primes from linear iterative schemes would be a rare occurrence.

The next stage of the investigation was to look at non linear iterations. More questions arose. Would it be harder or easier to find a long sequence of primes? What form would iterations take? Could I guarantee a 1,3,7 or 9 in the units column as before?

By now the General Studies course was ended but it was pleasing to see the investigations continue. The enthusiasm demonstrated by the students and the progress made left me feeling elated and excited: feelings I am afraid to confess that are not common to my normal teaching. This particular group and this particular investigation using DERIVE produced a partnership that allowed the boundaries of the group's mathematical knowledge to be stretched. What was just as pleasing was that there were as many questions still unanswered as answered. As far as I was concerned I experienced at first hand a glimpse of the potential that present technology had in both the teaching and learning processes. How much more can be discovered by approaching problems in an unfettered way where the tedious tasks are handed over to the computer. The main advantage of the new technology is that one is free to ask the most difficult questions and yet have a good chance of having them answered.

On reflection I think that what really made the investigation come to life was the fact that both teacher and students had the time and opportunity to allow the questions to keep bubbling into the open and, using DERIVE, were rewarded by answers. Having obtained some answers the impetus of the investigation was maintained. The boundaries of DERIVE were also being explored. It was just as important for the students to be aware of the limitations of packages like DERIVE as well as their potential as a mathematical tool.

What made this investigation so different to others I have been involved with was that I was as excited as the students by our success. The General Studies course consisted of twenty 40 minute lessons which will remain in my mind as the most fruitful lessons for a long time. The added bonus, of course, was that colleagues were now beginning to see the potential of the new technology in their own teaching. As a result of these investigations and interest among other students the school has now made available ten copies of DERIVE and has use of both MATHEMATICA and MAPLE.

Discussions with colleagues revealed another method of generating a sequence of primes. The following formula for example generates a sequence of forty primes: $T_n = n(n+1) + 41$

350

The following DERIVE procedure lists the first n values for a formula u beginning at n=1

$$PMARJ(n,u,x) := VECTOR(u,x,1,n)$$

Using this procedure PMARJ(40,x(x+1)+41,x) simplifies to [43, 47, 53, 61, 71, 83, 97, 113, 131, 151, 173, 197, 223, 251, 281, 313, 347, 383, 421, 461, 503, 547, 593, 641, 691, 743, 797, 853, 911, 971, 1033, 1097, 1163, 1231, 1301, 1373, 1447, 1523, 1601, 1681] where 1681 is 41^2.

The resulting collegiality was an added bonus.

Acknowledgements

Mr. A. Beck and Mr. O. L. C. Toller for their encouragement and support. F. Tang, R. Price and A. Shah my General Studies students for their enthusiasm and considerable skill with DERIVE. Mr. T. Marjoram for providing information on the 'formula' method. DERIVE version 2.3

19 Using Supercalculators to Affect Change in Linear Algebra

Donald R. LaTorre

High-level programmable graphics calculators, notably the HP-48G/GX and TI-85 supercalculators, are rapidly becoming commonplace on our university campuses. Although their formal use has been concentrated mainly within courses in calculus, introductory courses in linear algebra offer special opportunities for supercalculator enhancement. This chapter briefly outlines several of the ways that supercalculators can be used effectively in teaching courses in matrix-oriented, introductory linear algebra. It is based upon the author's experiences in teaching such courses over the past 5 years.

Introduction

Technology is helping affect major change in introductory linear algebra. Not only in terms of what we teach and what students learn, but also in terms of how we teach and how students learn.

Technology makes the power of matrix methods accessible to an ever widening range of disciplines. Thus, our courses are adopting new points of view. Topics such as matrix conditioning, the role of determinants and matrix inverses in a computational environment, and elementary iterative techniques are now being included in many introductory courses. And Gaussian elimination and the Gram-Schmidt orthonormalization algorithm are routinely being interpreted as LU and QR factorizations.

Because students learn best when their thinking is accompanied by doing, one of the prime benefits of technology in linear algebra is to facilitate a high level of active involvement with the course material. Activities that encourage students to explore, to conjecture, and to "discover" some of the very ideas we would have them learn are especially important in helping them to develop the independent thinking required for more advanced work.

Nowhere is this change more evident and more personalized than with hand-held technology in the form of the high-level programmable graphics calculators produced by Hewlett-Packard and Texas Instruments: the HP-48 and the TI-85. Equipped with sophisticated, built-in routines for matrix operations, high capacity expandable (in the HP-48) storage, and capable of communicating with other calculators and microcomputers, these devices are quite accurately termed supercalculators. They enable teachers to present linear algebra from a modern, matrix-oriented point of view and free students from unproductive hand computations - giving them a chance to concentrate more effectively on concepts. For both teachers and students, they promote and facilitate an increased level of mutual interaction, and serve to redirect the focus of teaching and learning from passive acceptance to active involvement. In this chapter I shall briefly outline several of the ways in which supercalculators can be used effectively in teaching courses in matrix-oriented, introductory linear algebra. My experiences are based on having taught such a course each term since 1989, in classes where each student has their own HP-48 calculator. A list of related publications providing further details is appended.

Systems of Linear Equations

Gaussian Elimination with back substitution is the most popular method for solving systems of linear equations in introductory linear algebra. It is a classic example of a simple algorithm that leads to a wealth of other concepts and ideas. Early on, I allow my students to use their calculator's built-in keystroke commands to carry out the elementary row operations of adding a multiple of one row to another row and to interchange rows. Later on, they use a slightly more automated procedure by executing a simple calculator program ELIM which, in a single keystroke, eliminates all entries below the desired pivot entry. I also require that they employ partial pivoting throughout the reduction process because it is widely incorporated into professional codes for linear systems. Partial pivoting quickly becomes unwieldy when done by hand because of the rational number arithmetic involved; but it is easy to implement with technology. After producing a row-echelon form for, say, the augmented matrix of a square linear system, students then perform back substitution one-step-at-a-time with program BACK. Thus, they control the entire process from start to finish by deciding when and where to pivot, which row interchanges to make, and then effect back substitution interactively. The calculators carry only the computational burden.

The same elimination procedure is applied to general linear systems that may include both pivot variables and free variables, the only difference being in back substitution. Since program BACK will not do symbolic back substitutions, students must backsolve by hand. However, I quickly make available to them a short program PIVOT that pivots on the specified pivot position to convert the pivot entry to 1 and then produces zeros elsewhere in the pivot column. PIVOT is thus effective in helping students generate the reduced row-echelon form (RREF) of the augmented matrix, from which all solutions can be readily obtained.

Applications to Vector Spaces

My students have always had difficulty understanding the vector space concepts of linear combinations, independence and dependence, and basis and dimension. But by concentrating on three vector spaces naturally associated with a matrix - the row space, the column space, and the null space - I have been more successful with these concepts than before. Supercalculators can be of real benefit here, for they facilitate students seeing these notions in the context of linear systems.

Consider, for example, the following matrix A:

$$
A = \begin{bmatrix}
4 & 1 & 8 & -6 & -3 \\
5 & -4 & 6 & -14 & -5 \\
8 & 4 & 6 & 2 & 6 \\
4 & -3 & 2 & -8 & -1 \\
-8 & 3 & -2 & 8 & -3
\end{bmatrix}
$$

Students use their calculator's RREF command to produce the reduced row-echelon form
E of A:

$$
E = \begin{bmatrix}
1 & 0 & 0 & 0 & 1 \\
0 & 1 & 0 & 2 & 1 \\
0 & 0 & 1 & -1 & -1 \\
0 & 0 & 0 & 0 & 0 \\
0 & 0 & 0 & 0 & 0
\end{bmatrix}
$$

All solutions to Ax=0 thus appear as

$$x_1 = - x_5$$
$$x_2 = -2x_4 - x_5$$
$$x_3 = x_4 + x_5$$

where $_4$ and x_5 are freely chosen. Since Ax=0 expresses the zero vector in R^5 as a linear combination of the column vectors A_1, A_2, A_3, A_4 and A_5 of matrix A, we have

$$-x_5 A_1 + (-2x_4 - x_5) A_2 + (x_4 + x_5) A_3 + x_4 A_4 + x_5 A_5 = 0.$$

Choosing $x_4 = 1$ and $x_5 = 0$ gives $-2A_2 + A_3 + A_4 = 0$, so that $A_4 = 2A_2 - A_3$ and A_4 is a linear combination of A_2 and A_3. Now choosing $x_4 = 0$ and $x_5 = 1$ gives $- A_1 - A_2 + A_3 + A_5 = 0$, so that $A_5 = A_1 + A_2 - A_3$ and A_5 is a linear combination of A_1, A_2, and A_3. A quick check of the first three columns of E shows that the first three columns of A are linearly independent. Students can now look back at matrix E and see how its columns reflect this analysis.

Since row operations are used to convert A to E, students readily understand that A and E have the same row space and that, indeed, the non-zero rows of E form a basis for the row space of A. But the column space of A is a bit different. Clearly A and E do not have the same column space because none of the columns of A (which lie in the column space of A) lie in the column space of E (since all vectors in the column-space of E have zeros for their last two components). Now think about finding a basis for the column space of A. Our earlier arguments show that columns 1, 2 and 3 of A are linearly independent; and, indeed, they span the column space of A because columns 4 and 5 are linear combinations of these first three. Thus columns 1, 2 and 3 of A, the columns that correspond to the pivot columns in E, form a basis for the column-space of A. Finally, consider the null space of A. All solutions to Ax = 0 can be expressed as

$$x = \begin{bmatrix} -x_5 \\ -2x_4 -x_5 \\ x_4 + x_5 \\ x_4 \\ x_5 \end{bmatrix} = x_4 \begin{bmatrix} 0 \\ -2 \\ 1 \\ 1 \\ 0 \end{bmatrix} + x_5 \begin{bmatrix} -1 \\ -1 \\ 1 \\ 0 \\ 1 \end{bmatrix}.$$

The two vectors on the right hand side are clearly independent (notice the placement of 0's and 1's), clearly span the solution space, and are, themselves, solutions (the first one comes from setting $x_4 = 1$ and $x_5 = 0$, while the second one comes from setting $x_4 = 0$ and $x_5 = 1$). Thus these two vectors form a basis for the null space. Notice, further, that the number of basis vectors (the dimension of the null space) is precisely the number of free variables.

This example serves to illustrate well the point we made earlier: computational capability (in this case, to obtain the RREF quickly) enables students to focus on the important vector space notions in the context of matrices.

Orthonormal bases and QR-factorizations

Of the many concepts that students encounter for the first time in linear algebra, perhaps none is so deceptively simple, yet so powerful, as that of orthogonality. Orthogonal projections, orthogonal bases, and orthogonal subspaces all play a key role in many of the more advanced topics in linear algebra: from least squares solutions of inconsistent systems to the more sophisticated theory and algorithms that depend upon factorizations with orthogonal matrices.

At the introductory level, supercalculators can be used to help students effectively engage the orthogonality concept that sets the stage for everything that follows: the construction of an orthonormal basis for a subspace W of R^n by the traditional Gram-Schmidt process. We begin with a simple calculator program, PROJ, that calculates the projection of a vector u onto a vector v in R^n:

$proj_v u = \dfrac{u \cdot v}{v \cdot v}$ program PROJ is then put to good use to help with the

Gram-Schmidt process as follows. The goal is to convert an existing basis $\{X_1, X_2, \ldots, X_k\}$ for W to an orthonormal basis $\{Q_1, Q_2, \ldots, Q_k\}$. Students begin by storing the vectors $X_1, X_2, \ldots, X_k$ in their calculators and building Q_1 as the normalization of X_1: $\quad X_1 \; X_1$ ABS /. Here, we are showing four keystroke commands in the reverse Polish notation used by the HP-48's. Recall that, having obtained orthonormal vectors $Q_1, \ldots, Q_j$ we obtain Q_{j+1} by normalizing X_{j+1} - (the sum of the projections of X_{j+1} onto $Q_1, \ldots, Q_j$). In terms of calculator use, after storing vector Q_1 my students proceed to construct Q_2 with a simple, one-line program:

$$\ll \; X_2 \;\; X_2 \;\; Q_1 \;\; PROJ \; - \; DUP \;\; ABS \; / \; \gg$$

So Q_2 is X_2 - $Proj_{Q_1} X_2$, normalized. Then, after storing Q_2, they construct Q_3 by

$$\ll X_3 \;\; X_3 \;\; Q_1 \;\; PROJ \; - \; X_3 \;\; Q_2 \;\; PROJ \; - \; DUP \;\; ABS \; / \; \gg$$

Thus Q_3 is X_3 - $Proj_{Q_1} X_3$ - $Proj_{Q_2} X_3$, normalized.

They continue in this way until all of $Q_1, \ldots, Q_k$ have been constructed. By composing and executing a simple program at each step, students reinforce their understanding of the algorithm and are able to review their work before execution.

Before supercalculators, only the best of my students were able to master the Gram-Schmidt process. The tedious hand calculations that were required ultimately prevented them from gaining sufficient experience with the algorithm. Now, all are successful, and go on to consider the process as a QR-factorization. The QR-factorization A=QR of an mxn matrix A of rank n is obtained by applying the Gram-Schmidt process to the column vectors of A. Here, Q is the matrix whose columns are the orthonormal vectors constructed by the Gram-Schmidt process, and R is the invertible upper triangular matrix whose non-zero entries are given by $r_{ij} = Q \cdot X_j$. I initially require that students build R using the DOT command on their calculators; but they inevitably discover that they may capitalize upon the orthogonal structure of Q and calculate $R = Q^TA$. We then apply our QR-factorization to obtain a least squares solution to Ax=b by backsolving $Rx = Q^Tb$ for x.

Eigenvalues and Eigenvectors

A study of eigenvalues and eigenvectors is essential for many of the more important applications of linear algebra, e.g., for Markov processes and systems of first order linear differential equations. And although it is simple enough for students to understand the defining matrix equation Ax = λx, x _ 0, they typically experience considerable difficulty with the subsequent material because of their inability to compute, by hand, the required quantities.

Supercalculators can be of considerable benefit in this regard by helping students to calculate such things as characteristic polynomials, eigenvalues and associated eigenvectors, and to construct diagonalizing matrices P. My students use a calculator routine CHAR that calculates the coefficients of the characteristic polynomial det(λI-A) of an nxn matrix A by using traces of powers of A. For example, using

$$A = \begin{bmatrix} -19 & -12 & 13 & 20 & -15 \\ 48 & 29 & -26 & -40 & 30 \\ 22 & 11 & -8 & -19 & 14 \\ -16 & -8 & 10 & 21 & -14 \\ -18 & -9 & 10 & 17 & -10 \end{bmatrix}$$

CHAR returns the list {1 -13 45 49 -490 600} which tells us that the characteristic polynomial of A is $\lambda^5 - 13\lambda^4 + 45\lambda^3 + 49\lambda^2 -$

$490\lambda + 600$. The calculator's built-in polynomial root-finder then shows the eigenvalues to be $\lambda = 5, 5, 4, -3,$ and 2. To find a basis for the eigenspace associated with $\lambda = 5$, students use the RREF command to obtain the reduced row-echelon form of A - 5I:

$$
\text{RREF} = \begin{bmatrix}
1 & .5 & 0 & 0 & -.5 \\
0 & 0 & 1 & 0 & 1 \\
0 & 0 & 0 & 1 & -2 \\
0 & 0 & 0 & 0 & 0 \\
0 & 0 & 0 & 0 & 0
\end{bmatrix}
\cdot
$$

Hence, all solutions to (A - 5I) x = 0 are given by

$$
\begin{aligned}
x_1 &= -.5x_2 + .5x_5 \\
x_3 &= -x_5 \\
x_4 &= 2x_5
\end{aligned}
$$

where x_2 and x_5 are freely chosen. Choosing $x_2 = 1$ and $x_5 = 0$ produces the solution

[-.5 1 0 0 0] and choosing $x_2 = 0$ and $x_5 = 1$ produces the solution [.5 0 -1 2 1]. These two vectors are the desired basis. Each of the other eigenvalues appears only once, so their eigenspaces are one-dimensional. Choosing bases as above and then aligning the five basis vectors as columns of matrix P, we have

$$P = \begin{bmatrix} -.5 & .5 & 1 & 1 & .5 \\ 1 & 0 & -2 & -2 & -1 \\ 0 & -1 & -2 & -1 & -.5 \\ 0 & 2 & 2 & 1 & 1 \\ 0 & 1 & 1 & 1 & 1 \end{bmatrix} \cdot$$

A single keystroke reveals that

$$P^{-1} = \begin{bmatrix} 0 & 1 & -2 & -2 & 2 \\ 2 & 1 & 0 & 0 & 0 \\ -2 & -1 & 0 & 1 & -1 \\ 4 & 2 & -2 & -3 & 2 \\ -4 & -2 & 2 & 2 & 0 \end{bmatrix}$$

and two multiplications show $P^{-1} AP = \text{diag} [5 \; 5 \; 4 \; -3 \; 2]$.

After students master the above process, I permit them to call upon their calculator's built-in routine for finding eigenvalues and eigenvectors in a few keystrokes. The routine evokes the QR-algorithm and returns both the eigenvalues and a matrix P of eigenvectors (but, for this example, the P is not nearly so nice as the one above). Thus, for more advanced topics, eigenvalues and eigenvectors can be easily found.

Student Reactions

We do not have objective measures of the learning that is taking place in the presence of supercalculators. For as we noted in the Introduction, this technology is a driving force for changing *what* we teach as well as how students learn. But, our students are experiencing their study of linear algebra in new ways: in an active, constructive environment. And they are demonstrating that they can achieve understanding of certain concepts better than before. For example, before we began using the technology our students often were unsuccessful in constructing orthonormal bases or diagonalizing real symmetric matrices simply because the hand computations quickly overwhelmed them. But the calculators have changed all that; they carry the computational burden and allow students to concentrate on what we value most in their learning, controlling the process by careful thinking. Grade distributions are not substantially different, but student attitudes towards linear algebra show noticeable improvement. Although we no longer track changes in attitudes, the chart below summarizes student responses across six different classes to an evaluation of the role of the calculators in the learning process, conducted by an outside evaluator during 1990-1992. While the response to question 2 might have been expected, the strong positive response to question 4 came as a surprise.

Questions	% of students who agree or strongly agree (n=156)
1. The graphics calculator helped me to understand the course material	76%
2. The graphics calculator was useful in solving problems	98%
3. The graphics calculator allowed me to do more exploration and investigation	83%
4. The graphics calculator helped me to have better intuition about the material	62%

Since this is likely to be the only course in linear algebra that non-mathematics majors ever take, it is important that the students experience linear algebra as the truly dynamic and exciting subject that it is, without having the technology dominate the course. Supercalculators help us do that. Indeed there is an atmosphere of enthusiasm that surrounds their use, which can only be interpreted in an overall positive sense.

References

LaTorre, Donald R.(1995) *Linear Algebra   Teaching Code for the HP-48G/GX,* Academic Press

LaTorre, Donald R., (1993) 'Iterative Methods in Introductory Linear Algebra,' *The College Mathematics Journal,* Vol. 24, No. 1, January 1993, 79-88.

LaTorre, Donald R., (1992) 'Using Graphing Calculators to Enhance the Teaching and Learning of Linear Algebra, Symbolic Computation' in *Undergraduate Mathematics Education,* MAA Notes No. 24, The Mathematical Association of American, 109-120.

LaTorre, Donald R., (1992) 'Explorations in Linear Algebra', in John G. Harvey and John W. Kenelly (Eds.), *Explorations with the Texas Instruments TI-85,* Academic Press.

20 Models of Technology

Spreadsheets and the Language of applicable Mathematics

Chris Bissell

Spreadsheets offer the possibility of carrying out quite complex mathematical modelling with a minimum of mathematical formalism. This raises a number of questions concerning the role of traditional applicable mathematics for engineering and technology. A number of examples of the spreadsheet approach are presented, together with a discussion of some of the ways spreadsheets can be used to enhance and enrich the mathematical experience of students of technology.

Introduction

Traditional engineering and technology curricula have laid great stress on mathematics. Budding engineers and technologists, it seems, require a thorough grounding in vectors, matrices, calculus, differential equations, and so on. The traditional mathematics service course in colleges and universities has thus devoted a lot of time to solving 'problems' on these topics: generating solutions to differential equations; reproducing the derivations and results of classic mathematical models; working through 'examples' of vector analysis or matrix manipulation - all too often, in other words, the computation of 'right' answers to tried and trusted academic questions! Practising engineers, on the other hand, rarely - if ever - need to carry out such tasks. True, they need to select and use appropriate standard models, to scale and transform known results, to manipulate notations and patterns, and - increasingly - to use mathematically-based computer tools with insight and understanding. But the context in which such 'mathematical' activities are carried out is often so far removed from conventional engineering mathematics that it makes more sense to think of engineers and technologists speaking a language essentially different from that of mathematicians.

To give just one example, consider the topic of linear differential equations with constant coefficients - a 'core' topic of engineering mathematics. Differential equation models of this type are widely used in electronics and control engineering, for example. Yet the electronics or control engineer will not be heard talking about particular integrals and complementary functions, and rarely be seen solving differential equations by hand or by computer (contrary to the fond belief of many mathematics teachers!). The language used by the engineer is far more likely to be that of standard step response curves (Figure 1), frequency-response curves (Figure 2) or other, related, pictorial representations such as pole-zero plots in the complex plane - all using the physically more convenient measures of damping ratio and natural frequency rather than differential equation coefficients, and all characterised by a striking visual presentation of information. The engineer learns to think in terms of such models and patterns, and to manipulate them in a way which is very different from standard 'engineering maths'. The engineering representations take on a life of their own: engineers might talk of particular poles in the complex plane 'causing' a particular class of behaviour, of the need for an electronic amplifier or compact disc control system to 'shape frequency response' in particular regions, and so on. Over the years, in fact, engineers have put enormous effort into developing techniques designed precisely to reduce or eliminate the need for traditional mathematics - techniques which typically involve pictorial and graphical representation, tables, heuristic rules, and so on. What is striking about the developments of the last few years is that the computer-aided tools now likely to be running on the modern engineer's desktop or laptop computer *also* tend to use this technologist's language - rather than the traditional language of mathematics.

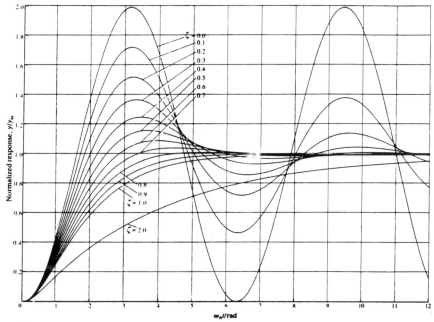

Figure 1

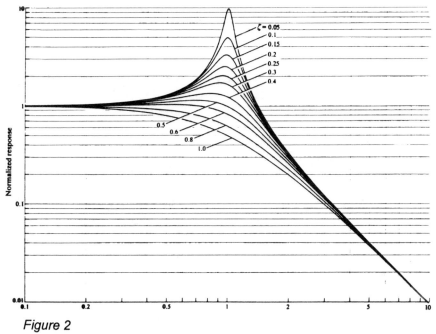

Figure 2

The argument for the essential difference between these mathematical and engineering languages is developed at somewhat greater length elsewhere (Bissell & Dillon, 1993). This paper addresses the use of spreadsheets as a tool both for investigating and understanding such models in their technological context, and for becoming fluent in this new language.

What's so good about spreadsheets?

In his book *The visual representation of quantitative information*, Tufte (1983, p 40-41) remarks that Figure 3 'may well be the best statistical graphic ever drawn', noting that 'six variables are plotted: the size of [Napoleon's] army, its location on a two dimensional surface, the direction of the army's movement, and temperature on various dates during the retreat from Moscow'. The size of the army and the direction of its march is indicated by the breadth and orientation of the shaded areas; other data is added to that plot or beneath the main graphic.

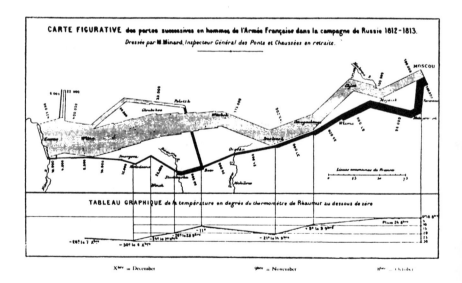

Figure 3

368

Tufte classes Figure 3 as a 'narrative graphic of space and time' in which the artist has added 'spatial dimensions to the design of the graphic, so that the data are moving over space (in two or three dimensions) as well as over time'. Furthermore, 'multivariate complexity can be subtly integrated into graphic architecture, integrated so gently and unobtrusively that viewers are hardly aware that they are looking into a world of four or five dimensions.'

This 'higher dimensional' quality of good graphical representation of data is one of the great advantages of the spreadsheet. As a paper at TTM'93 put it:

> The spreadsheet is not unlike a piece of paper on which one might have notes in shorthand, quotes in copperplate, squiggles in pencil and a sketch on the side. There are rows and columns, but they are not all only spatially interconnected: the user might have generated three or four conflicting algorithms, for use side by side and simultaneously with the same data, in order to test a conjecture. There is a sense in which we want to attribute higher-dimensional use to this piece of software. Perhaps it is *three* dimensional. (Nevile, L. & Mason, J., 1993)

Or perhaps a good spreadsheet - like Minard's graphic, or many traditional engineering design charts - is of even higher dimension. The reader is invited to bear in mind this 'higher dimensionality' when considering the examples of the following sections. Each example is presented in the context of a number of challenges to teachers using traditional approaches to applicable mathematics:

- Is conventional mathematical formalism (a) necessary and (b) desirable when teaching maths to students of technology and engineering?

- Precisely what degree of facility in the 'language of mathematics' do student engineers and technologists require?

- How might teachers of disciplines other than the ones represented here use spreadsheet examples to enhance and enrich their teaching?

From modelling assumption direct to spread-sheet cell

The classic models of technology are generally based on some set of (more or less realistic) modelling assumptions which are then translated into mathematical expressions. To the novice, however, the manipulation of the mathematical expressions can prove much more difficult than understanding the underlying assumptions, and problems with the mathematics may easily obscure the whole modelling process. Spreadsheets can provide valuable experience in working with particular models and exploring the effects of different modelling assumptions - even for students possessing only an elementary knowledge of mathematics. Models of growth and decay are a case in point. Classic continuous models of this type are based on modelling assumptions such as 'growth rate is proportional to population' or 'fractional growth rate declines linearly with increasing population' - assumptions which have the dual advantage of being fairly realistic (in certain situations) while also leading to tractable differential equation models. Traditionally, therefore, a knowledge of calculus has been required before tackling them. All this changes, however, with a spreadsheet.

Consider, for example, the population model known as the logistic. This is particularly interesting to explore by spreadsheet in its discrete form - that is, we assume that the annual fractional population increment r declines linearly with increasing population, reaching zero at some equilibrium population P_e. Algebraically, we might write this modelling assumption as:

$$r_n = r_I (1 - P_n/P_e)$$

where r_I represents some 'initial' fractional increment for very low values of population.

Then successive annual populations can be calculated iteratively using the expression

370

$$P_{n+1} = P_n (1 + r_n)$$

Each of these modelling assumptions can be converted directly into a spreadsheet formula. Once entered into the first row, they are simply copied down the appropriate columns, the spreadsheet package taking care of the row-by-row modifications. Figure 4 shows the formulae used, while Figure 5 gives numerical results and a bar-chart plot showing the characteristic logistic rise to a limit. (Note that the INT function is used to round to a whole number of individuals. The $ sign signifies an absolute cell reference, unchanged on copying. The starting and equilibrium (maximum) population values are set parameters of the particular model.)

	A	B	C	D	E
1	year	population	annual fractional	initial r	max pop
2			growth		
3	0	100	=$D$3*(1-B3/$E$3)	0.35	1000
4	1	=INT(B3*(1+C3))	=$D$3*(1-B4/$E$3)		
5	2	=INT(B4*(1+C4))	=$D$3*(1-B5/$E$3)		

Figure 4

Students investigating the effect of different initial annual fractional increment r_I will soon come across oscillatory or chaotic behaviour. Figures 6 and 7 show the result of changing the value of r_I held in cell D3 to 2.1 and 3.0, respectively. Further exploration of such models quite naturally raises important issues about model accuracy, validation and the computational process, as well as those relating to the internal characteristics of the models themselves.

Note that in this example the spreadsheet has simply permitted, without any great mathematical prior knowledge, the exploration of the consequences of a particular modelling assumption. Whether the teacher should express this as a 'non-linear discrete mapping', or even relate it to 'Euler's numerical solution of the continuous logistic equation' is another question altogether - and the answer will depend very much on the type and philosophy of the particular course and teacher (see Stone's comment on the approach presented here and Bissell's reply, 1994). One thing is certain, however: the exploration of such models (and many others) by spreadsheet is perfectly feasible for students without a knowledge of calculus, differential equations, partial fractions or even the exponential function. And this fact calls into question many of the dearly held assumptions about the teaching of such models to non-mathematicians.

Figures 5 - 7 illustrate clearly the 'higher dimensionality' quality of spreadsheets referred to in the previous section. In each case, the user sees a two-dimensional array of numerical values together with a two-dimensional graphic. A simple keystroke or mouse operation displays the formulae used to calculate the values instead of the values themselves. And by rapidly changing the r_I value held in cell D3, the user sees the three figures in quick succession, enabling immediate comparisons to be made. Alternatively, storing the results for different starting conditions in separate columns allows comparisons to be made on a single plot.

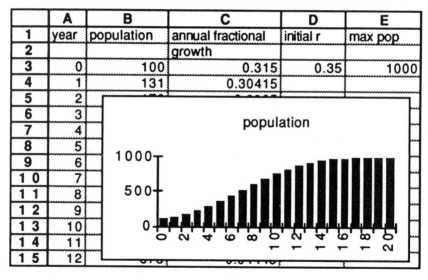

	A	B	C	D	E
1	year	population	annual fractional	initial r	max pop
2			growth		
3	0	100	0.315	0.35	1000
4	1	131	0.30415		
5	2				
6	3				
7	4				
8	5				
9	6				
10	7				
11	8				
12	9				
13	10				
14	11				
15	12				

Figure 5

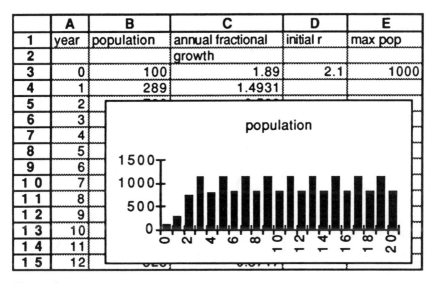

Figure 6

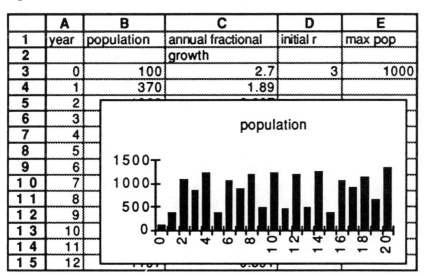

Figure 7

	A	B	C	D	E	F	G
1	time	capacitor	resistor	voltage	applied	growth	freq
2	step	volts	volts	increment	voltage	factor	factor
3						T/RC	
4	0	0	0	0	0	0.05	10
5	1	0	0.1	0.00499	0.0998		
6	2	0.005	0.194	0.00968	0.1987		
7	3						
8	4						
9	5						
10	6						
11	7						
12	8						
13	9						
14	10						
15	11						
16	12						
17	13						
18	14						
19	15						
20	16						
21	17						
22	18	0.4426	0.531	0.02656	0.9738		

Figure 8

Models similar to the above can also be constructed for lumped-parameter physical systems: RC-circuits, mass-spring-damper systems, and so on. Figure 8 shows an example simulating a lowpass resistor-capacitor circuit. Instead of following the standard approach, which is to derive a first order differential equation relating the time-varying voltage at the input and output, the spreadsheet offers an intuitive discretisation. Reasoning from Ohm's Relationship, each increment in capacitor voltage is computed by assuming it to be proportional to the difference between its present value and the instantaneously applied voltage - an assumption valid for sufficiently small time increments. Column E holds the 'driving function', the voltage applied to the input of the circuit. In the case shown it is filled with values from the spreadsheet SIN function, to simulate the filter's response to an applied sinusoidal voltage. The chart shows this applied sinusoidal voltage together with the

374

resulting voltages across the capacitor and resistor: note yet again the 'higher-dimensional' quality of the final display. Changing the contents of column E to represent a new driving voltage - a step change, a linearly increasing 'ramp' function, or a rectangular waveform, for example - leads immediately to the calculation of the new voltages across resistor and capacitor. As before, the modelling assumption has been built into the spreadsheet directly, without recourse first to a differential equation and then to explicit discretisation. In this case, though - and in constrast to the discrete logistic example, where the discrete nature of the model may be a function of the breeding cycle for certain organisms - the discretisation is *implicit* in the modelling assumption. Again, depending on educational aims, teachers may choose to relate the spreadsheet to the traditional differential equation model (and its numerical solution) - *or they may not*. In this case the challenge to teachers of electrical and electronic engineering is the dual one of examining critically their approach both to circuit modelling and to the mathematical content of their service courses. Attempting to model higher-order systems in this way is possible, but can rapidly become complex and confusing. An initial stage of modelling the flows of material, energy or information diagramatically, as proposed in a recent guide produced by Armstrong & Bridges (1993) for the Mathematical Association, can perhaps ease the construction of more complex spreadsheets. But the great virtue of student spreadsheet use is that very simple examples are adequate for exploring the *generic* properties of many important technological models. Students can begin to get a feel for the factors which imply oscillatory, rather than non-oscillatory behaviour, for example, or a slow rather than a fast response of a dynamic system to a given input; they can follow the evolution in time, row by row, of their simulation; they can easily investigate the general effects of various changes to the model. In the logistic example above, for instance, it is easy to explore in general terms the recovery of a population after a disaster, or the effect of a predicted step change in growth rate as a consequence of environmental protection.

An important point - and one which teachers of mathematics and technology are sometimes unwilling to accept - is that modelling

375

exercises of this type do not necessarily have to produce accurate numerical answers to a particular problem to be a useful learning experience. In order to understand the basic properties of RC circuits it is more important that the student should appreciate the *general* features of the phase and amplitude relations of the curves of Figure 8 than for the spreadsheet to generate precisely correct numerical values by using a 'good' numerical algorithm. (If accurate numerical values are required in a circuit simulation, an electronics engineer would turn to a reliable professional package such as SPICE within which the relevant circuit equations are solved with a high degree of robustness and accuracy.) Simple, but detailed, generic investigations are surely more productive for student engineers and technologists than using a spreadsheet for, say, Runge-Kutta numerical solution of classical continuous models (typical of some of the applications of spreadsheets recently reported in the engineering education literature). And perhaps the greatest advantage of using spreadsheets in this way is that students can construct (in both the everyday and the psychological sense) the model for themselves.

Topology, strategy, and structure

Recursive models and recursive numerical solutions to standard continuous models are food and drink to engineering mathematics. Moreover, in electronics and control engineering the physical systems themselves often involve feedback, and recursive solution of dynamic response is then closely related to system topology. In such cases, spreadsheets offer a direct approach to understanding system behaviour, and can encourage in the learner a qualitative feel for the mathematical description. Once again, it is the spreadsheet array which allows system structure or solution strategy to be closely linked to the spreadsheet structure. Three examples will be outlined briefly.

Figure 9 shows a simple scrambler of the type sometimes used to 'randomise' digital data before transmission over a network - in certain classes of modem, for example (Bissell & Chapman, 1992). The scrambling process is achieved by carrying out logical exclusive OR

Self-synchronising scrambler with tap polynomial 1 + x^(-1) + x^(-3)						
Input	Stage 1	Stage 2	Stage 3	Feedback	Output	
1	0	0	0	0	1	
1	1	0	0	1	0	
1	0	1	0	0	1	
1	1	0	1	0	1	
1	1	1	0	1	0	
1	0	1	1	1	0	
1	0	0	1	1	0	
1	0	0	0	0	1	
1	1	0	0	1	0	
1	0	1	0	0	1	
1	1	0	1	0	1	
1	1	1	0	1	0	

Figure 9

(XOR) operations on the input, output, and versions of the output delayed by certain numbers of time intervals. The columns of the spreadsheet correspond to input, output, delayed versions of the latter (note the simple diagonal shifts across the spreadsheet corresponding to time delay), and the results of the two XOR operations. The particular spreadsheet used by the author (Microsoft Excel) also allows the incorporation of simple diagrams to aid spreadsheet construction and general understanding: the correspondence between the diagram and the spreadsheet columns is clear. (Depending on the precise functions offered by the spreadsheet, the result of the XOR operation can be computed using built-in spreadsheet logic or modulo-2 arithmetic.) The 'randomisation' of the scrambler is apparent from the way the input 1,1,1,1,1,1,1,1,1,1,1,1, ... is converted to the output 1,0,1,1,0,0,0,1,0,1,1,0, ... Other logic devices can be simulated in a similar way. Once again, the great advantage of the spreadsheet approach is that students can construct simulations themselves, observe the time evolution, and experience the various classes of behaviour. (What happens, for example, if the scrambler input becomes the output sequence of Figure 9? ... Is there a scrambler input sequence which gives an output 1,1,1,1,1,1,1,1,1,1, ... ? Will this be true for any scrambler? Is the last question one which can be answered easily by spreadsheet, or do we need a different model? ...

A similar approach may be applied to recursive systems or simulations involving arithmetic, rather than logic operations. Figure 10 shows how the relationship between the z-plane pole positions (difference equation eigenvalues) and the unit step response sequence of a discrete linear system may be explored using a spreadsheet. First and second differences correspond to diagonal shifts across the spreadsheet in a simulation of the damped oscillatory behaviour so often observed in technological systems. In this particular example an iterative solution is computed to the second-order difference equation

$$y[n] = x[n] + 2c \cos\theta \, y[n-1] - c^2 \, y[n-2]$$

defined by an arbitrary pair of complex poles at $z = c \exp(\pm j\theta)$. This particular difference equation could model a simple, recursive digital filter, in which case the filter realisation might be directly related to the spreadsheet structure in the same way as in the scrambler example. Or it might represent a numerical approximation to a second-order differential equation - in which case the recursion is a function of the numerical algorithm rather than the system. In both cases, the spreadsheet allows easy exploration of the relationship between system variables (complex-plane poles and zeros, say, or digital filter coefficients) and physically important effects such as overshoot, oscillation frequency, or settling time. Depending on

378

student requirements the model can be kept generic, or built and investigated directly using the language of the appropriate discipline.

	A	B	C	D	E	F	G
1	input			output	pole positions		
2	x[n]	y[n-2]	y[n-1]	y[n]	c	theta (degrees)	
3					0.88	32	
4	1	0	0	1			
5	1	0	1	2.4926			
6	1	1	2.4926	3.9459			
7	1	2.4926	3.9459	4.9593			
8	1	3					
9	1	4					
10	1	5					
11	1	5					
12	1	4					
13	1	3					
14	1	3					
15	1	2					
16	1	2					
17	1	2					
18	1	3					
19	1	3.5632	3.8361	3.9664			

Figure 10

It should be clear by now that spreadsheet layout is a key factor in pedagogical applications in technological modelling. As a final example, in which the spreadsheet layout reflects a particularly important mathematical algorithm, consider a rather specialist application from the field of signal processing: the Fast Fourier Transform (FFT). The normal Discrete Fourier Transform (DFT) involves a computational effort proportional to N^2 when computing the transform of N data points. By a cunning re-ordering of the computations this can be reduced to an effort proportional to $N log_2 N$ in what is known as the Fast Fourier Transform. These re-orderings (given the rather fanciful names of butterflies, twiddles, shuffles, and so on) are generally explained in textbooks by reference to two-dimensional arrays and diagrams. These diagrams are perfect for spreadsheet illustration, in which the modified computation order becomes a corresponding rearrangement of the spreadsheet cells, as indicated in outline in Figure 11 (Chapman, 1993). Advanced students can, with some help, develop such a spreadsheet for themselves; at lower levels the spreadsheet can be provided for students to work with in order to facilitate understanding of a rather complex idea.

Each of the above three examples - the scrambler, the discrete linear system, and the FFT - comes from an area of information engineering traditionally considered to be highly 'mathematical'. In each case the spreadsheet has significantly modified the way in which the 'mathematics' can be presented, manipulated and interpreted. How such novel approaches affect students' understanding calls for detailed investigation.

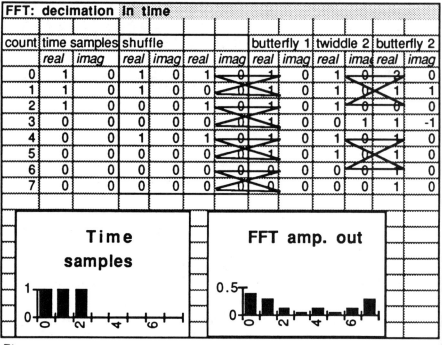

Figure 11

Discrete event simulation

Examples so far have involved the use of spreadsheets for rather conventional applicable mathematics. There are, however, numerous possibilities for applications in combinatorial mathematics and discrete event simulation. For example, the Open University Course T247 *Working with Systems* has used spreadsheets for simulating a tennis game; the 'bunching' of buses on a crowded route; pest control; and scheduling problems. Spreadsheets can also be used to model the queueing effects characteristic of Markov processes - as might occur in a digital telecommunications packet-switching buffer, on a motorway, or at a post-office counter. Such applications require more detailed contextualisation than can be given here. The general principles are well-illustrated, however, by the more abstract example of Figure 12, in which the binomial distribution is demonstrated by exploiting the spreadsheet cell array as a branching tree structure.

	A	B	C	D	E	F	G	H	I	J
1	rand	0.66	0.06	0.88	0	0.4	0.43	0.78	0.43	
2		down	up	down	up	up	up	down	up	total
3									0	0
4								0		
5							0		0	4
6						0		0		
7					0		0		0	7
8				0		0		0		
9			0		0		1		1	20
10		0		0		1		1		
11	1		1		1		0		0	16
12										
13									0	8
14										
15									0	7
16										
17									0	2
18										
19									0	0
20										64

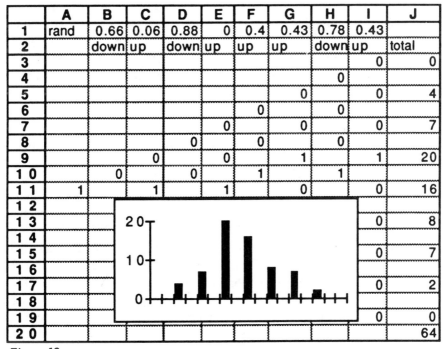

Figure 12

The demonstration begins with a 1 half way down the leftmost column. Spreadsheet logic functions are then used to insert a 1 into successive diagonally neighbouring cells, depending on a random 'up' or 'down' decision. Repeated running of the spreadsheet shows how the characteristic shape of the binomial distribution gradually builds up as the 1s 'trickle through' into the appropriate position. A similar approach might be used to model the probabilities of particular failure combinations, or the propagation of various physical phenomena in two dimensions. In such cases the spreadsheet structure renders possible new types of representation; yet again the representation is characterised by its direct links to modelling assumptions, rather than the use of mathematical formalism.

Conclusion

In this paper I have argued that what counts as doing mathematics *in* technology is often very different from what is taught as 'maths *for* technology'. Furthermore, the discrepancy is becoming increasingly noticeable now that new computer tools allow technologists to make more direct links between systems and models, and thus bypass much mathematical formalism. An important part of mathematics and technology education is therefore to learn to use a range of maths-based computer tools effectively and with understanding.

Spreadsheets are one such tool, particularly appropriate for engineering and technological applications. They form a highly approachable intermediate option between a high-level computer language on the one hand, and a more sophisticated generic package such as Mathematica, Mathcad, Matlab, Derive, or Maple, on the other. It is precisely *because* spreadsheets have not been designed specifically for the solution of mathematical problems by experienced mathematicians that their use can greatly enhance a student technologist's understanding! Spreadsheets are ideal for the expression of both standard models and more open-ended modelling exercises directly in the language of engineering and technology, rather than the language of mathematics. The examples here are taken primarily from the author's own specialist areas of information engineering, but a similar approach to spreadsheet use is possible in many other areas of technology.

The approach presented in this chapter has grown out of the use of spreadsheets in Open University courses such as T102 *Living with Technology* and T247 *Working with Systems*. Similar activities have been carried out with students at a number of conventional institutions, although experience has often been disseminated informally, published papers on spreadsheets being much less common than those addressing the use of more sophisticated and powerful tools such as Derive, Mathcad and Mathematica. Spreadsheets should not, of course, be pushed beyond their natural limits, nor should they be used when another computer-based tool (high-level language, numerical analysis library, generic maths package, dedicated engineering package, etc) is more appropriate. Enthusiasts can become tempted to 'overload' cells with mathematical functions or logical expressions, and such overload can result in excessively-complex spreadsheets very difficult to debug or understand. Readers should not conclude from this paper that spreadsheets are ideal for *any* mathematical exploration or simulation - nor should they be seen as a replacement for all traditional skills! A number of recommendations can, however, be made to teachers/educators:

- Encourage students to build their own spreadsheets wherever possible.
- Exploit spreadsheet layout in striking visual ways.
- Include graphs and dynamic simulations.
- Try to relate modelling assumptions or system topology directly to spreadsheet formulae or structure.
- Make use of the random number generation function of spreadsheets for stochastic models.
- Don't assume that explicit discretisation of a standard continuous model is necessarily the best approach.
- Don't build a spreadsheet to do something that is better done by a different computer tool.
- Don't restrict spreadsheets to textbook examples, or to teaching standard textbook material in a conservative way.

383

Those of us involved with the education of engineers and technologists, and particularly with their *mathematical* education, are faced with a multitude of sometimes conflicting requirements. We must ensure that our students are conversant with a wide enough range of the conventional tools of applicable mathematics for them not to disgrace themselves in polite mathematical society. At the same time, our students also need to be fluent in the 'engineering dialect(s)' of mathematical language in order to operate properly in their profession. Furthermore, as educators we need to have the confidence to allow our students to explore for themselves the nature of the mathematics, and especially the *models*, they are expected to use. We must be willing to accept that the considerable time students need for such exploration, constructing their own models and explanations, is a valid educational activity. Spreadsheets are comparatively easy to learn, are well-integrated with other office packages such as text processors, and have excellent graphics. Furthermore, modern spreadsheet and text-processing packages offer many 'authoring' features which enable complete CAL sessions to be developed completely within such environments (Cuttle et al, 1992). And it should not be forgotten that the general spreadsheet skills developed by students as they explore technological modelling will stand them in good stead beyond their academic career.

Acknowledgements

This paper is based on Bissell (1993), but thas been thoroughly revised and extended. The author wishes to thank his colleague David Chapman for many useful discussions, and specifically for the FFT and scrambler examples. Figures 1 and 2 are reproduced with permission of the Open University; versions of figures 11 and 12 appeared in Bissell (1994a), for which IEE copyright exists. An expanded version of this chapter including additional examples and suggestions is available as Bissell (1994b).

References and Further Reading

Armstrong, P. and Bridges, R. 1993, *Spreadsheets: Exploring their Potential in Secondary Mathematics*, Leicester: Mathematical Association

Benson, H. & Kopp, J. 1991, *Spreadsheet Physics: Study Units for University Physics*, London: Wiley

Bissell, C. C. 1992, 'Maths for technology, What do we need and how do we teach it?' *Theta*, 6(2), 40-43

Bissell, C. C. 1993, 'Technology for maths for technology: the role of spreadsheets', in *Proceedings of International Conference on Technology and the Teaching of Mathematics,* Birmingham, UK, 17 - 20 September, 1993, 141-148

Bissell, C. C. 1994a, 'Spreadsheets in the teaching of information engineering', *IEE Science & Education Journal*, 3(2), 89-96

Bissell, C. C. 1994b, *Spreadsheets in Engineering Education: Mathematics, Modelling and Simulation*, Systems Architecture Group Internal Report SAG/1994/RR37. Faculty of Technology, The Open University, Walton Hall, Milton Keynes, UK

Bissell, C. C & Chapman, D. A. 1989, 'Modelling applications of spreadsheets', *IEE Review*, July/Aug, 267-271

Bissell, C. C. & Chapman, D. A. 1992, *Digital Signal Transmission*, Cambridge University Press, Appendix C, 308-311

Bissell, C. C. & Chapman, D. A. 1993, 'Spreadsheets as a learning aid in engineering education', in *Proceedings of CAEE '93* (International Conference on Computer-aided Engineering Education), Bucharest, Romania, 22-24 September, 1993, 277-82

Bissell, C. C. & Dillon, C.R. 1993, 'Back to the backs of envelopes', *The Higher*, 10 September, 1993, 16.

Chapman, D. A. 1993, 'Spreadsheet demonstration of discrete and fast Fourier transforms', *Int. J. Elec. Eng. Ed.*, 30(3), 211-215

Cuttle, M. L., Young, C. P. L. & Heath, S. B. 1992, *A Practical Introduction to Creating Courseware with Microsoft Excel*, CTI CLUES, University of Aberdeen
(e.mail: clues@uk.ac.aberdeen)

Healy, L. & Sutherland, R. 1992, *Exploring Mathematics with Spreadsheets*, London: J Murray

Kral, I. H. 1991, *Excel Spreadsheets for Engineers and Scientists*, London: Prentice Hall

Nevile, L & Mason, J. 1993, 'Looking at, through, and back at: useful ways of viewing mathematical software', in *Proceedings of International Conference on Technology and the Teaching of Mathematics*, Birmingham, UK, 17-20 September, 1993, 391-98

Open University 1989, *T102 Living with Technology*, Milton Keynes: Open University Press

Open University 1991, *T247 Working with Systems*, Milton Keynes: Open University Press

Stanton, B. J., Drozdowski, M. J., & Duncan, T. S. 1993, 'Using spreadsheets in student exercises for signal and linear system analysis', *IEEE Transactions on Education*, 36(1), 62-68

Stone, J.A.R. 1994, 'A Note on the Logistic', *Theta*, 8(1), 54-9 (includes Bissell's reply)

Tufte, E. R. 1983, *The Visual Display of Quantitative Information*, Graphics Press, 40-41

21 Using Technology In Examinations

Jim Tabor

> *Following the tremendous increase in the power and availability of personal computers, we consider the implications of the routine use of computer assistance in teaching a final year engineering module, and how this implies the use of computers during the examination. Finally, suggested guidelines are given for conduct of an examination using computers.*

Coventry University has recently become fully modular. The scheme revolves around an eight module year. Each module has two contact hours per week. Students are assumed to undertake a further three hours private study for each module, leading to a 'student week' of 40 hours. An optional final year module offered by the Division of Mathematics is *Computational Mathematics for Electrical Engineers*. The syllabus contains material as follows:

> *Numerical integration, solution of systems of nonlinear equations, systems of linear equations, finite element analysis, eigenvalue problems and systems of ordinary differential equations.*

In the teaching, the emphasis is on the solution of practical engineering problems and methods for large sparse matrices which would not be dealt with using full matrix methods. The assessment mark for the rest of the modules on the Electrical Engineering course was to be 20% for coursework and 80% for the exam, with an aggregate pass mark of 40% . The author did not wish to deviate from this scheme. In particular, there was no thought that the module should be assessed entirely on coursework.

The Division of Mathematics has quite extensive computing facilities, including an open access PC facility and PC teaching laboratories, both licensed for Matlab (1). For the first run of the module in October 1991 there were 18 students. The relatively small size of the first class prompted the author to experiment with the means of assessment. The class was timetabled for the first hour in one of the Division's teaching laboratories. The second hour was timetabled in a classroom, in the same block, four floors down. This arrangement placed constraints on the way the module was lectured, and cut down on the total time available because of the time wasted in moving rooms. It became clear that individuals in the class had very different levels of preparedness, particularly as far as Matlab was concerned. There were three distinct groups. Some of those previously on course had undertaken a year in industry. They had attended a different second year mathematics course from the group moving straight into the final year. There were also a small number of direct entrant students from France. Given the relatively large number of queries throughout the week, mainly concerning Matlab, the author decided to timetable an optional hour in a teaching laboratory. Attendance varied from a few individuals up to most of the class in the last few weeks of the course. On balance, the extra optional timetabled hour probably saved the author time since out of lecture queries were reduced dramatically.

For the second run in October 1992 there were 28 students. It was not possible to timetable a teaching room for either of the two hours of the module. It has been possible exceptionally to use a teaching laboratory for both hours. Due to the physical arrangement of the laboratory, there are few disadvantages to lecturing in the teaching laboratory, as opposed to a teaching room. There is a slight time advantage in that there is no need to move rooms. It is also possible to vary the time allocated between lecture and laboratory class. The main objection is that the laboratory is being used inappropriately for the 50% of the time devoted to the lecture. There is a similar spread of preparedess in the students for the second run. The class composition broadly follows that for the first run with the addition of a large component of Israeli direct entrants, who have taken the same second level mathematics course in Israel, but who have not met Matlab. It has again been necessary to allocate an optional hour in the laboratory in order to help level up.

Teaching the Course

This substantial section has been included in order to show how significantly the routine use of an excellent computer aid can affect the teaching process throughout the year. The level of insight and adventure that I hope to show is demonstrated by the students just cannot be achieved with static and pre-prepared material. I apologise in advance to those who will find the material on signal processing hard going, but claim that perseverence will be rewarded.

It is immediately clear, given the audience for the module, that a highly mathematical numerical analysis based course would be inappropriate. The author takes the view that engineers, particularly electrical engineers, should know a lot of mathematics. Professional electrical engineers still seem in accord with this view. Accepting that the primary function of engineers is to make things work, it is still the contention that this is not best done by simple application of formulae. Unless the engineer has a good understanding of the underlying ideas, the formulae will be wrongly selected or applied beyond reasonable limits.

As a basis for the planned module, the author felt that the practical *application* of the numerical methods should be emphasised, but with sufficient theory to allow the engineering students to assess strengths and weaknesses and limits of applicability. In order to retain their respect for the module it would be necessary to enable them to address problems of real interest. This can only really be done by using industrial strength packages in the classroom.The Matlab suite was chosen for its wide use in industry, its extensive library of pre–prepared funcions relevant to electrical engineering and its easily learned programming langue which allows modifications and extensions to existing functions to be readily made.

Almost the whole of the course could be addressed using Matlab, though the finite element component would use the ESDUfine (2) finite element package, itself partly written by the author.

Because of the lack of previous experience for most of the students it was apparent that some considerable effort and class time would have to be devoted to teaching the Matlab language to the students. There is a good on-line help facility, detailing both the range of commands available and giving detailed help on each, but in the first contact hour the students were given a 'Matlab Primer' by Kermit Sigmon (Sigmon). For the second run of the module, the 'Student Edition of Matlab' (3) had been published and should have been of great assistance. The author received a review copy, but the MS-DOS edition of the book was sold out for the first term.

All the topics except finite element analysis in the module have been dealt with using Matlab, even when there were purpose written teaching tools (Jaques & Judd) available. This was to maintain the momentum and ensure enough practice in Matlab. The Division also has significant experience with the Derive 'computer algebra for the masses' and Maple packages. Knowledge of one of these packages in particular and computer algebra in general would be a significant advantage to the students in their later careers, but the author felt that the time constraints were too great to allow introduction of another major package in concert with Matlab.

One of the significant benefits of a sensibly powerful implementation of Matlab is that the more able students find it enjoyable to experiment beyond the set work. The first real 'hook' for the students was the fast Fourier transform, which formed the last part of the section on numerical integration. It is not the purpose of this article to discuss signal processing, but the way the module can be *taught* has a bearing on how the module should be *examined*. One of the student exercises was to analyse for frequency components the signal defined in the following equation ,

$$s(t) = \sin(50 \times 2\pi t) + \sin(120 \times 2\pi t) + \text{random noise}$$

where a number of different sampling rates were used. One instance of the signal is plotted in Figure 1.

The result of the Fourier analysis should be δ-functions at 50 and 120 representing the pure frequency components of the underlying signal, with the noise represented by a relatively even frequency distribution. About a dozen Matlab commands are sufficient to produce a plot of the signal as shown in Figure 1 and the power density spectrum $Pyy(\omega)$ as shown in Figure 2. These commands were supplied to the students in the form of a Matlab M-file, which they could then modify.

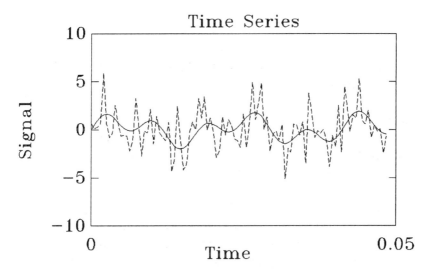

Figure 1. Underlying signal + noise

Figure 2 shows two power density spectra, where 128 and 256 samples of the signal have been taken over a second. The time between samples dictates the maximum frequency that can be analysed (Nyquist frequency), so for 128 samples per second, frequencies up to 64 Hz

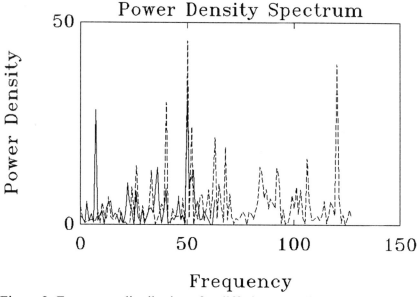

Figure 2. Frequency distributions for differing sampling rates. Power density spectrum vs. frequency

can be considered. Thus the component with a frequency of 120 Hz is out of range. At 256 samples per second both the components at 50 and 120 Hz appear as strong peaks. However, it is no accident that for the 128 samples per second analysis there is a strong peak at 8 Hz. At the slower sampling rate, the 120 Hz signal would yield the same sample values as an 8 Hz signal. This phenomenon is called *aliasing*, and works by *foldback* where the frequency line is folded like a concertina back onto the allowed frequency range. It was very easy to illustrate aliasing, not by words, but by allowing the students to experiment using Matlab.

Normally, the sampling rate cannot be freely varied, and if frequency components up to 120 Hz are under consideration, a sampling rate of 256 makes sense. While working on aliasing, several students remarked on the very wide variation of the heights of the peaks that occurred on different attempts with the same underlying signal but different random noise. This was worse with 128 samples, but was still quite evident at 256 samples per second. The obvious question was how to improve the the reliability of the signal to noise ratio with the same sampling rate of 256 samples per second. One student was very firm that a number of distinct 1 second data collection runs should be averaged since this would average out the noise. The author suggested that running for longer at the same sampling rate and letting the mathematics do the work would be beneficial. The student set out to average eight one second runs, while the author modified the Matlab M-file to run for eight seconds. The results appear in Figure 3.

The maximum frequency that can be resolved is the same since the time interval between samples is the same, but more points in the frequency domain are calculated when more points in the time domain are used. It would be normal to plot these figures in semi-log format since this is the preferred method of display when signal to noise ratios are being considered. Such plots are perfectly possible in Matlab, but the effect of the linear plots is more striking to the untrained eye.

The group was very rapidly able to distinguish between the two possibilities for improving the power density spectrum.. For eight one second runs, the noise is averaged out nicely, but the peaks are smeared out quite badly. The spectrum for the single eight second run is very uneven, with some high spurious peaks, but the real signals are very distinct and sharp.

Examining with Computers

Having determined that the module should be substantially based on appropriate use of Matlab we have the problem of how to make a fair assessment. The use of ESDUfine can be best assessed by the coursework element. There is no need to detail how this was assessed. For the remainder of material of the course it was clear that *not* to allow the use of Matlab in the examination would be a cause of unfairness to the students. The author also found it much easier to envisage the setting of a sensible examination paper using Matlab.

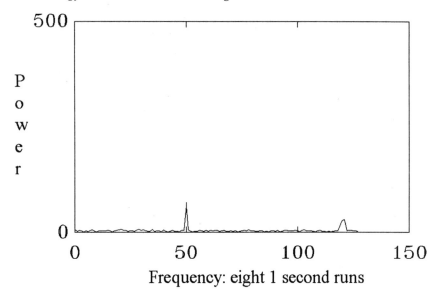

Frequency: eight 1 second runs

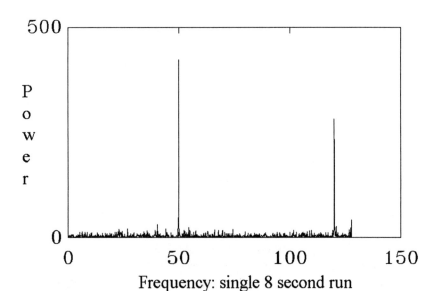

Frequency: single 8 second run

Figure 3. Comparison of calculated power density spectra.

During the examining period, there is greatly reduced pressure on the teaching laboratories, so there is little difficulty in timetabling them to be used in the examination.

There has been some previous experience of using computers in an examination situation within the Division. However, this has been at a lower level, where single numbers have been filled into spaces left in a form. At final year level it is necessary to be able to consider an overall performance and the author would not have embarked on a computer aided examination unless marks for method could be awarded.There are several attributes of Matlab that make it sensible to use as an aid in an examination at this level.

- It is a *very* robust package. The only way the author knows of causing a catastrophic error is to write a recursively defined M-file which never terminates so that the program stack tries to grow out of memory.

- There is a robust diary facility where the diary file is readable after most (but not all) disasters. Quitting and re-starting the program is fail-safe since a subsequent session appends to the diary file by default.

- Even PC-Matlab has a scroll back facility so that previous commands can be retrieved and edited easily.

- Most use of Matlab requires writing M-files which persist after quitting the program.

- The edit/run cycle for M-files is so rapid that writing and de-bugging M-files in an examination situation is feasible.

There were few problems encountered during the examination. One serious problem occurred after two hours when a student made the discovery of how to hang Matlab. Given the capabilities of the student, the author concluded that the discovery was due to incompetance rather than design. Re-booting the machine seemed the only way out, but the diary file had been left open and could not be accessed. All M-files were still intact. The student continued, completing the examination. It was possible to assess on the contents of the examination booklet, the M-files for the whole of the three hour period and the diary file for the last hour's work. There was only one part of a question where the marks for method were in some doubt. A small allowance of marks was made in lieu of the contents of the diary file. Fortunately the student passed, so the author still has no experience of the examination appeals process.

The Examination Paper

It is not feasible to reproduce more than one question from the examination paper for June 1992 in any detail within the constraints of this article. The examination was over three hours, and complete attempts to 4 questions out of 7 would attract full marks. In general the marks were distributed 50/50 between theory, which had to appear in the examination booklet, and Matlab applications. A section of the rubric which will be discussed later is as follows:

You are required to use Matlab to assist in sitting this exam.

You are required to produce a DIARY file and, depending on question choice, several M-files which will form part of your exam script, and which will be collected along with the exam booklet at the end of the exam.

Note that no printing is required.

Question 7.

7. Show that the second order Taylor scheme for the equation

$$y' = \frac{4t}{y} - ty \quad \text{for } y(0) = 1$$

is given by

$$y_{i+1} = y_i + h\left(\frac{4}{y_i}\right)\left[t_i + \frac{h}{2}\left(1 - t_i^2 - \frac{4t_i^2}{y_i^2}\right)\right]$$

Use Matlab to calculate the solution for $0 < t < 2$ using stepsize $h = 0.1$ both for the above Taylor method and for Euler's method.

Assuming the exact solution is given by

$$y = \left(4 - 3e^{t^2}\right)^{\frac{1}{2}}$$

plot the two numerical solutions against the true solution and plot the absolute error ($\times 10^3$) for both methods.

Example Solution to Question 7

$$y' = f(t, y) \quad \text{where} \quad f(t, y) = \frac{4t}{y} - ty$$

From Taylor's theorem we have, if $y' = f(t, y)$,

$$y(t + h) = y(t) + hf(t, y(t)) + \frac{h^2}{2}f'(t, y(t)) + \frac{h^3}{3!}f''(t + \theta h, y(t + \theta h))$$

where $0 < \theta < 1$. For the 2nd order Taylor method, we neglect the term in h^3.

Finding $f'(t, y)$ we have

$$f'(t, y) = \frac{4}{y} - y - \frac{dy}{dt}\left[t + \frac{4t}{y^2}\right]$$

$$= \frac{4}{y} - y - \left(\frac{4t}{y} - ty\right)\left[t + \frac{4t}{y^2}\right]$$

397

$$= \left(\frac{4}{y} - y\right)\left[1 - t^2\left(\frac{4}{y^2} + 1\right)\right]$$

Taking $t_i = ih$ and $y_i = y(t_i)$ we have

$$y_{i+1} = y_i + h\left[\frac{4}{y_i} - y_i\right]t_i + \frac{h^2}{2}\left(\frac{4}{y_i} - y_i\right)\left[1 - t_i^2\left(\frac{4}{y_i^2} + 1\right)\right] + O(h^3)$$

$$= y_i + h\left(\frac{4}{y_i} - y_i\right)\left[t_i + \frac{h}{2}\left(1 - t_i^2 - \frac{4t_i^2}{y_i^2}\right)\right] + O(h^3)$$

as required.

10 marks

Diary file.

```
>> for i=1:21;
t(i)=0.1*(i-1);
yt(i)=sqrt(4-3*exp(-t(i)^2));
end
>> h=0.1;
>> y(1)=1;
>> e(1)=1;
>> for i=1:20;
   e(i+1)  =  e(i)  + h*(4/e(i)-e(i))*(t(i));
   y(i+1)  =  y(i)  + h*(4/y(i)-y(i))*(t(i)+...
            h*(1 - t(i)^2*(1+4/y(i)^2))/2);
end
>> plot(t,y,t,e,t,yt)
>> erre=1000*(yt-e);
>> errt=1000*(yt-y);
>> plot(t,erre,t,errt)
```

The plots should appear to the student as those below.

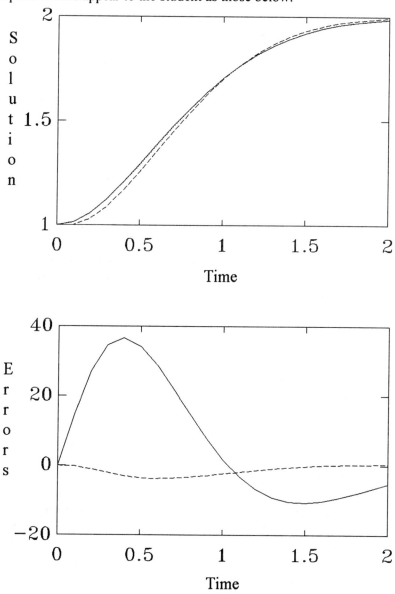

15 marks

As in the case of an examination with solutions purely in the form of written text, full marks would be awarded for equivalent programming.

Since this examination is for a final level module, it was externally moderated, and with the permission of the moderator and that of Coventry University, the comments are reproduced.

321MTE Computational Mathematics for Electrical Engineers.

This is an excellent paper, and a valuable subject for the engineering student. The syllabus describes the usual kind of material, but the teaching is imaginative, and the use of MATLAB in the examination gives the opportunity to students to demonstrate their ability to solve representative engineering problems. The presentation of the paper is as good as the content, and thanks are due to the examiner for the beautifully prepared model solutions.

These are very positive comments.

Assessment

As indicated in the rubric, for the first examination there were three components to the examination 'script', namely

1. The written examination booklet.
2. The collection of M-files.
3. The diary file.

All three components were required to be assessed to award the final mark. It was convenient to print out the M-files and diary files for each candidate since they could be annotated and presented to the external moderators, if required. Only one candidate ended up writing a huge diary file of 600 Kb by sending vast amounts of irrelevant intermediate results to the screen. The other students' diary files ranged from 20Kb to 70Kb. Producing these printouts for the group

400

of 18 took about 3 hours. The author takes about half an hour to mark a conventional examination script. About the same time was required for marking these 'scripts', so the printing out time was a significant overhead.

As an experienced Matlab user, it was quite possible to read and assess the diary file and the M-files without recourse to running them through Matlab for the vast majority of question attempts. A point which appears important is the observation that those students who 'roughed out' M-files in the examination booklet seemed to spend less effort, as indicated in the diary file, in correcting and de-bugging them.

For the second examination the rubric was changed to allow the examiner to assess on the contents of the examination booklet alone. The students were instructed to provide the text of any M-files written in the examination booklet. This is not a great burden since the M-files requested would always be relatively short. Indeed, a student who produced a correct written version of an M-file in the examination booklet, with no evidence of the production or running of an M-file and no evidence of the diary file, would still have been awarded full marks. Most of the questions did require use of Matlab to obtain full marks, but on average 75% of the available marks in the examination could be obtained purely on paper working. The ability to run Matlab during the examination provides on-line help and the possibility of testing M-files and using the visible results as a clue to errors. The students were made aware of the way marks were to be awarded well in advance of the examination date.

The problem of awarding partial marks is little different in the case of a computer examination, if the diary file and M-files, or similar facilities, are available to resolve ambiguities.

Experienced lecturers are used to having their judgements of students broadly confirmed by final examination results. The odd surprisingly good performance is welcome, but if several 'good' students perform badly the fairness of the method of assessment has to be questioned.

401

For both the June 1992 and 1993 examinations the average mark was close to 55% and several marks were above 70%. In both years the order of merit roughly confirmed both the author's impressions and correlated reasonably with performance in other modules. It was possible to conclude that the examinations did form a fair and reasonable assessment of the groups.

Health and Safety

The University *recommendation* for CRT use is to limit continuous use to 50 minutes, with a 10 minute break. Longer periods with longer breaks are permitted, for example 2 hours with half an hour break. This recommendation was not explicitly observed during either examination. However, there is an argument that the students were not using the CRT *continuously* during the exam, since they were spending significant periods writing in the exam book. There is a good reason, apart from health and safety regulations, for enforcing breaks after an hour or so. Such breaks would allow invigilators to take copies of the students files. This would be best performed on a network, but could be adequately performed using floppy discs.

Should it be necessary to explicitly observe the CRT usage recommendations during a three hour examination, there would be some advantage to requiring the students to perform work only in the examination booklet, with the CRTs switched off, for the first hour. Since each question contains about 40% - 60% of effort that does not require use of the computer, and the writing of M-files and the rest of the computing is probably best thought out in advance, this should not prove restrictive.

Conclusions

A large fraction of the University examination regulations is designed to ensure that *all* the materials produced by a student during the sitting of an examination are collected in an orderly manner and delivered intact to the examiner for marking. In particular, the regulations, if followed, are supposed to prevent a student from removing an examination script from the examination room and claiming that the script was lost after the student correctly handed it in. In the case of an examination using a computer package we have the problems of a written examination script, but we also have the much more serious problems associated with any computer files that form part of the examination 'script'. Were a student to rip up all or part of an examination script, there would be no hesitation in holding the student to be responsible since there is no reasonable way for the damage to have occurred accidentally. If a computer file is deleted, or worse still, overwritten, there would be some sympathy for a student claiming an accident.

It has to be said that the rubric for the examination for June 1992 left many opportunities for loss of the computer based part of the examination 'script'. In the case of examination booklets, because it is obvious that attempts at cheating are unlikely to be successful, almost no such attempts are made.The rubric for June 1993 made it clear that only material appearing in the examination booklet is necessary for the full assessment of the examination. M-files and diary files were still be collected, but would only be used to resolve ambiguities in the same way rough work within an examination booklet may be taken into account.

To summarise, if it is desired to use computers in an examination situation, the following guidelines may be useful.

- The examiner should be immediately available to deal with any minor or major emergencies;

- no printing should be required where it will cause disruption;

- spare computers should be available in case of machine failure;

- each student should have a desk with computer individually assigned. If a floppy disk is to be used this should be labelled prior to the examination;

- if computer files are to be used in any way in the assessment, copies should be taken at one or two intervals during the examination;

- all work to be assessed should appear in a written form in an examination booklet. Computer files should have the same status as rough work in the booklet-they *may* be taken into account to resolve ambiguities.

Notes

1. *High Performance Numeric Computation Software,*
The MathWorks, Inc., Cochituate Plasce,24 Prime Park Way, Natick, Mass 01760, U.S.A.
2. *ESDUfine - Personal Modular Finite Element Suite*
ESDU International plc., 27 Corsham Street, London N1 6UA,
3. *The Student Edition of Matlab* ©1992 The MathWorks Inc., Prentice Hall, Inc., Englewood Cliffs, New Jersey 07632, U.S.A.. ISBN 0--13--855982-1
4. Jaques, I. & Judd, C., *Numerical Analysis*
Division of Mathematics, School of Mathematical and Information Sciences, Coventry University, Priory Street, Coventry, CV1 5FB.
5. *DERIVE – A Mathematical Assistant for your Personal Computer,* Soft Warehouse Europe, GmbH,b Schloss Hagenberg, A-4232 Hagenberg, AUSTRIA.

References

Sigmon, K. 1990, MATLAB Primer, Department of Mathematics, University of Florida, Gainsville, FL 32611, U.S.A.

Section Six: Innovative Uses of Technology

In this final Section, four chapters report on different ways in which technology, whether film or video, or computer based, can enhance the learning of mathematics. We are particularly pleased at the breadth these chapters show in their choice of technology but also note the consistency between authors despite being based in different countries and dealing with learners of very different ages. All four chapters (seven authors) see the need for mathematics to be developed within a context and for the meaning of the mathematics to be negotiated by the learners themselves. The resonances with chapters in earlier Sections will not be lost on the reader.

In Chapter 22, *Mathematics and Technology*, Michele Emmer sets the scene by reviewing how technology can aid the visualisation of aspects of mathematics, not only by supporting students' acquisition of mathematical knowledge but, possibly more importantly, by emphasising and supporting their mathematical intuition. He describes the use of computers in Calculus and Advanced Calculus courses for students of mathematics, chemistry and biology. Films are presented as one way of enabling a mathematical perspective to inform on particular artistic images which carry scientific meaning; a mathematical exhibition, *The Eye of Horus: a journey into mathematical imagination* which was directed at the general public, offers yet a different experience. The approach is consistently one of deriving through use leading to formalisation, rather than the more frequently encountered inverse.

In the final three chapters, our attention shifts to the learning of mathematics by those in the first phase of schooling. Taking a similar view to Michele Emmer with respect to the passage from experience to formalisation, Geoff. Sheath describes, in Chapter 23, *Children Making Maths Videos*, a project which sharpened primary children's conceptual understanding of both content and process in mathematics. The children made for parents a video which explained their approaches to a mathematical investigation. This emphasis on communication highlighted the gap between doing, and reflecting upon the meaning of what has been done.

In Chapter 24, *Planning for Portability*, Janet Ainley and Dave Pratt report on the impact which the introduction of portable computers made on the use and understanding of data handling techniques by primary-aged children. Important to this innovative project were the activity-based pedagogy, and the cross-curricular thematic planning, enabling mathematical ideas to be encountered in a range of contexts.

Finally, in Chapter 25, *Arithmetic Microworlds In a Hypermedia System for Problem Solving*, Rosa Maria Bottino and her colleagues complete the Section by describing work that they did with children aged 7-12 years using a piece of complex computer software designed to facilitate pupils attempting to solve arithmetic word problems. Included as part of the software are data-bases, visual representation microworlds and an environment to build problem solutions. The authors conclude that the system allows users to insert the problem-solving process into a framework which is both operational and relational in that it structures and supports the children's actions and fosters the construction of socially shared meaning of their outcomes.

As with chapters in the other Sections of this book, the authors of these four chapters underline how the classroom environment, its ethos and expectations, and its consistency with a view of mathematics as derived by learners, for their own purposes, is central to the quality and quantity of mathematics which is learned and the attitudes which are held towards the discipline. The strength of technology does not lie in its being an up-to-date way of doing more of the same mathematics. It can truly provide a bridge linking the purposes of the teacher with the desired outcomes of learning mathematics.

22 Mathematics and Technology

Michele Emmer

Most people in industrialized countries recognize that mathematics is a fundamental discipline touching nearly all branches of science and technology. Even the possible cultural role of mathematics starts to be recognized. Technology has played an important role for this recognition, in particular the new (and old) visual techniques. With so many images constantly surrounding us today, one of the problems we are confronted with is that of finding the right images to visualize situations in a wide range of scientific, in particular mathematical fields.

Introduction

One of the most interesting aspects of the development of computers and network architectures is the way in which these technologies are modifying the general scientific panorama, in particular that of mathematics, reducing the differences in methodological terms between mathematics and the experimental sciences. Modern computers are so powerful that they can deal with numerical simulation and graphical representation of complex mathematical models. This is true even for the smaller personal computers which are so widespread in schools and universities. In particular, computer graphic techniques have been used not only just to visualise already known phenomena but in a more interesting way to understand how to solve problems only partially solved. In some specific cases, such techniques have provided a new way of proving results in mathematical research.

Computer graphics work not only as a pure visualisation of well-known phenomena but also as a new way of studying mathematical problems, in particular geometrical ones. It can be said that a new branch of mathematics has been developing in the last few years that can be called *Visual Mathematics* (Emmer, 1991[1], 1992, 1993). In 1987 a group of mathematicians at Brown University, including Thomas Banchoff, realised a computer animation movie showing the Hypersphere. Two of them wrote:

> The great potential of computer graphics as a new exploratory medium was recognized by mathematicians soon after the relevant technology became available. As display devices and programming methods grew more sophisticated so did the depth and scope of applications of computer graphics to mathematical problems. (Koçak & Laidlaw, 1987, 8)

A more interesting example from the point of view of mathematical research is the discovery of new types of minimal surfaces by William Meeks and David Hoffman. Hoffman and colleagues have pointed out very clearly the possible interest of mathematicians in picture. The reasons given for this interest are:

> Computer-generated images allow new, often unexpected, mathematical phenomena to be observed.
> Richer, more complex examples of known phenomena can be explored.
> On the basis of exploration of examples and phenomena, new patterns are observed.
> Easier and more fruitful connections can be made with other scientific disciplines. (Callahan et al, 1988, 648)

So mathematicians today are often able to make use of experimental data on which to base the formal approach to the solution of problems. The mathematical community is becoming more and more aware of this significant innovation. All this has had important consequences for mathematical activity both in the applied/industrial sectors and in basic research. (see Concus et al, 1991)

Of course this was bound to have an effect on the university training of mathematicians, and has made it necessary to rethink ideas about the content and organisation of courses, as well as the best use of newly available instruments and equipment, including the possibility of developing numerical experiments and simulations using even very small computers. (see Falcone, forthcoming).A number of authors in the last few years have discussed the limits and benefits of the use of computers, and in particular of personal computers for teaching different subjects. (see for example Banchoff et al, 1988) As we pointed out in 1988 the problem is: what are the new methods which we might adopt to use in full the possibilities offered by personal computers? What must be changed in the content of traditional courses in view of the technical revolution of the last few years?:

> We are assuming here that something must be changed since, in our opinion, courses based on the use of new technologies can offer to the students more than traditional 'chalk & blackboard' courses do. Probably this assumption is not generally accepted but it is clear that the spread of informatic tools throughout our society cannot leave the school apart, so in some sense we are forced to introduce computers in schools either as a new subject of study or as new teaching instruments. What is not clear at the moment is the following: are these tools really useful when teaching traditional subjects, or is their use in this context due only to a current fashion? (Capuzzo Dolcetta et al, 1988, 637).

The Mathematics Laboratory at the University of Rome

Our activities concerned the use of personal computers in mathematics courses for undergraduate students, and in particular, given our own mathematical background, in courses of Calculus and

Advanced Calculus. Experiments started in 1983 along two different lines: on the one hand by direct involvement in courses held in the departments of Mathematics and Physics at Rome University La Sapienza (in collaboration with Italo Capuzzo Dolcetta, Maurizio Falcone and Stefano Finzi Vita), leading to changes in the structure of the courses themselves and bringing about the setting up of a true Mathematics Laboratory; on the other hand, by comparison with similar experiences underway in the universities of Paris Sud Orsay (France) and Leeds (UK), both of whom are partners in an ERASMUS project funded by the European Commission. In this phase it was possible for the teaching staff involved in the project to enlarge their knowledge and their experience in the other member states and for some students to spend a recognised and integrated part of their courses in another participating university.

As everyone knows, both *Calculus* and *Advanced Calculus* are fundamental courses devoted to developing logico-deductive skills while presenting mathematical techniques and results requested in the subsequent courses. Traditionally, the organisation of these courses consists of general lectures and practical exercise sessions.The introduction of computers induced some modification in both, since computers have been used to present and visualise mathematical phenomena as well as to solve problems from a numerical point of view.

In the first three years of our experiment we developed, with the students' help, our own demonstration programs to visualise and illustrate some aspects of mathematics using the graphic facilities offered by personal computers. No commercial software has been used. (Capuzzo Dolcetta & Falcone, 1992) The main point regarding programs developed for demonstrations in mathematics, in our opinion, is that they must not be so sophisticated to make typical computer errors disappear. Student must be aware of limitations connected with the use of computers in scientific investigations and programs must not cancel mathematical difficulties. It is also very important for the student to have a sufficient background to distinguish personal errors from computer errors. These programs

410

constitute the basis for the Mathematics Laboratory in which a number of different experiments can be presented and discussed: functions, sequences, solution trajectories of ordinary differential equations, curves and surfaces can be displayed to students showing them the wide variety of behaviours and situations hidden in a mathematical definition. The main target is to drive students rapidly towards an active use of computers postponing the refinements to the discussion of particular problems. In this way students can start programming from the very beginning, trying by themselves to inplement numerical methods in order to solve concrete mathematical problems. In this active work of programming, which was done using 20 Olivetti M-24 personal computers, the emphasis is mainly on mathematics rather than on programming. The choice of a programming language is not very important in our opinion and BASIC is probably the best choice to minimise the number of prerequisites, considering that many students already have familiarity with it on their home-computers.

The visualisation of a series of examples helps students to absorb concepts that can be quite difficult to grasp in their abstract mathematical formulation, specially in the first two years of their curriculum. From this point of view the use of a computer is very practical and effective since, if your software allows an interactive modification of the data, you can easily modify the parameters in a given example showing immediately the changes due to the new choice. Students can therefore use this *dynamic blackboard* to make investigations, verifying their knowledge and the validity of their intuitions.We have discussed in detail the experience of the first three years of the mathematics laboratory. (Capuzzo Dolcetta et al, 1988, 637)

Mathematics Outside of Mathematics Departments

Starting from 1987 my personal experience of teaching was in departments of Chemistry and Biology in small universities. The situation of computing facilities for students in chemistry and biology, even in large universities, is quite different from students in physics and mathematics. Moreover there are only few mathematics courses in the chemistry curriculum in the first two years covering topics such as calculus and advanced calculus, linear algebra and analytic geometry. Due to the number of hours each student has to spend in the chemistry laboratories, it is not possible to increase the number of hours devoted to mathematics; so it is almost impossible to introduce a mathematics laboratory. The situation is even worse for the biology students. They have only one or two mathematics courses in the first two years. Biology students' interest in mathematics is not very high. It is interesting to quote from a paper by S. A . Garfunkel and G. S. Young on *Mathematics outside of Mathematics Departments.*. They sent many non-mathematics departments a letter asking for the opinions of the staff on the mathematics courses. They classified the comments into five categories:

1)   The mathematics faculty does not know or appreciate applications. Some typical quotations:

- There is an inability of mathematicians to come to grips with a difference between pure and applied mathematics. Mathematicians of the pure strain look down their noses at the other strains. Thus science and engineering departments feel that their students will not be adequately serviced by math departments.

- Mathematics departments have become so abstractly oriented that their courses are not given any applied content. Our engineering school tried to convince our math department to teach some applied courses but eventually gave up.

- The math department courses "turn off" most of our kids and it's up to us, back in the professional departments, to turn their enthusiasm back on again.

2) Mathematics faculty teach mathematics as an art with full abstraction, not as a tool.

- The content of most math courses focuses on theoretical development. This is not 'bad' per se but leaves most students wondering about the 'what, when, where, and why' of applications.
- Applied departments use math as a tool; math departments often become more interested in its description and generation of the 'tool' itself.
- Engineers find math to be a need, not a love. Mathematicians are out of touch with the real world and are more like mathematical artists than real world scientists.

3) Topics span too many mathematics courses.

- There is not room for every student to take the separate courses in differential equations (ordinary and partial), vector and tensor analysis, complex variables, Fourier series, probability. All these are covered in a one-year course by omitting the detailed proofs and generalization.

4) The mathematics departments have not kept up with new applied mathematics.

- Mathematics departments generally do a lousy job with mathematics their faculty has no training for. Specifically, Shannon's information theory; automata; transformation geometry; graph theory; algebraic coding theory; polynomial rings and finite fields; computer ability; and, one suspects, probability and statistics. (Garfunkel & Young, 1990, 408)

My personal experience is that in teaching courses for non-mathematics students one way of trying partially to avoid all these problems is to introduce the use of personal computers from the very beginning. It is important to visualise and to give graphic examples of mathematical phenomena and at the same time introduce the students to the use of simple software already prepared. I have always used the non-commercial software prepared in the Laboratory at the University of Rome, described in the previous section. I do not believe that in teaching students in mathematics and physics, all details and formal proofs must be given while, for non-mathematics students, courses in mathematics should be a list of properties and results with almost no explanation. Non-mathematics students also need a precise idea of the fundamental mathematical concepts and methods, preferably with full details. At the same time it is important to look for examples, possibly using the graphics facilities of your department, in which it is easy to show connections with biology, medicine, chemistry, etc. A typical example is the following: the function $\exp(-x2)$ is very important in biology, in particular in genetics; it is possible to obtain experimentally with the students the graph of the function making simple experiments or using experimental data known to the students from other courses (genetics, chemistry,....). Then it is possible to show that the function is integrable on the real line, but it is not possible explicitly to calculate the value of the integral. Of course this example is very useful to convince students of the necessity to introduce improper integrals, an otherwise very abstract definition. Using a personal computer it is possible to show live how to obtain numerically a good approximation of the value of the integral. In this way, many mathematical ideas (graph of a function, integrability, numerical approximation) can be introduced in an interesting way, showing all possible links with the other non-mathematics courses. It is also important not to give an idea of mathematics as a simple list of techniques and formulas. So formal definitions, theorems and numerical approximations must be well balanced during the course of mathematics.

Finally, I agree with the observations of Falcone in his paper *Calculus Revisited* , that:

> As everyone knows, the range of applications of mathematical science has greatly increased: engineering and physics are the traditional fields of application but also biology, physiology and economics now use sophisticated techniques of modern

mathematics. A presentation of simple models, their simulation on a computer and a discussion of the mathematical models necessary to study them is probably one of the best ways to motivate students who look at mathematics more as a difficult topic than as a useful science. This approach would be particularly valuable especially for students attending degrees other than Mathematics. (Falcone, forthcoming)

New and old technologies in mathematics: films and videos

With so many images constantly surrounding us today, one of the problems we are confronted with is that of finding the right images to visualize situations in a wide range of scientific fields. In mathematics, we are dealing with ideas that are often abstract and difficult to grasp; obviously, we are not always able to find images that effectively clarify the question. In order to use cinema techniques for mathematical subjects, the two most important aspects (the scientific facts and the images used to illustrate them) must arise from the same source. One cannot hope to decide on the subject first, and then to search for the images with which to visualise it.

In my personal experience, the decision to link mathematical subjects to the visual arts (to architecture as well as to physics, chemistry and biology) seemed quite natural. One of the aims of artists is to make *visible the invisible;* why not use the images that artists have created, starting from a more or less scientific base with the addition of a personal element, to talk about mathematics? My main idea with regard to producing math-movies is that of creating cultural documents seen through the eyes of a mathematician. In other words, as I have written elsewhere:

> The movies are attempts to produce works which are, at the same time, vehicles of information of a scientific and artistic nature on various mathematical subjects, and also to stimulate the observer towards further

> investigations of these same topics. The possibility
> offered by cinema techniques are fully exploited. My
> intention was to use the full language of images and
> sound. The problem, of course, is to maintain a
> balance between entertainment and informative
> popularization in such a way that one aspect does not
> dominate the other. (Emmer, 1986, 249)

The movie should not be a lesson with pictures, but rather a new language that integrates the two ingredients. Compared with other media, the cinema has the great advantage of being able to provide a large quantity of information in a limited period of time.

Another major advantage is that the language of the cinema is universal. The language of images in movement is understood by people of all ages and all cultural backgrounds. This aspect has to be fully exploited in movies concerning mathematics, which have to catch the interest of an audience ranging from primary school children, to university students and even the general public. My own series of movies *Art and Mathematics* is based on these principles. The principal aim of my cinema experience was, and still is, to create *cultural documents* seen through the eyes of a mathematician who therefore gives priority to his discipline. The choice of the subject becomes a problem of finding a balance between scientific requirements and artistic values, between scientific explanation and movie sequences. The aim of this vast undertaking, more than 20 films and videos of 27 minutes each have been produced in the last 15 years, was to present an interdisciplinary model more or less inspired by scientific ideas for the teaching of several aspects of mathematics in connection with the other sciences and art movements. My wish to show how mathematics plays an important role in our *culture,* was a truly educational ambition even if not strictly related to a specific didactic curriculum. This is the main reason why the films have been translated in English, French, Spanish and are still distributed all over the world. For the project I have asked and obtained the cooperation of many famous mathematicians like Roger Penrose, H.S.M. Coxeter, Fred Almgren,

Jean Taylor, Thomas Banchoff. Due to the universality of their language the films have also been used for more strictly didactic purposes. (see for example Emmer, 1989a, 213)

Mathematics on show

A fairly new development, over the last few years, is the itinerant exhibition of mathematics. There are many examples of permanent exhibitions devoted to mathematics (in the USA, France, UK). But, over the last few years, we have seen the appearance of mathematical exhibitions linked to recent theories and results. Images play an essential role in all these topics, especially those created by computers with sophisticated graphic capabilities. Then another interesting aspect is that these events have become not just scientific shows but also artistic displays.

Here too, as I mentioned for films, the idea of building a bridge between scientific and artistic images enables one to deal, in a comprehensible manner, with a history of mathematics parallel to the artistic and scientific events with which the audience is more familiar (Emmer, 1991b, 23). It is clear that I am referring in particular to the Dutch graphic artist M.C.Escher, (Coxeter et all, 1986) but the question does not end there, as can be seen in the exhibition *The Eye of Horus: a Journey into Mathematical Imagination* held in various Italian cities in 1989.

This exhibition was an attempt to put together all my personal experience in films, videos, computer graphics, art exhibitions, educational experiences in order to make a contribution towards spreading the *culture of mathematics* as widely as possible. The exhibition was sponsored by the *Musée des Sciences et de l'Industrie de la Villette* in Paris. My idea was to organise a show that was informative not just for mathematics students and their teachers but also for the general public, enabling them to *see, hear and even touch* the exhibits while enjoying themselves. It was an exhibition dominated by the visual aspects of mathematics, by images. Knowing that we had to make a few basic mathematical concepts interesting and entertaining for the public (and I say a *few* because I am convinced that there are many subjects that cannot be made accessible to non-experts in an exhibition), we attempted to make these ideas visible by enlisting the aid not only of mathematicians but also of artists who might be interested in such terms through their work. I wanted to reinforce the idea that this was not a purely didactic exhibition on mathematics but rather an exhibition with several didactic sections. The exhibition was arranged in thematic sectors, each of which contained explanatory panels, one or two tables from la Villette's exhibition *Horizons Mathématiques,* some films and videos (most from the series *Art and Mathematics*) and works by Italian and non-Italian artists, including paintings, sculptures and computer graphics. The first section had the same title as the exhibition itself: *The Eye of Horus.* The reason for choosing this subject was that the eye of Horus had a precise mathematical significance for the Egyptians, closely linked to the myth of Isis and Osiris. Other thematic sectors were focused on symmetry, including quasicrystals and Penrose tilings, Platonic Solids, Knots, Labyrinths, Moebius Band, Fourth Dimension and more. The exhibition was visited by more than 100,000 people, including many pupils from elementary schools. The Italian State Television made a video of the exhibition in the series *The Great Exhibitions of the Year.* A large catalogue was also produced with original papers by mathematicians and artists. (Emmer, 1989b; Emmer, 1990, 89)

Final comments

In conclusion, I would like to say that in the case of films and exhibitions it is the image that plays the essential role, awakening interest and stimulating the imagination. The same is true with computer graphics. It is no doubt that visualisation of mathematical phenomena significantly increases the rate of assimilation; the students acquire basic knowledge more quickly and develop mathematical intuition. The great majority of students said that lessons conducted with the aid of computers had clarified aspects that purely theoretical lessons had not made clear. Of course it is important to recall that the emphasis on applications and on the relevance of visualisation does not mean that the abstract and rigorous approach to mathematical problems should be abandoned in favour of simulations and heuristic reasoning.

I agree with the opinion of Falcone who wrote:

> The visualization of concepts in mathematics is a revolutionary event which will change completely our way of teaching, increasing the role of geometry in the whole learning process. Not only will it be possible to refresh traditional lectures giving "real time" examples and motivations, but it will also be possible to treat new subjects and give to students some ideas of the new frontiers of current research in mathematics. This seems to be a very important point to motivate students from the very beginning of their university career and, if this approach is also developed in secondary schools, probably the number of students choosing to study mathematics will increase, inverting the tendency in most industrialized countries where the best students move to engineering and computer science departments. (Falcone, forthcoming)

The aim is to share experiences on computer-aided teaching and also to point out some topics where current mathematical research takes decisive advantage of the facilities offered by a computer, in the spirit of a strong interaction between research and didactics that we believe to be the distinguishing feature of university teaching, even at the undergraduate level. The purpose of using various media and technologies over the last 20 years of my personal experience (films, videos, computer laboratories, books, exhibitions) was to improve the teaching of mathematics and to enlarge the diffusion of *mathematics culture*. In doing so, as the mathematics culture is diffused, I assume also that the teaching of mathematics will receive an important benefit.

This paper was partially supported by a grant MPI 40%.

References

Banchoff, T.F., Capuzzo Dolcetta, I., Dechamps, M., Emmer, M., Koçak, H. & Salinger, D., 1988, eds., *ECM/87, Proceedings of the International Congress on Educational Computing in Mathematics.*, Amsterdam: North-Holland.

Callahan, M.J., Hoffman, D. and Hoffman, J.T., 1988, 'Computer Graphics Tools for the Study of Minimal Surfaces', *Comm. ACM*, 31 n. 6, 648-661.

Capuzzo Dolcetta, I., Emmer, M., Falcone, M. & Finzi Vita, S., 1988, 'The Impact of New Technologies in Teaching Calculus: a Report of an Experience', *Inter. J. Math. Educ. Sci. Technol.*, 19 n.5, 637-657.

Capuzzo Dolcetta, I. & Falcone, M., 1992, *Calculus, un software per il laboratorio di Matematica*, Rome: Gea; see also the textbook: Capuzzo Dolcetta, I. & Falcone, M., 1990, *L'analisi al calcolatore: il personal computer nel laboratorio di Matematica*, Bologna: Zanichelli.

Concus, P., Finn, R. & Hoffman, D.A., eds., 1991, *Geometric Analysis and Computer Graphics*, MSRI Series n. 17, Proceeding of the workshop, Berkeley, 1988, Berlin: Springer-Verlag. A new workshop took place Coxeter, in October 1992.

Koçak, H. & Laidlaw, D., 1987, 'Computer Graphics and the Geometry of S^3 , *The Mathematical Intelligencer*, 9 n. 1, 8-11.

Emmer, M., 1986, 'Movies on M.C. Escher and their Mathematical Appeal' , in Coxeter, H.S.M., Emmer,

M., Penrose, R. & Teuber, M., eds., *M.C. Escher: Art and Science,* Amsterdam: North-Holland, 249-262.

Emmer, M., 1989a, 'Art and Mathematics: an Interdisciplinary Model for Math Education', in Blum, W., Huntley, I., Kaiser-Messner, G. & Profke, L., eds., *Applications and Modelling in Learning and Teaching Mathematics*, Chichester: E. Hordwood Ltd., 213-218.

Emmer, M, 1989b, ed., *L'occhio di Horus: itinerari nell'immaginario matematico*, Roma: Ist. Enciclopedia Italiana.

Emmer, M,, 1990, *Mathematics and the Media.*, in *The Popularization of Mathematics*, ICMI Series, Cambridge: Cambridge University Press , 89-102

Emmer, M, 1991a, *La perfezione visibile*, Roma: Theoria.

Emmer, M, 1991b, 'Art and Mathematics: a Series of Interdisciplinary Movies' *ZDM* 89/1, 23-26.

Emmer, M, (1992) ed., , 'Visual Mathematics', special issue, *Leonardo,* 25 n. 3/4.

Emmer, M, (1993) ed., *The Visual Mind: Art and Mathematics*, Cambridge: The MIT Press.

Falcone, M., *Calculus Revisited.*, forthcoming

Garfunkel, S.A. & Young, G.S., 1990, 'Mathematics outside of Mathematics Departments', *Notices AMS*, 37 n.4, 408-411.

23 Children Making Maths Videos

How making videos can aid children's mathematical communication and reflection

Geoff Sheath

This paper describes how children made short videos to report mathematical investigations and analyses the contribution of the making of the videos to their mathematical communication and reflection. It was carried out in a class of 30 ten year olds at a school in the Docklands area of London, UK. The work was part of the Parental Involvement in the Core Curriculum project.

Review of previous work

Television has not been used as much as the computer to aid learning in mathematics. However, in Britain there is a long tradition of broadcast schools' television mathematics programmes which goes back at least 25 years, and in the United States there are mathematics programmes such as Square One. Television has also been used quite widely in Britain for teacher training, particularly by the Open University.

The making of video by children in British primary classrooms has usually been in support of media education in which children explore how messages are shaped by the media that transmit them. The work described in this paper is not ostensibly connected with media education but it is of note that the 'four key areas of investigation which underpin what is taught in media education at primary level' which Craggs (1992) identifies:

1. Selection and construction
2. A sense of audience
3. Representations of reality
4. Narrative technique

all had to be considered by the children when making videos themselves.

The production of videos by pupils on mathematical themes does not appear to have been widely explored. There has, however, been plenty of work with primary children in other curriculum areas. Emerson (1993), for example, describes primary children making their own adverts and gives suggestions to teachers embarking on such a project.

The investigation Frogs (also called Leapfrog or Frogs and Toads) on which the children's video is based is described in Burton (1984) and Brissenden (1988). It is often introduced by taking a strip of seven squares, placing three red counters at one end and three green at the other with an empty square in between and then inviting the children to swap over the red and green counters by a process of sliding counters or leapfrogging opposite-coloured counters using the empty square. Children may be asked in how few moves this can be done. There is then ample scope for children to look at similar problems and to find rules.

The background, scope and aims of the investigation

In the period 1991 to 1993 the author spent one day per week working in a school for 3-11 year olds in the Docklands area of London as part of the University's programme of professional renewal for academic staff involved in teacher training. The school has many pupils from ethnic minority backgrounds (predominantly from Bangladesh) and a high proportion of poor children. 27 out of 30 children in the class in which this project was undertaken were entitled to free school meals, one measure of poverty in the U.K.

Besides teaching, the author worked within the school on a number of projects designed to support development of parental involvement. This was part of the wider Parental Involvement in the Core Curriculum (PICC) project (Hancock et al, 1993, 1994).

The school had already undertaken many initiatives to foster home-school links. Despite this range of initiatives it was felt by the school that there were certain things that staff regarded as very important that had been difficult to communicate to parents. One of these was the school's emphasis on problem solving and process learning. Although parents could see what their children were learning (for example by looking at the products of that learning in exercise books, written accounts, drawings or models), it was often difficult for parents to appreciate how their children were learning or what they were learning about their learning.

It was hypothesised that video would enable evidence that teachers used to gain insight into children's problem solving and learning strategies (for example answers to questions and class presentations) to be captured and communicated to parents. It was also hypothesised that in discussing how to present their work on video children would be enabled to reflect more deeply on their work and the work of others. Finally it was hypothesised that identifying parents as the audience would help to give focus to the making of the video.

Descriptions of the procedures used

The class of ten year olds with which the author was working were introduced to the process of making video through an introductory activity in which each member of the class participated as part of a group of six. Emerson's (1993) list of suggestions for children making videos was used in planning this activity. Each group was invited to choose an aspect of computer work around the school about which they could interview someone on video. These interviews would be turned into a report by editing the interview, adding extra video and a commentary as necessary. As examples, one group interviewed the headteacher about the overall IT strategy in the school, another group concentrated on the use of a computer program by 8 year olds, and a third group prepared a report about work they had done building computer-controlled Lego models.

The first task for each group was to set up an interview related to their report. Each group of six planned their questions, rehearsed them within the group, modified them as necessary, rehearsed them with the subjects on audiotape, modified the questions as necessary, and finally recorded the interview on videotape. Each group member took a turn at camera operator, sound operator, floor manager (responsible for continuity, cueing and clapper board), interviewer and director.

Further planning was then done using mainly story-boards (a sequence of small sketches used to represent the video sequences to be taken) and scripts to identify the further shots needed. These were then taken, commentaries written and recorded and the author edited the final report to the group's instructions.

The planning activities were easy to integrate into the normal work of the class as the class regularly worked in small groups. The videoing was generally done at playtime, lunchtime or during school assemblies, partly because of the need for direct supervision out of the classroom, but mainly to overcome the problem of ambient noise. Later some groups were able to video out of the classroom without supervision.

After the introductory activities the author and the class teacher chose certain pieces of work that children had done and invited small groups to make videos to describe them. The results, reported on here, refer to a video made about the mathematics investigation Frogs.

426

Data was collected as field notes and children's work. As the author was working with the class as a teacher and only on one day per week the data is sometimes second hand or incomplete.

Statement of results

As part of their mathematics work the class teacher had asked a number of children to investigate Frogs. The children worked on their investigations four or five times a week over a period of about two weeks. During this time they kept a record of their work in their exercise books and some children shared their work with the rest of the class by giving oral reports.

On the completion of the introductory video activities Frogs was identified by the class teacher as a suitable piece of mathematics work for presentation as a video and the class teacher selected a group to work on this. The group comprised four children who had worked on the investigation (Akosua, Kung, Vinh An and Forida) and one child, Makhon, who had not. The group was selected to include a range of accomplishment. Makhon, who had not done the investigation, was to act as an interviewer. It was made clear to the children at the outset that the principal audience for the video was to be their parents.

All four children had written accounts of their investigation in their exercise books. These were in the form of results, tables and diagrams with minimal explanations. Akosua had also produced a summative written account of her results in the form of a poster. As an example of her style (which was fuller than any of the accounts in exercise books) here is part of what Akosua included about the rules she had found:

> I know that if I have 5 cubes each side I would start with a slide, then do 1 hop then a slide, then two hops, then a slide and three hops. You carry on doing that up to 5 and then, when you get to 5, you begin to go backwards for instance five then slide, four then slide etc. It's the same

with 6 but you do six hops then you go down. An easier way [to calculate the] moves you make on each amount of cubes you just add two to the number before for instance:

3

5   +2=7

8

7   +2=9

15

9   +2=11

24

11

35

The poster also included diagrams which showed how Akosua had recorded her results and which she had used to spot symmetry and patterns. However the account was descriptive and included few details of how she had discovered the results.

The first meeting of the group took about 30 minutes. The group first explained Frogs to Makhon by answering his questions. Makhon was shown the initial puzzle using three red and three green cubes on a seven-square track and asked questions in a fairly informal way.

Drawing on their previous experience of making videos the group used Makhon's questions to draw up a list of questions for the formal on-camera interview. Both questions and points that answers would need to include were written down. Initially the pupils were going to explain the puzzle by showing cubes being moved on the video, but they finally decided that six children on seven chairs would show it more clearly and be more interesting for their parents. Akosua was chosen to be interviewed for the first segment. The interview was practised several times without being videoed. Each time the group offered suggestions for its improvement. Finally the group videoed the interview. During the first take Akosua forgot one of the points she was supposed to include and so a second take was made. Later in the day the group were able to view the takes.

During the following week the group had to decide exactly how to explain the solution to the basic puzzle. The second sequence of the video, which explained the solution to the basic puzzle was videoed during playtime using six other children from the class. This took about 15 minutes.

At this point the group had cohered and established a way of working which included defining question and answer sessions in note form on paper. The next planning session was crucial because it had to knit together all the results that the pupils had discovered.

The author suggested that Makhon interview each of the four pupils in turn and ask them to explain their results. The pupils used their original notes in their maths exercise books as a prompt. However a number of points came out in these interviews that were neither in the children's exercise books nor on Akosua's poster. Akosua and Vinh An talked about the struggle they had to find solutions, and talked about looking for simpler examples once they had solved the initial puzzle. All children talked about the need to record systematically and explained the systems they had used. All were looking for patterns and described finding both geometric and number patterns. Akosua and Vinh An explained how they used the patterns to predict further results and to verify them. Vinh An explained how he looked for and found algebraic generalisations for some of his results. During this discussion statements and questions showed that each child was attempting to understand each of the others' results and to relate their own ways of working to those described. Very little of this process information had come out in writing and yet the pupils talked about it easily.

By the end of the discussion it was clear that some results had been obtained by all the group, some by only one or two. A list was drawn up of who was to explain what. A formal script was not written. Instead questions to prompt each pupil to explain the important features of their work were identified. The question of what to say was frequently related to the parental audience. Statements about how the children had worked and why they had chosen to move in the directions they had were seen as necessary in their explanations to provide the links between results.

When the main structure of questions and answers had been agreed storyboards were used to prompt discussion of the visuals that would help illustrate the points being made. In most cases either close ups of manipulations of cubes or close ups of diagrams or tables of results from exercise books were used.

This whole planning session took about 45 minutes. The video making provided both focus and motivation in this session.

During the following week the group finalised the questions and practised them and on the author's next visit to the school the interviews were run through and videoed. Following a viewing one interview was retaken because the group felt that it was not clear enough. The following is an excerpt from the soundtrack.

> [After a description of the solution to the basic puzzle Akosua says] When I'd got an answer [for the three cubes] I looked for an even similar (sic) problem with one cube and two cubes. When I drew it [the pattern of moves] out I found it was symmetrical ... There's one cube, and two cubes and three cubes [shows patterns in book] ... But then I looked ahead of three and went straight from four to five, and I know that if I had five cubes each side I would start with slide and one hop [continues explaining pattern] and you can continue with six or seven up to about seventeen ... [Akosua then goes on to explain the method based on differences as shown on the poster. Vinh An then takes up the explanation] ... The difference between 3 and 8 is 5, between 8 and 15 is 7, between 15 and 24 is 9 so my hypothesis is that it goes up in odd numbers. [Makhon asks, "Have you noticed any other patterns?"] Yes, like this, 1×3 is 3, 2×4 is 8, 3×5 is 15, 4×6 is 24 ...

Once the explanations were completed satisfactorily the other shots, for example of cubes being moved and diagrams from exercise books, were videoed and the author edited up the completed video to the group's instructions.

The video was shown to parents who, in discussion afterwards, commented particularly on how motivating their children had found making it. It was not possible to research their views systematically, but most parents found the video itself impressive. However they did find parts of it rather difficult to understand, particularly where the children discussed sequences that they had found.

Discussion

The class had had a lot of experience of investigations. In particular the members of this group showed evidence of being aware of the value of recording systematically, trial and improvement, simplification (by starting with one red and green cube on three squares), looking for patterns, generalising and validating results. All of these were things that the class teacher and the author stressed in their teaching

The children were all aware of the 'vocabulary' of television through their own viewing and the author found, like Emerson (1993), that, 'basic techniques and procedures were not a noticeable problem except for steady camera operation'.

The children had completed their investigations into Frogs before it was decided to use it as the basis of a video, so their written and video accounts drew on the same pool of results. In comparing the written accounts and the video accounts of the children's work it was apparent that the videos were fuller and richer. There are several reasons that might account for this.

The first was the motivation that making the video produced in the children. Of itself this may have been a consequence of the interest of the author and the class teacher, but almost without exception the children wanted to be involved, liked using the equipment and wanted to produce their best work for recording.

The second was that the sense of their parents as audience enabled the pupils to focus on the video as a real communication task. On a number of occasions during the planning a discussion of whether parents would understand a particular answer led to a fruitful exchange on the effectiveness of the communication.

The third was that the necessity of working as a team and sharing out the presentation forced meaningful discussion of the content in a way that compiling a composite written record would not have done. Papert (1980) asserted that debugging routines in LOGO enabled children to reflect on their own thinking. In a comparable way, planning and implementing the videos necessitated children reflecting on the content and modes of their mathematical communication.

The fourth was that oral presentation to camera cut out much of the need to write down what was to be expressed. This was most graphically shown in another video where a boy in the class who had special educational needs and could not write at all was able to make a real contribution to a video by answering and asking questions in his own words.

The fifth was that the video account could be very detailed without requiring the time and sustained effort needed to *write* a long account. Although many of the children were fluent writers and experienced at editing their work in longhand or on the computer, no written account could be as easily edited as a video sequence where the whole process of taping a sequence, reviewing it on television, discussing it and then recording another take might only take ten minutes. The contrast between Akosua's written and spoken account may be due to this effect, particularly as no new investigative work was undertaken for the video.

Conclusions

Three hypotheses were made at the beginning of this paper. For the first, the video did capture children describing how they had worked on the problem in a way that had not been captured on paper. This included evidence of their problem solving and learning processes. It is not the author's contention that the children were incapable of

432

writing about their problem solving and learning processes but that the making of the video made it easier to communicate them and raised their significance.

The second hypothesis, that making the video deepened reflection, is less easy to substantiate. Certainly each child articulated their reflections on their work both during planning discussions and on the video and they also had to try to understand how the other members of the group had worked. In the author's view therefore there was at least a limited increase in reflection for each member of the group.

The final hypothesis, that the parental audience would focus the video, was borne out by the planning discussions where the children's intimate knowledge of the proposed audience enabled them to judge whether an explanation would be understood by their parents or not.

In conclusion the author believes:
- that most similar classes could make mathematical videos, but that the teacher would have to have a good grasp of video techniques to support the group in the early stages;
- that most children would find the making of videos motivating and that this would result in high quality discussion both of processes and results and some increase in reflection;
- that identifying parents as the audience would focus the discussion and communication.

References

Burton, L. 1984, *Thinking things through.* Oxford: Blackwell.

Brissenden, T. 1988, *Talking about Mathematics.* Oxford: Blackwell.

Craggs, C. 1992, *Media education in the primary school.* London: Routledge.

Emerson, A. 1993, *Teaching Media in the Primary School.* London: Cassell Educational.

Hancock, R., Sheath, G., Smith P., Beetlestone F., 1993 *PICC Annual Report.* London: Faculty of Education, The University of Greenwich.

Hancock, R., Sheath, G., Smith P., Beetlestone F., 1994, *Getting started on the PICC Project* **in Dombey, H. and Spencer, M. (ed.)** *First Steps Together* Stoke-on-Trent: Trentham.

Papert, S. 1980, *Mindstorms.* Brighton, Harvester.

24 Planning for Portability

Integrating Mathematics and Technology in the Primary Curriculum

Janet Ainley and Dave Pratt

This chapter reports on the early stages of the Primary Laptop Project, in which we are exploring the effects of access to portable computers on mathematical learning. We describe the initial planning, where our aim was to set up cross-curricular activities which would take advantage of the portable technology and also contain opportunities for mathematics. We suggest that in such an environment immediate access to computers can produce significant changes in children's learning, leading to what we have called portable mathematics. We also consider the support needed by teachers to take advantage of the opportunities offered by this environment.

Background

This paper reports on aspects of the work of the Primary Laptop Project, based at Warwick University and at Brookhurst Combined School, Leamington Spa, U.K.[1]. The project has the long-term aim of studying the effect on children's and teachers' learning and attitudes when offered high levels of access to personal technology[2], with particular emphasis upon mathematical understanding. Laptops offer a way of creating a computer-rich environment *within* the primary classroom, since they are small enough to be placed on a table alongside other resources. Furthermore, they can be easily moved from room to room or from school to home, allowing children and teachers to use them in very flexible and powerful ways which would simply not be possible with desktop machines.

The first phase of the project was used to help a group of three teachers become familiar with using a portable computer and a commercial software package. For a period of about a term, the teachers each had a laptop for their personal use. During this time they received regular support, but we made no specific demands as to what they did with the machines, except that they were asked to keep a journal, using the word processor, throughout the project. These journals gave the teachers a purposeful and substantial task using the laptop, and encouraged them to explore and experiment. They also provided a valuable source of data about when and how the teachers used the laptops, and the successes and difficulties which they experienced: events we could not witness directly. As their confidence increased, the teachers found it valuable to look back at and reflect on their journal entries, and eventually they were able to relate the difficulties which children encountered to their own early experiences.

From this initial, preparatory phase, the project has moved into a second phase in which we have studied three classes of Y7 (11/12 year-old) and Y5 (9/10 year-old) children working on activities with a data handling component, chosen because the teachers felt a lack of expertise in this area. During this second phase, each class has had access to one laptop between two children for about three weeks. Six laptops have remained in the school throughout the year.

Throughout both phases of the project, data has been collected by means of field notes made by the researchers, supplemented by the journals kept by the teachers and examples of children's work (collected on disc). A series of individual interviews and structured discussions were conducted with the teachers. These were recorded on audio tape, and also by means of notes taken by the researchers.

These experiences have led us to a number of insights about the mathematics children tackle in this computer-rich context, and about the support needed by teachers both at the planning stage and during the classroom activities in order to use the technology effectively for mathematical development. In this chapter, we wish to focus on these issues.

436

Planning for Portable Technology and Integrated Mathematics

In setting up the Primary Laptop Project we started from the position that the machines must primarily be used as *portables*, operating on battery power whenever possible. The flexibility offered by a portable machine proved to be very important, not just in terms of convenience, but also in terms of the ways in which it could be integrated into the activity of the classroom. The teachers involved in the project planned work in their classrooms around themes, which spread over several weeks. Within these themes, activities often crossed many curriculum areas. The children were used to being given responsibility for planning and evaluating their work within frameworks provided by the teacher, and class discussions about their work were a regular part of the school day. This fitted well with the ways we envisaged the laptops being used. *Our* main interest was in mathematics, but we wanted the children to feel that they had control over the use of the machines, and that they were free to use them whenever they felt it would be appropriate.

The project teachers asked us to make some input on data-handling, an area they felt they were not yet addressing in their use of IT. We worked with them to plan activities with data-handling potential, which fitted into existing topic work being undertaken by the three classes. We aimed to design these activities so that they would engage the children's interest, involve the use of a range of IT skills, and also have the potential to introduce mathematical ideas in contexts which would make them accessible to the children.

The Y7 children were already working on medieval life at Kenilworth Castle as a theme. This group of children quickly realised the potential for using the laptops within their topic, and produced large quantities of work making use of the way in which writing and drawing can easily be integrated within the software. The theme offered a number of data handling opportunities, including extracting information from historical sources such as the Domesday Book and Lay Subsidy Rolls, which give records of taxes paid in various years.

437

Despite the confidence which these children developed with the hardware and software, it was clear that they did not really get to grips with the mathematical ideas which we hoped a data handling project would raise. There was a tendency for the children to use the power of the technology to produce quick but meaningless graphs, which indicated that they had no real grasp of the nature of the data with which they were working. For example, Adam had entered information from the lay subsidy rolls about the taxes paid in one district into a spreadsheet, using one column for shillings and another for pence. (currency which is no longer in use.) Another year's taxes were entered into two more columns. Finally, the difference between these amounts was entered into two further columns. In the graph below (Fig. 1) Adam's intention was to present tax changes but, in fact, he graphed all six columns of his spreadsheet. Adam was happy with his graph even though it was quite meaningless.

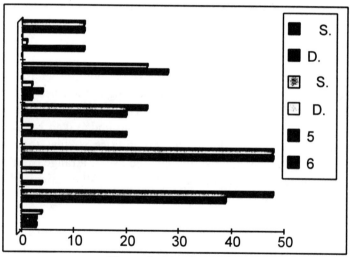

Figure 1 Adam's graph

The children were reaching levels of facility with the computers which were superficially impressive but which also revealed the limitations of their mathematics. Their teacher, who was new to using computers in this way herself, was pleased with the speed with which the children had become confident with the software, but slightly bewildered when they began to discover facilities which she did not know about, such as he three-dimensional effect in Adam's graph. This made it difficult for her to focus on the mathematical problems in the children's work which were apparent to us because of our greater experience in both mathematics and IT.

When the project team met to discuss these experiences, the Y7 teacher said that she felt the children's problems arose partly because of the context in which the work was set. The children were working from secondary data sources which needed considerable analysis from a historical perspective before the data could be translated into a suitable form to be entered into the spreadsheet.

As a result of these discussions, the two Y5 teachers decided to concentrate on more obviously mathematical and scientific activities, which seemed to offer more straight forward opportunities for data handling. They felt that it was important for the children to be involved in data collection, so that they would have a better understanding of the data, but the teachers found it harder to get a sense of the uses to which the data might then be put. One expressed dissatisfaction with her original ideas because they involved what she called *dead data*; that is, data whose collection was the end-point of the activity, rather than what we might call *hot data*, which is collected for a clear purpose as part of a wider activity. Even though she recognised this distinction, her own inexperience with this sort of work made it difficult for her to translate her ideas into more realistic and engaging activities. At our suggestion, she reformulated her activities in the form of problems which the children had to solve, and which required experimentation and data handling techniques.

The class were working on the topic of wheels, and the two problems which were identified for investigation were *what affects the way your toy car rolls?* and *can you predict how far a wheel will roll in one turn if you know its diameter?*

The children worked on the first of these problems by setting up experiments rolling toy cars down slopes, and began by considering a very wide range of variables. These included a number which were later abandoned, such as the colour of the car, or how many doors it had. We envisaged that this experimentation would lead the children to explore relationships between independent and dependent variables, and to use graphs to identify trends and make predictions.

Work on the second problem involved children measuring the diameter and circumference of as many wheels as they could find around the school. They then continued their investigations at the British Transport Museum, where they were allowed access to many more vehicles. This activity also offered the opportunity for children to use graphs as well as the numerical and symbolic possibilities of the spreadsheet to explore relationships and find an approximate value for Pi. As the activity progressed, another mathematical possibility emerged as the children used the spreadsheet to find the average of their results.

Movement in the Classroom

When they began working on the question *what affects how your car rolls?* the portability of the machines meant that children were able to spread out from the classroom, making use of space in corridors in which to set up their slopes. They could enter their results directly into a machine which was beside them on the floor thus helping them to maintain the momentum of the activity. They could take their machines to someone else to ask for help rather than having to persuade someone to come to them, and put their machines together to compare or copy results. For example, some groups shared data on the dimensions of the cars, in order to cut down on the time taken for the experiments.

This movement and sharing in the classroom contributed to what we have come to think of as *fluidity of data*. Data handling is essentially the process of reorganising and re-presenting information. Information which is stored in electronic form can be reorganised and re-presented much more easily than anything in 'hard copy', and a good computer environment almost invites this sort of activity. The children naturally combined text, graphics, tables and graphs in writing reports of their work. Many children would normally find this kind of factual writing difficult and unmotivating, but the ability to create an illustrated document with data presented in several forms seemed to make the task much more purposeful. Thus when we talk about *movement* in the classroom, we are thinking of the way in which data flows between different components of the software and between children as well as the more visible physical movement of children and machines.

A third sense in which we use the term *movement* is that of children moving on in their mathematical understanding. Rory's data on the weight of the toy car and the distance it travelled was first entered into two columns of a spreadsheet. He used scatter graphs during his experiment to guide the direction of his project. When he was satisfied that he had reached some conclusions, he wrote his report, which was created largely by cut and paste techniques from his previous tables and graphs. Figure 2 below is an extract from his report, which also included drawings and tables.

> We put loads of plasticene on the car to make it heavier To see if It made it go further or not. We put all the data on the car on the spreadsheet and saved it on a disk. We found out that The car went further when we adid more weight. The line showes the trend on the scatergraf. On the scatergafe it showes you that the hever the car the ferver it goes.

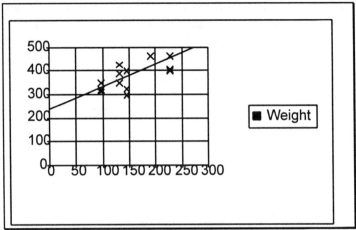

Figure 2 Rory's Report

Rory's teacher was surprised and delighted at his progress with the software and with the development of his mathematical ideas. She recognised that the machine seemed to be freeing him to some extent from his difficulties with language and allowing him to express his real potential in science and mathematics.

Teacher intervention was an important factor in exploiting the potential for children like Rory to draw new ideas from the activity. Whilst they worked with the toy cars, we introduced the facility to draw scatter graphs as a way of helping them to see trends in their results. Because they were able to do this immediately, beside their experiment, interpreting the graph was relatively easy for them, even though they had never used this kind of graph before. They were able to relate points on the graphs to particular cars and events, and realised, at an intuitive level at least, the significance of the points lying roughly in a straight line. Some of the children, like Rory, spontaneously picked up a line from the graphics tools and dropped it onto their graphs.

Looking at the graphs *during* their experimental work, rather than as part of a final reporting stage, meant that they could use them in a formative way, identifying results which needed to be checked because they seemed to fall outside the expected pattern, and

442

designing new experiments to test hypotheses and to fill gaps in their range of data. It was very exciting to see children learning to use graphs as active analytical tools, rather than seeing them in the more passive presentational role they are usually given in primary classrooms.

Another example of movement occurring in the classroom was when the children were measuring the diameters and circumferences of wheels. Having entered the data into

Diameter	Circumference	
12	38	3.17
9	24	2.67
30	96	3.2
20	65	3.25
5	17	3.4
10	30	3
32	100	3.13
....		
....		
....		
101	310	3.07
	Average	3.16

Figure 3   *Wheels data*

two columns, they were encouraged to think about connections between them. One girl was struggling to express the fact that there was a relationship between them: "It's as if that's the mummy (pointing to the diameter) and that's the baby (pointing to the circumference)". They were shown how to use formulas to divide the circumference by the diameter.

The third column contained numbers around 3. They were then shown that the computer had a trick called the average. These children had not learned how to calculate the average but they were able to use the computer's facility to generate it in their spreadsheet. These children were beginning to understand average as a good estimator, at least in one experimental situation. As in the previous example, the teacher intervention in a context where the children were working with fluid data had enabled the pupils to encounter two new and important mathematical ideas; there is a fixed relationship between the diameter and the circumference of wheels and that an average is a better estimator than individual measurements.

Portable mathematics and integrated technology

We began with portable technology and teachers who liked to work with cross-curricular themes. By working closely with the teachers, we were able to plan activities which we predicted would integrate mathematics and science (and other fields of knowledge) in a process where the children themselves would be carrying out research design, implementation and evaluation. The technology gave the data a fluidity which allowed the children to present and re-present their ideas very easily. As well as helping the children to gain new understanding, this property also seemed to make evident the children's misunderstandings and hence offered opportunities for teacher interventions.

We conjecture that this process allows the children to make their mathematics *portable*: the support of the computer, in the context of the teaching approach employed, is enabling the children to gain insights into the mathematics that underlies particular situations, in a form which enables them to transfer the mathematics to new contexts.

We acknowledge that this is an ambitious conjecture. It arises from observations which we feel provide evidence that there are a number of features of the computer-rich environment created within the project which can be particularly supportive enabling children to 'port' mathematical understanding into new contexts. In the following section we give examples which illustrate this idea. Exploring the nature of the support offered by the learning environment, which is schematised in Figure 4, will form a major focus for future research within the project.

Observations of Portable Mathematics

When children who had used scatter graphs in the toy cars experiment began to explore the diameter and circumference of wheels, some of them used scatter graphs in a very similar way with little further encouragement from the teacher. They quickly recognised that the points fell roughly into a line, and used the graphics tools to add this. These children seemed to have understood something about the use of a scatter graph which they had managed to re-apply to a new context. The children have never drawn a scatter graph themselves but the computer gave them direct access to the concept without needing to go through the process of learning

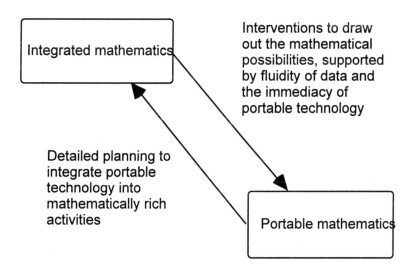

Integrated mathematics

Interventions to draw out the mathematical possibilities, supported by fluidity of data and the immediacy of portable technology

Detailed planning to integrate portable technology into mathematically rich activities

Portable mathematics

Figure 4 A model of 'portable mathematics'

how to draw such a graph, which might have proved an obstacle. We think that the children developed their understanding in a way which enabled them to port this concept for a number of reasons.

- The children had encountered the idea in a meaningful context. The graph was re-presenting their own data, with which they had already become very familiar, through presenting it in other forms (i.e. numerical in this case). The nature of the technology allowed this easy movement between different representations.
- They learnt about scatter graphs as a technique to be carried out on the computer. In the first instance, this powerful idea was to them merely a simple technique they learned to carry out which generated something concrete and useful.
- The children were encouraged to share their ideas with each other. Thus learner soon became teacher. The children naturally shared these ideas because of their experience of peer-tutoring, and their curiosity to learn about what the computer could do.
- Careful planning gave the children a chance to try out this new technique in a different context.

The children also shared the technique of using the computer to produce an average. It became common knowledge that there was this special number that you multiply the diameter by to get the circumference (although no-one called it Pi), and that the trick of finding the average gave you a better estimate. By the time the children went to the transport museum to measure further wheels, they were able to predict the circumference of very large wheels, beyond the range of numbers they had looked at so far.

These children are at a curious stage where, for them, average is not what you get when you add up all the numbers and divide by how many there are. For them, average is a black box with a rather subtle statistical meaning to do with estimation. In the future, they will meet other meanings of average, such as its role as an indicator of the size of a set of numbers. No doubt at some stage, they will open up the black box and be shown the process of calculating the mean average. A real test of their ability to port the mathematics will come when they have to apply the idea of average as a better estimator in a

446

quite different context. At the time of writing, further activities are planned to explore this.

In this chapter we have presented some ideas about the special ways in which immediate and continuous access to computers can have a direct impact upon children's mathematical thinking and learning. We have illustrated how integrated software has the potential to encourage children to reflect upon and evaluate their work, and presented examples of children using graphs in new and analytical ways. Their ability to port these skills into new contexts suggests a developing powerful understanding.

However, we are aware that in order to achieve these results, much more than access to personal technology was needed. It seemed to us that the support given to teachers at the planning stage helped them in a fundamental way to appreciate how activities could be set up which would have the potential for this learning to take place. Furthermore it was not always obvious to the teachers when there were opportunities to intervene in order to help the children to move on in their thinking. This was to do with their limited mathematical expertise and appreciation of the potential for the technology to facilitate this learning. However, we have also seen dramatic developments in the teachers' confidence in both these areas, and feel that their sharing of expertise and reflections in their journals have contributed significantly to this.

In future phases of this project, we wish to explore in more depth the ways in which the technology can assist the porting of mathematics in other contexts and also to come to a better understanding of the kinds of support teachers need in order to take advantage of these facilities.

Notes

1   This project has been supported by Warwick University Research and Innovation fund and by UFC InSET funding.

2   For this project we have used Apple Macintosh Powerbook 100's with *ClarisWorks*, a software package which includes word processor, graphics, spreadsheet and database.

25 Arithmetic Microworlds in a Hypermedia System for Problem Solving

Rosa Maria Bottino, Giampaolo Chiappini,
Pier Luigi Ferrari

In this paper we refer to the ARI-LAB system, an educational computer-based system which combines hypermedia and communication technologies in order to allow the user to build her/his own solution to a given arithmetic problem by navigating through different integrated environments. In particular, we focus on the arithmetic microworlds which are included in the system. The discussion will be carried out taking into account both pedagogical and technical aspects.

Introduction

This work is concerned with arithmetic education at the ages of 7-12. We consider the opportunities offered by technology and we refer, in particular, to the ARI-LAB system, a computer based tool designed and implemented to assist pupils in solving arithmetic word problems.

The ARI-LAB system is an educational tool which combines hypermedia and communication technologies in order to allow the user to build her/his own solution to a given arithmetic problem navigating through different integrated environments. The environments available include microworlds, databases containing sets of solved problems and a communication environment which allows the user to cooperate with other students, or even to interact with the teacher, through a local or a remote network.

The ARI-LAB system is implemented in HyperCard 2.1 using HyperTalk programming language. It runs on Apple Macintosh computers with at least 12 inch monitors.

The evaluation of the system was carried out in 1993 with four deaf children attending the third class of primary school; another experiment has been carried out in 1994 in a ordinary second class of primary school (7 years old pupils).

In this paper we give only a sketch of the whole system (for a more complete presentation see Bottino et al., 1994), but focus on the arithmetic microworlds included in it. When discussing this aspect we try to point out our view of what we mean by 'arithmetic knowledge' and of the role that technology can assume in its acquisition.

In particular, we consider the graphic representation systems enclosed in the microworlds and the way in which they affect the acquisition of problem solving skills. Even if in the following we discuss an example derived from the experiments, the analysis we perform here is an a-priori one. With this analysis we want to justify the visualization features of the ARI-LAB system and their cognitive significance according to results from research in mathematics education and psychology of learning.

Arithmetic Knowledge and Problem Solving

In this section we explain the general ideas underlying the design of ARI-LAB; in particular we discuss what we mean by 'arithmetic knowledge' and some findings from mathematics education research about how to develop arithmetic learning. Arithmetic knowledge includes a lot of procedural skills, but cannot be identified with the ability to perform calculations. It should be related to the ability to use arithmetic concepts and procedures in order to solve problems. The problem solving process requires a suitable representation of the problem situation and the design of a resolution strategy. A crucial point is that a correct resolution process requires the student to grasp the sense of the problem situation and to recognise some relationships among the data.

Arithmetic symbolism and written computation algorithms are no doubt important tools in order to concisely represent the problem situation and some steps of the resolution process, provided that pupils are already able to handle numbers and to plan informal strategies within concrete problem situations. If pupils have not yet developed these abilities, they might use symbols without being able to give them any meaning and thus a sort of *contract* is implicitly stipulated between the pupils and the teacher, according to which to solve an arithmetic problem means to guess the correct operation and to perform the computations involved. Therefore the task for an educational project should be to allow pupils to deal with a wide range of problem situations and to construct specific strategies to obtain a correct answer, not only starting from their mathematical knowledge but also from their knowledge on the problem situation.

Symbolic systems play a crucial role at different stages of the development of mathematical ideas and processes. In this regard, symbols that are not completely arbitrary but preserve some analogical link with the related objects (such as geometrical figures, histograms, lines of numbers, coins) have proved very important (Lesh, 1981). Languages of this kind could act as mediators between the problem situation and its meanings and the mathematical ideas, relationships and processes involved, as discussed in Dreyfus, 1991. Only a close contact with the meanings of the problem situation can help pupils consciously to find a resolution procedure and prevent them from some dysfunctional behaviours, such as providing stereotyped answers which by-pass the interpretation of the text, or being unable to perform any step to solve them. Moreover, recent studies (see, for example, Cobb et al., 1992) have pointed out that the inclusion of problem solving activities within a context of social interaction may become fruitfully related to both the opportunity of using different representation systems in order to solve a problem and the validation of the strategies performed. By 'a context of social interaction' we mean all the kinds of cultural exchange between the solver and the environment (teachers, other students, ...) even through the mediation of various devices (books, computers, board, ...).

The outlined ideas have guided us in the design of ARI-LAB.

The ARI-LAB System

A general insight

ARI-LAB is a hypermedia system which consists of a structured and connected set of environments which offer the users various options in order to solve a given arithmetic word problem. Two different kinds of users are expected: the student who has to solve a given problem and the teacher who can configure the system according to the needs of her/his students. The student interacts with four main environments: the *visual representation environment*, the *strategy building environment*, the *communication environment* and the *database environment*. The teacher, in addition to the environments available to the students, can access another environment, *the teacher environment*, which allows her/him to set different layouts of the system.

The user interface of ARI-LAB, which is entirely mouse driven, is designed in order to make easy the passage from one environment to another, to automatically perform the dynamic increment of the system and to take account of the changes caused by users' actions. Starting from a main display frame, where the user can choose if she/he wants to solve a new problem or to see the solution of previously solved ones, different windows are presented according to the choice made. For example, if the student wants to solve a new problem, the system presents her/him with an interface which is structured (see Fig. 1) into three different display areas.

Each area corresponds to a different environment, namely the visual representation environment, the communication environment and the strategy building environment. Each environment has its proper icon commands and link buttons which can be activated by 'clicking' over them. The layout of the environments is thought to make it easier for the user to perceive the components of the systems, interpret them and make decisions about what to do next while building a solution for a given problem.

Let us consider now very briefly the different environments of the system. In the next session we discuss with more details those features of ARI-LAB which make it a suitable tool for the acquisition of additive and multiplicative strategies in arithmetic problem solving and we consider in particular the visual representation environment.

In the *strategy building environment* the user chooses an arithmetic problem, from a set prearranged by the teacher, and builds her/his solution step by step using different kinds of languages (written, visual, symbolic) according to taste. At each step of the resolution process she/he can access the visual representation environment, choosing one of the available microworlds to get a representation.

In the *visual representation environment* students can perform experiments through interaction with a number of available microworlds; within each microworld graphic animations are available. Through interaction with the microworld chosen, pupils can activate different animations until they get a visual configuration representing a resolution step they recognise as adequate to their purposes in relation to the problem situation. Pupils can also partially or totally paste their productions into the *strategy building environment*.

Visual representation environment

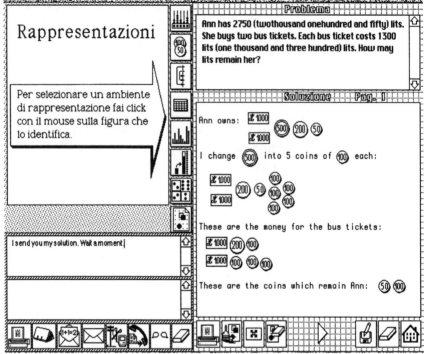

Communication environment   Strategy building environment

Fig 1: Interface of the problem solving environment with an example problem. This environment is composed of: a strategy building environment, a visual representation environment, and a communication environment.

At any moment of the interaction with the system the user can access the *communication environment* which allows her/him to insert the production of a solution into a social interaction process; in this environment different kinds of interactions are possible: the student can send or receive messages and solved problems (or partially solved problems) to another student who is in contact via a modem connection; or the student can be in contact with other students of her/his same class connected via a local network. The

454

communication environment is also accessible from the data-base environment.

Users can access the *data-base environment* when they want to see problems previously developed and classified by the teacher or problems they had previously saved (their own problems or problems received from other users). The user's interface of this environment is built so that teacher's problems, problems solved by the actual user and problems received by other students are organized in different databases. The classification of teacher's problems is done according to the visual representation in which their solution is presented, whereas problems in the other data-bases are stored according to a temporal criterion. In the teacher's problems data-base, if there are two or more solutions for the same problem (obtained using different microworlds), mutual links are given to connect them. In the user's interface links are visualized so that the student can choose to inspect alternative solutions by selecting the corresponding icon. The user can access the data-base environment at any time of the session.

The *teacher environment*, which is not accessible to the students, allows the teacher to set different layouts of the system according to her/his educational goals. In particular it is possible to add or modify problems of the teacher's data-base, to choose the microworlds the teacher wants to be accessible by her/his students and to choose local network and distance communication connections. This implies that the interface of the system is dynamically and automatically changed according to the teacher's options.

The Arithmetic Microworlds of ARI-LAB

The visual representation environment of ARI-LAB consists of a number of different visual representation microworlds which are provided in order to allow to the user to represent a wide range of problem situations and, at the same time, to point out different mathematical structures and perspectives. The user interface of the visual representation environment is shown in the up left window of fig.1. The user accesses this environment by choosing one of the available microworlds, clicking on the corresponding icon. The representation microworlds available at present are: 'abacus', 'coins', 'spreadsheet', 'calendar', 'histogram maker', 'measurement division', 'partitive division', 'art bits'.

The selection of each icon gives access to one or more cards. For example, in the 'abacus' microworld the user accesses only one card; in the 'calendar' microworld the user has at her/his disposal twelve cards (one for each month of the year); in the 'histogram maker' microworld there are three cards, the table, the units of measure and the histogram cards which allow manipulations according to different criteria (insertion of data in the table, ordering of data, setting of the units of measurement, drawing of the histogram, etc.). In the following figure 2 the user's interfaces of the main cards of each available microworld are shown.

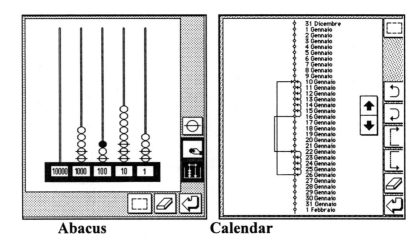

Abacus   **Calendar**

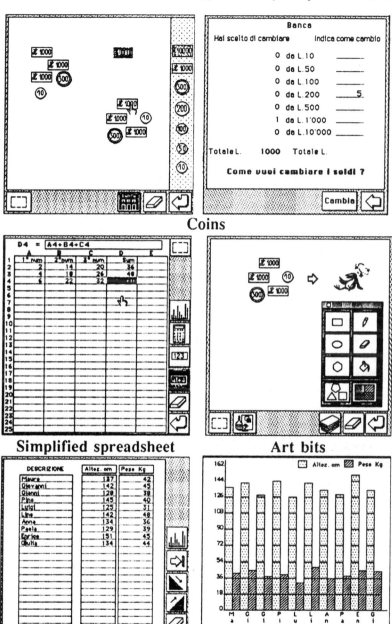

Coins

Simplified spreadsheet

Art bits

Histogram maker

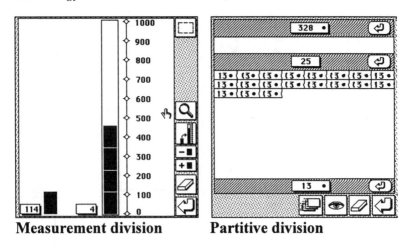

Measurement division   **Partitive division**

Fig. 2 Interfaces of the main cards of the microworlds available in the ARI-LAB system

Some microworlds are strongly characterized on the semantic ground ('coins', 'calendar'). Others ('abacus', 'histogram' and 'spreadsheet') allow content-independent representations of mathematical objects. Others ('integer division' microworlds) are of a mixed nature and have been planned on the basis of empirical observations of pupils' behaviour. The 'art bits' microworld has characteristics which are different from those of the other microworlds. Here the user may access a set of prearranged drawings and can produce new ones as well by means of a number of available tools in order to modify or enrich the representations obtained working within the other microworlds. The aim of this microworld is to offer the student the opportunity to produce visual representations which better fit her/his resolution process.

Even if there is no standard definition of 'microworld', the environments available in the ARI-LAB system have a number of characteristics which correspond to those usually indicated as necessary to qualifying them (see, for example, Laborde and Strasser, 1990). In particular, each of the microworlds available:

458

- provides the user a specific visual language which allows her/him to generate and manipulate objects in order to produce visual representations of a problem situation or of a solution step;
- offers a variety of ways to achieve a goal;
- embodies an abstract domain described in a model;
- in many cases allows the direct manipulation of the objects.

When the user selects one of the available microworlds she/he is provided with a language (with a specific set of rules) which allows her/him to generate objects and to perform actions (animations). For example, in the microworld 'coins' she/he can generate coins of different kind, change them with other ones of the same value (selecting the icon 'bank') and move them over the screen. In the microworld 'abacus' she/he can generate a ball in the abacus, change it against an equivalent set of others or mark it (in order to perform subtractions).

A number of actions are available in all microworlds, such as the opportunity of copying a visual representation produced by the user (or part of it) into the strategy building environment, of returning to the environment menu and cancelling. In this regard, we note that in any microworld, if some action does not yield suitable results, backtracking is always allowed in two ways: stepwise or automatically, according to the ways the action has been carried out.

Each microworld is designed so that it is possible, under suitable conditions, to get an automatic feedback to some actions, based on the knowledge of the system. Actually, in each microworld, the user's manipulations are controlled by a set of rules which prevent her/him from some specific incorrect step. For example, if some user tries to perform an incorrect change of coins, or of balls in the abacus, the system prevents her/him continuing and provides a specific error message. This feature can promote a reflection on some mathematical aspects (such as the decimal notation) even if the complete control of the strategic plan is left to the student.

After the first experiment in the classroom, we have added to the system another facility (the monitoring facility) we consider very useful in order to evaluate students' behaviour: the system now can automatically record anything a pupil has produced within a microworld in a working session, and then even anything she/he does not want to paste in the strategy building environment or she/he cuts before quitting. At the end of the session the teacher can review the sequence of the steps performed, in a sort of movie. This feature was widely used during the subsequent tests.

During the experiments three microworlds were used: 'money', 'abacus' and 'art bits'. Addition and multiplication arithmetic problems were proposed mainly concerning 'market' situations. The solution of these problems required different strategies to be performed (e.g. completion, total/part/reminder, containment, partition, ...).

Technology And The Representation Process

The ARI-LAB system works as a mediator in the transition from an iconic to a symbolic use of visual representations. To explain this statement let us take into account the following problem: "Four children want to buy a cake which costs 7200 lit.. How may lit. spend each child if all the children pay the some amount?". In educational practice it is well known that a drawing of the cost of the cake consisting of seven 1,000 lit. notes and one 200 lit. coins, which is often used when working with paper and pencil (also by pupils having no idea about the solution), hardly helps pupils to find a solution and most often induce them to grasp only some exterior feature of the problem situation described in the text, as being a static representation embodying no mathematical relationship or procedure.

The solution we report in figure 3, which was realized by a 8 year old student during the experiment we performed in a second class of primary school, shows how the visual and operation stimulae offered by the 'coins' microworld can foster the realization of representations useful to quantitatively solve the problem.

The frames reported in fig.3, which have been obtained by means of the monitoring facility of the system, show the steps through which the student achieved a solution to the previous problem. The subject, after the generation of the coins which are necessary to buy the cake (frame 1), tried to distribute the generated coins in four groups (frame 2). When she understood that this is not possible she cancelled a 1000 lit. note and generated ten 100 lit. coins (frame 3) instead. She tried to distribute these coins in the previous four groups (frame 4) but realised that this is not completely possible. She then decided to cancel two 1000 lit. notes and to generate different coins instead. Firstly she generated five 200 lit. coins (frame 5) and distributed them in the four groups (frame 6). Then she generated a 1000 lit. notes and changed it (through the 'bank' facility) with ten 100 lit. coins (frame 7). She distributed the obtained coins in the four groups so obtaining a solution for the problem (frame 8).

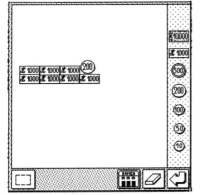

Frame 1

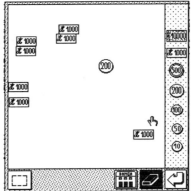

Frame 2

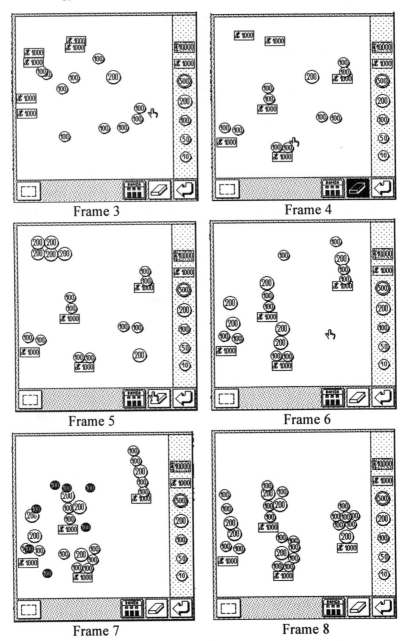

Frame 3

Frame 4

Frame 5

Frame 6

Frame 7

Frame 8

Fig. 3: Example of the steps of a solution strategy performed through the interaction with the microworld 'coins'

The example points out how the mediation of the computer has fostered the passage from an iconic representation, based only on rigid imagination processes (which can hardly be modified) to a representation based on a symbolic system more effective and flexible, which allowed the pupil to go beyond the actual perceptions and grasp connections through time and invariant properties in spite of superficial changes (see Bruner, 1966).

We note that the interaction with the microworld induced the pupil to perform mental experiments that helped her to solve the proposed problem. The availability on the screen of coins of different value, for instance, led her to represent the total amount to be shared in some more effective ways, and the availability of the icon 'bank' to change coins caused her to use this facility (even if in some cases she prefers to cancel and generate new coins instead). Moreover, the opportunity to move coins and group them on the screen in different fashions helped the student to build a resolution procedure. In this process suitable schemes that the student also developed in her everyday-life experience were activated (coins are usually changed, the same value can be formed with different coins, coins can be grouped in different ways, etc.). Hence the pupil gradually built a mental representation of the problem situation which is more efficient in order to quantitatively express the relations between the elements of the problem, and has performed actions (with the mediation of the system) which produced resolution steps based on visual representation.

The representation system of the microworld 'coins' can be used with different functions. Coins can be used both instrumentally to support counting activities and as symbols for the representation of the problem situation. Instrumental and symbolic functions involve cognitive abilities of different level which correspond to the different roles assumed by the graphic representation of coins. The interaction opportunities offered by the system together with the graphic feedback it offers to the actions performed give the possibility to exploit these opportunities, fostering the passage from an iconic use of the representation system to a symbolic one.

Technology And The Resolution Process

In problem solving, any written or graphic report produced by the solver during the process may assume different attributes, from being the trace of a sequence of actions performed (even mentally) by the subject up to being a whole piece of reasoning (not only a sequence of unrelated steps) which schematises some piece of information about the problem (Fischbein, 1994). These interpretations may be very different if we take into account the cognitive and metacognitive processes involved. The former generally embodies the semantics associated with the actions performed by the subject producing it, which are not necessarily related to the mathematical structure of the problem; the latter embodies its own semantics, modelled by the drawing (or the syntax of the written report produced) and depending on the specific features of the language used. Moreover, in the latter case, the solution is already regarded as an autonomous object which can be discussed and communicated. In our opinion these distinctions make sense since they point out a relevant feature of arithmetic thinking, that is the transition from strategies based on actions to incorporation within some relational model.

We believe that some features of our system may promote the development of some interplay between the procedural and relational aspects. In the conclusions some examples of such processes can be found.

In the system, microworlds are spatially and conceptually detached from the strategy building environment where the pupil produces a global interpretation and thus a solution of the problem. The resolution steps the pupil produces within different microworlds have to be copied in the strategy building environment and properly structured, so that the whole solution may be obtained, communicated and discussed. It is also possible to find and compare different solutions for the same problem, e.g. working in different microworlds in order to get different representations, which can even be combined to make up one solution. When a visual representation built within a microworld is partly or totally copied in the strategy building environment, the pupil is induced to detach her/himself from it (and thus from the actions which have contributed to its building and their links with time) and actually to use it as an object that takes part in the construction of a resolution procedure.

For communication purposes a graphic representation may be inadequate, in itself, to explain a whole resolution process, if not equipped with notes and comments. In this regard, natural language may be used both as a component of the graphic representation proposed and as a tool to describe it. In the first case natural language performs functions otherwise lacking (e.g. naming a graphic component or referring to it. In the second one, the subject may gain a new interpretation of the solution of the problem, in the sense that representations can be interpreted by means of natural language in the framework of the problem situation. This may lead to a chain of interpretations in which any representation (which embodies its own interpretation of the problem situation) is in turn interpreted by means of another symbolic system. The aspects described allow us to characterise the production of a solution in the ARI-LAB system in the following way:

- the actions the subject performs in the microworlds do not exhaust the whole resolution process because of the structural features of the system;

- the resolution process develops in a communication practice including the interactions not only between pupil and teacher but also among pupils; co-operation and negotiation processes are thus enhanced. This can lead to the discussion of the problem solving process and also of the solution; moreover, the interactions opportunities given by the system compel the student to use the different languages available also to communicate the results of her/his own thinking.

Moreover, processes based on analogy may be activated through the interaction with the data-base environment. These processes have not been thoroughly analysed in this paper, for details see Bottino et al. (1994).

Final Remarks

ARI-LAB allows the user to insert the problem solving process into a framework which is both operational and relational as far as it structures and supports her/his actions and fosters the construction of a socially shared meaning of their outcomes. To justify this statement we briefly discuss two further examples.

The first one is concerned with the difficulties of primary school students when dealing with problems involving some total amount, some part of it and the remainder (the so-called combine problems). Throughout the experiment, we became aware that these difficulties are mainly linked to the ability of giving different meanings to the symbols that have been introduced in order to represent the problem situation. By means of ARI-LAB it becomes possible to plan educational paths aimed at the construction of this kind of competence. For example, in market problems solvable within the microworld 'coins', we have observed that the opportunities of displacing coins over the working space support the process of planning which allows the user to attribute different meanings (e.g. component of the total and, at the same time, of a part) to the same symbol (e.g. a coin). This strengthens student's grasp of the relational structure involved (e.g. the inclusion between sets or the ordering of the numbers corresponding to the total and to the part). The opportunity of changing coins supports the achievement of a representation that preserves some semantic similarity to the initial one but allows the user to perform an appropriate strategy. The anticipation of ideas and steps of the strategy under development is supported also by the interface of the changing facility (see fig.2, frame 4) which allows the user to control at any time: the coins she wants to change, their amount, the coins that replace the previous ones and, after the change, their amount or, if the change is not correct, some warning message. This makes the acquisition of arithmetic relationships between the quantities involved richer as well.

The second example concerns the learning of the decimal notation for nonnegative integers and the difficulties of attributing to the grouping activities meanings that make easier the transition to the positional value of figures. During the tests already performed, the transition has been attained by integrating the work within the microworlds 'coins' and 'abacus'. In this case a feature of the system that has proved crucial is the opportunity of easily jumping from the microworld 'coins' to 'abacus' preserving the same resolution page within the strategy building environment. It is worth to note that the microworld 'coins' supports structured counting processes by means of the spatial arrangement and the grouping of coins on the screen. So the students were allowed to: draw some representation within 'coins', paste it into the strategy building environment, turn to 'abacus' (keeping the representation with coins on the screen), produce a corresponding representation within it, paste it on the strategy building environment and writing down the corresponding value as a decimal number.

These examples show how the students could exploit the opportunities provided by ARI-LAB in order to enrich the meaning of their actions, to transfer it to their products and to integrate it within their relational structure. Of course larger-scale testing is needed to point out features and limits of these processes.

References

Bottino R.M., Chiappini G., Ferrari P.L. 1994, 'A hypermedia system for interactive problem solving in arithmetic', *Journal of Educational Multimedia and Hypermedia*, AACE, to appear.

Bruner, J.S. 1966, *Studies in Cognitive Growth*, J. Wiley & Sons.

Cobb, P., Yackel, E., Wood, T. 1992, 'Interaction and learning in mathematics classroom situations', *Educational Studies in Mathematics*, 23, n.1, 99-122.

Dreyfus, T. 1991, 'On the status of visual reasoning in mathematics and mathematics education', *Proceedings PME 15,* 1, 33-48.

Fischbein, E. 1993, 'The theory of figural concepts', *Educational Studies in Mathematics,* vol.24, 139-162.

Laborde, J.M. and Strasser, R. 1990, 'Cabri-Géomètre: a microworld of geometry for guided discovery learning', *ZDM*, 90/5, 171-177.

Lesh, R. 1981, 'Applied Mathematical Problem Solving', *Educational Studies in Mathematics,* vol.12, 235-264.

The Contributors

Janet Ainley

Institute of Education, University of Warwick, Coventry, CV4 7AL, U.K.

Janet is a lecturer in mathematics education in the Mathematics Research Centre. Her main areas of work are in primary and early years mathematics, and in the use of computers in the learning and teaching of mathematics. As well as writing for a number of journals, she has edited *Micromath*, a journal of the Association of Teachers of Mathematics.

Christopher Bissell

Open University, Milton Keynes, MK7 6AA, U.K.

Chris is Senior Lecturer in Electronics in the Faculty of Technology, and has a long-standing interest in the use of computer-based tools to enhance engineering education. He has contributed to Open University courses in signal processing, control engineering and telecommunications, and is currently leading a team responsible for introducing computer-mediated communication into Technology Foundation Course.

Rosa Maria Bottino

C.N.R., Institute for Applied Mathematics, Via De Marini 6, 16149 Genova, Italy.

Rosa Maria is a mathematician who is a researcher at the Italian National Research Council. Her current research interests are in mathematics and computer science education and in educational computing.

Leone Burton

School of Education, University of Birmingham, Edgbaston, Birmingham, B15 2TT, U.K.

Leone is Professor of Education (Mathematics and Science), author of mathematics education books in such areas as the learning of

469

mathematics particularly from an investigative perspective, gender and assessment. Her passion is expended on demystifying and humanising mathematics in learning. Her son has been known to feature in her work.

Giampaolo Chiappini
C.N.R., Institute for Applied Mathematics, Via De Marini 6, 16149 Genova, Italy.
Giampaolo is a mathematician who is a researcher at the Italian National Research Council. His current research interests are in mathematics education and in the development of software for mathematics education.

Giuliana Dettori
I.M.A., C.N.R., Via De Marini 6, 16149 Genova, Italy.
Giuliana is a mathematician, working as a researcher at the Institute for Applied Mathematics of the National Research Council of Italy. Her main current research field is the didactic of mathematics and informatics.

Andrea A. diSessa
Graduate School of Education, University of California, Berkeley, CA 94720, U.S.A.
Andrea has a degrees in physics from Princeton University and MIT. He worked with Seymour Papert and the Logo Project at MIT in the development of Logo. Since 1985 he has been Professor of Education in mathematics, science and technology at the University of California in Berkeley. His recent research includes a focus on new methods of learning with computational media and conceptual development in science education.

Michele Emmer
Dipartimento di Matematica, Università Ca' Foscari, Dorsoduro 3825/E, 30123 Venice, Italy.
Michele is Professor of Mathematics, a filmaker and writer. He is currently president of the Italian Association for Scientific Media.

Pier Luigi Ferrari

University of Torino at Alessandria, Department of Advanced Science and Technology, Via Cavour 84, 15100 Alessandria, Italy.
Pier Luigi is Professor of Mathematics Education. His research interests are in logic as well as mathematics education.

Rossella Garuti,

I.M.A., C.N.R., Via De Marini 6, 16149 Genova, Italy.
Rossella is a teacher of mathematics and science in an intermediate school.

Mark Hunter

John Smeaton High School, Leeds, LS15 8TA, U.K.
Mark is Faculty Director of Mathematics and Business Studies. He has been teaching in secondary school for 10 years and has research interests in students' algebraic understanding and the use of computer algebra systems.

Barbara Jaworski

University of Oxford, Dept. of Educational Studies, 15 Norham Gardens, Oxford, OX2 6PY, U.K.
Barbara lectures in mathematics education. Her current research interest is the development of mathematics teaching through the enquiry of teachers into aspects of their practice or of students' learning. She is the author of *Investigating Mathematics Teaching: A Constructivist Enquiry* (Falmer Press, 1994).

Colette Laborde

Laboratoire IMAG-LSD2, Université Joseph Fourier, Grenoble, France.
Colette is a professor in "didactique des mathématiques" and head of a doctoral program in science and mathematics education in Grenoble.

Jean-baptiste Lagrange
Institut Universitaire de Formation des Maîtres, 35043 Rennes Cedex, France.
Jean-baptiste teaches mathematics education. His Ph.D. was in "Didactique de I'Informatique".

Donald R. LaTorre
Clemson University, Clemson, South Carolina, U.S.A.
Don is Professor and Director of Undergraduate Studies in Mathematics Sciences at Clemson. An enthusiastic advocate of reform in the teaching and learning of mathematics, he has helped Clemson achieve national prominence in the widespread use of high-level graphics calculators.

Enrica Lemut
I.M.A., C.N.R., Via De Marini 6, 16149, Genova, Italy.
Enrica is a mathematician, working as a researcher at the Institute for Applied Mathematics of the National Research Council of Italy. Her main research interest is in mathematics and technology education.

Eric Love
Open University, Milton Keynes, MK7 6AA, U.K.
Eric is a lecturer in Mathematics Education, with extensive experience in school as mathematics department head and, later, as advisory teacher supporting the professional development of colleagues. His particular interests are in the use of software to support mathematics teaching and learning, and the analysis of teacher accounts of their professional activities as narratives.

Paul Marshall
Leeds Metropolitan University, Leeds, LS6 3QS, U.K.
Paul is a lecturer in mathematics education in the Faculty of Cultural and Education Studies before which he was a secondary school teacher and department head for 18 years. He has research interests in students' learning and the use of computer algebra systems.

John Mason
Open University, Milton Keynes, MK7 6AA, U.K. and Sunrise Research Laboratory, R.M.I.T., Melbourne, Australia.
John is a Professor of Mathematics Education, Director of the Centre for Mathematics Education, and Visiting Sunrise Professor at RMIT.

John Monaghan
CSSME, University of Leeds, Leeds, LS2 9JT, U.K.
John is a lecturer in mathematics education. He was a secondary school teacher and department head for 10 years. He has research interests in students' learning, advanced mathematical thinking, new technologies in mathematics education and the use of computer algebra systems.

Ljuba Netchitailova
Department of Mathematics and Analysis, RUDN University, Moscow, Russia.
Ljuba is a mathematician who worked as a Fellow in the Institute for Applied Mathematics of the National Research Council of Italy.

Liddy Nevile
Sunrise Research Laboratory, RMIT, Melbourne, Australia.
Liddy is Director of the Sunrise Research Laboratory. She has worked extensively over many years with children and teachers using LOGO, and has spearheaded the use in Australia of the BOXER computational environment.

Adrian Oldknow
West Sussex Institute of Higher Education, Bognor Regis, PO21 IHR, U.K.
Adrian recently directed, with Nuffield Foundation support, a project involving 10 schools with 1000 graphic calculators with pupils aged 14-16. He chairs a government steering group which is supporting IT-mathematics work.

Pat Perks

School of Education, University of Birmingham, B15 2TT, U.K.

Pat is currently a lecturer in mathematics education. Prior to that she worked for nine years as an advisory teacher alongside teachers in primary and secondary classrooms.

Hilary Povey

Sheffield Hallam University, Sheffield, S10 2NA, U.K.

Hilary is a lecturer in mathematics and mathematics education. In her teaching, research and writing her concerns lie in promoting active learning and in creating contexts for learning which encourage autonomy and independence. She has one daughter.

Dave Pratt

Institute of Education, University of Warwick, Coventry, CV4 7AL, U.K.

Dave is a lecturer in mathematics education in the Mathematics Education Research Centre. His work relates to the learning and teaching of mathematics across all age phases with a particular interest in how the use of computers can support mathematical learning. Dave regularly contributes to professional and academic journals and has been involved in the design and implementation of a range of Logo microworlds.

Phil. Rippon

Open University, Milton Keynes, Mk7 6AA. U.K.

Phil. is a senior lecturer in pure mathematics. His research interests range from potential theory to complex analysis to iteration, where he makes extensive use of computer experiments. He also likes to lecture using computer demonstrations on Acorn machines.

Tom Roper

CSSME, University of Leeds, Leeds, LS2 9JT, U.K.

Tom is a lecturer in mathematics education. He was a secondary school teacher and department head for 17 years. He has research interests in students' understanding of mechanics, the assessment of

474

'using and applying mathematics' and the use of computer algebra systems.

Andrew Rothery
Worcester College of Higher Education, Henwick Grove, Worcester, WR2 6AJ, U.K.
Andrew teaches mathematics and is also the Academic Director for IT.

Kenneth Ruthven
Department of Education, University of Cambridge, 17 Trumpington Street, Cambridge, CB2 1QA, U.K.
Kenneth is a university lecturer in mathematics education and a fellow of Hughes Hall. He currently advises curriculum projects in the United Kingdom and the United States of America on calculator issues.

Geoff. Sheath
University of Greenwich, Eltham, London, SE9 2PQ, U.K.
Geoff is Head of Mathematics Education at the University of Greenwich where he teaches both mathematics and mathematics education. He has taught in both primary and secondary schools, and has long experience of working with children to make films and videos.

Mary Margaret Shoaf-Grubbs
Division of Mathematics and Natural Sciences, College of New Rochelle, New York 10805, U.S.A.
Mary Margaret's research interests include the role that spatial visualization plays in the leaning and understanding of mathematics and the integration of technology into all levels of mathematics.

Teresa Smart
University of North London, London, N7 8DB
Teresa is a senior lecturer in mathematics education. Previously she taught in a London secondary school. Her research is around the issue of gender and technology in the mathematics classroom.

Rosamund Sutherland

Department of Mathematics, Statistics and Computing, Institute of Education, University of London, London, WC1H 0AL
Rosamund has worked at the Institute of Education, University of London for the last ten years. During this period she has directed a number of research projects concerned with the use of computers for teaching and learning mathematics, using software such as Logo, Excel, and Cabri-Géomètre. Her particular interest is in researching the ways in which teaching with computer-based symbol systems supports students to learn mathematics and make links between mathematics in paper-based and computer-based setting.

Jim Tabor

Division of Mathematics, Coventry University, Priory Street, Coventry, CV1 5FB, U.K.
Jim was educated at Royal Holloway College, University of London from 1970 to 1976 taking a B.Sc. in Mathematics and a Ph.D. in High Energy Physics. From 1976 to 1978 he worked for Rolls Royce and Associates Ltd. and since then has lectured at Coventry University.

Patrick Wild

Merchant Taylors' School, Sandy Lodge Lane, Northwood, Middlesex, HA6 2HT, U.K.
Patrick at Merchant Taylors' School, Northwood teaches mathematics to pupils in the age range eleven to eighteen years. After a short spell as a civil engineer he entered teaching seventeen years ago. His recent M.Sc. in Applicable Mathematics sparked off his interest in computer algebraic packages.

Names of Panel of Reviewers:

Johnston Anderson
Michelle Artigue
Mike Askew
Ruth Atkinson
Barbara Ball
Mike Beilby
John Berry
Neil Bibby
Pam Bishop
John Bradshaw
Laurinder Brown
Diana Burkhardt
David Burghes
Bob Burn
Dave Carter
John Costello
Kathy Crawford
Alan Davies
Sandy Dawson
Janet Duffin
Judy Goldfinch
Alan Graham
David Green
Susie Groves
Angel Guttierez
Chris Haines
Robert Harding
Lulu Healey
Dirk Hermans
Dave Hewitt
Joel Hillel

Keith Hirst
Christine Hopkins
Celia Hoyles
Glyn James
John Jaworski
Sylvia Johnson
Betty Johnston
George Joseph
Christine Keitel
Dudley Kennet
Daphne Kerslake
Carolyn Kieran
Jeremy Kilpatrick
Ann Kitchen
Lydia Kronsjo
David Le Masurier
Steve Lerman
Kevin Lord
Lynn Marston
Graham McCauley
Jean Melrose
Eleni Nardi
Hugh Neil
Richard Noss
Adrian Oldknow
Tony Orton
Derek Peasey
Stefano Pozzi
Steph Prestage
Phil Ramsden
Melissa Rodd

David Saunders
John Searl
Christine Shiu
John Silvester
Fred Simons
Kaye Stacey
Jan Stewart
Malcolm Swan
David Tall
Andrew Tee
Anne Watson
Joe Watson
John Wood
William Wynne-
    Willson

Subject Index

A-level students
  access to graphic calculators 101,
  102, 197-8
  and Prime Iterating Number
  Generators 326, 345-51
activities *see* problems
affine mappings 30-3
algebra
  algebraic symbolism and
    conception of objects 289-304
  algebraic understanding and
    Derive 258, 307-23,
  algebraic understanding and grap-
    hic calculators 217-27
  linear algebra and supercalcula-
    tors 326, 353-64
  with Logo 258, 275-85
  with spreadsheets 257, 258, 261-
  72, 275-85
animations 124, 129, 132, 159-60
applicable mathematics 365-84
ARI-LAB 449-67
arithmetic
  algebraic and arithmetical think-
    ing 263, 264, 276, 279-80, 318,
  323
  with ARI-LAB 449-67
  and calculator use 177-90
Art and Mathematics (film series)
  416-17

artistic and mathematical images
  415-18
assessment 400-2
attitudes to calculators 174-5, 176,
  191-2, 201, 207-10, 245-6, 363-4
averages 443, 446-7

BASIC 71, 72, 257, 293, 411
biology, computer mathematics in
  411-15
black box tasks 52-4
Booleans, pupils' difficulties with
  296-9
Boxer 6, 7, 12, 69-92, 125
Brookhurst Combined School (Leamington Spa) 435
Brooks-Matelski sets 24, 27-8
Bunny Hops *see* problems

Cabri-géomètre 12, 40-65, 158
Calculator and Computer Pre-
  Calculus Project (C2PC) 216
Calculator Aware Number (CAN)
  project 173-4, 196-7, 238, 249
calculators
  compared with computers 232-4
  overview 172, 231-50
  use by primary children 171, 173-
  93

see also graphic calculators;
personal technology
Calculators in Primary Mathematics
Project (Australia) 175-6, 238-9
calculus
curricular issues 325-6, 329-42,
370-2,
using personal computers 409-11,
413-14
Card Rotation Test 219
CAS *see* Computer Algebra
Systems
Chelsea Diagnostic Mathematics
Tests 219, 309
chemistry, computer mathematics in
411-15
Chocolates problem *see* problems
'chunking' (Boxer) 77-8
City of Birmingham calculator
project 174-93
class box (Boxer project) 89-91
Cockcroft Report 6, 235
code-sharing (Boxer) 73-4, 81-4
collaboration 143-4, 190, 432, 446
with Boxer 12, 69-92
graphic calculators encouraging
196, 201-2
groups making videos 425-31
learning groups 139-40
combine problems *see* problems
communication environment (ARI-
LAB) 454-5, 465
Computer Algebra Systems (CAS)
226
and the curriculum 329-42
see also Derive
computer programming *see*
programming
computers
compared with calculators 232-4

in examinations 327, 393-403
as problem-solving environments
35-7
see also laptop computers; palmtop computers; personal computers; personal technology
concepts, and calculator use 185-7
cones, maximisation 333-6
confidence, and calculators 176-7,
205-7, 246
consideration, need for 5, 105-7, 239
control
dimensions of 159-64
learners gaining 142-3
teachers relinquishing 224-6
counting problems *see* problems
Coventry University 387
creativity, enhanced by calculators
188-90
critical pedagogy, development of 9-
10, 96, 135-50
curriculum 8, 165
benefits of calculators 177-88
controversy about calculators
246-9
impact of Information Technology (IT) 104
implications of software 113-14,
325-6, 329-42
need for review 106

data-base environment (ARI-LAB)
455
data handling with laptop computers
436-47
decimal numbers
and ARI-LAB 467
and calculators 181-2, 186, 242
DERIVE 300, 382, 383, 390

and optimisation problems 325-6, 333-42
and Prime Iterating Number Generators 326, 345-51
and quadratic functions 307-23
diagrams 124, 129
diaries and journals
as records of learners' learning 141-8, 149-50
as records of teachers' learning 436
didactic tension 9, 122
didactic transposition (*transposition didactique*) 115, 121, 122
didactical contract 63
dimensionality of software 7, 159-64, 369, 372
and increasing opportunity 164-5
discrete event simulation 381-2
discrete linear systems 378-9
drag mode *see* mouse mathematics
drawings
Cabri-drawings 40-65
compared with figures 37-40

eigenvalues and eigenvectors 360-2
empirical combinatorial strategy 58
empowerment 136, 155
engineering
applicable mathematics andspre-adsheets 365-84
electronic engineering and Matlab 387-404
enjoyment, of calculators 176-7, 246
enthusiasm 4-5, 195, 201, 174, 191-2, 210, 349-50
equivalence problems *see* problems
ERASMUS project 410
EUCLIDE 300-2
Eureka 157, 160

everyday mathematics 110, 116, 155, 231
examinations
using computers in 327, 393-404
using graphic calculators in 101, 102, 208 -9
exhibitions of mathematics 417-18
Eye of Horus: a Journey into Mathematical Imagination, The (exhibition) 405, 417-18

Fast Fourier Transform (FFT)
spreadsheet illustration 379-80
using Matlab 390-3
female learners, use of graphic calculators 195-210, 213-27
figures (geometrical)
Cabri-figures 40-3
compared with drawings 37-40
'Fill your screen' tasks *see* problems
films and videos
children making 405-6, 423-33
dimensionality of control 159-60
integrating art and mathematics 415-17
fractals 11, 15-33, 89-91
Frogs *see* problems
functions
generality 115
introducing 267
understanding 217, 317-18
see also quadratic functions
funding 101, 103-4

Gaussian Elimination 354-5
gender issues 5
benefits of graphic calculators to girls 200-2, 216

spatial ability in males and females 214-15
generalised arithmetic 318
generality 9, 115, 120, 122-3, 153, 275-6, 277-9, 291-2, 299-300
  comparison of generalities 163-4
  generalisation of problems (algebra) 268-71
  generic properties of models 375-6
geometrical strategies and visual strategies 55-63
geometry
  Cabri-géomètre 40-65
  drawings and figures 37-40
  EUCLIDE 300-2
  geometric construction, dimensions of control 161-3
  Pythagoras' theorem 157-8, 166-7
Gram-Schmidt process 358-60
Grants for Educational Support and Training (GEST) 104
graphic calculators 232, 239, 241-2
  supercalculators and linear algebra 326, 353-64
  use by female students 171-2, 195-210, 213-27
  see also personal technology
graphical iteration 17
graphing adventure game (Boxer) 79-84
graphs
  families of curves 336-40
  graphical interpretation of derivatives 340-1
  graphical understanding and Derive 258, 307-23
  graphical understanding and graphic calculators 202-10, 217-27
  meaningless 7, 438

see also scatter graphs
GRID Algebra 157, 160-1
growth and decay models 370-3

health and safety 402
higher education students
  developing critical pedagogy 9-10, 96, 135-50
  using *Derive* 325-6, 332-42
  using graphic calculators 172, 217-27
  using Matlab 327, 387-404
  using spreadsheets 326-7, 365-84
  using supercalculators with linear algebra 326, 353-64
history project, using laptop computers 437-9
HOELL problem *see* problems

I-You and I-It 9, 121
imagery 127-31, 159
  artistic and scientific 407-20
  see also visual representation
investigations *see* problems
INVITE problem *see* problems
iterative strategies 244-5

'jigsaw' situation *see* problems
journals *see* diaries and journals
Julia sets 24-6, 29-31
Junior Calculator Pilot Study (Durham) 173

language
  cinematic 416
  engineering and mathematical 366
  of mathematics, and calculator use 187-8
  natural, in resolution process 465
laptop computers

Primary Laptop Project 435-48
using *Derive* 307-23
large numbers, and calculators 177-80, 242
learners
  construal of tasks 35-6, 44, 125
  control of learning 142-3
  didactic transposition 115, 121, 122
  didactical contract 63
  involvement in learning 199-200, 227, 354
  learners, users and mathematicians 7, 109-17, 154-5
  'scaffolding' 331-2
  working *on* and *through* 121, 163
  see also primary pupils, secondary students, higher education students
Leeds University 410
linear algebra and supercalculators 326, 353-64
linear differential equations, engineers' approach , 366
Logo 54, 71, 111, 112, 115, 125, 165-6, 257
  and algebra 258, 275-85
  compared with Boxer 72-3, 74, 92
  learning, and development of critical pedagogy 96, 135-50
looking
  at and *through* screens 120, 124, 157
  at, through and *back at* software 96, 153-68
lumped-parameter physical systems 374-5

'magic', of calculators 207-10

Mandelbrot sets 24-8
Maple 7, 115, 125, 382, 390
market problems *see* problems
Mathcad 125, 382, 383
Mathematica 7, 115, 125, 300, 382, 383,
mathematical facilities (personal technology) 98-100
mathematicians
  mathematicians, learners and users 7, 109-17, 154-5
  non-mathematicians' views of 412-13
mathematics, nature of 137-8, 146-7
Matlab 327, 382, 387-404
matrices 353-64
Measurement of a Field task *see* problems
microworlds (ARI-LAB) 455-60
'milieu didactique' 36
Mindstorms - Children, Computers and Powerful Ideas (Papert) 136-7
modelling
  with spreadsheets 370-84
  with *Derive* 333-42
Morton Brown problem *see* problems
mouse mathematics 123-4, 160-1
  drag mode 41, 63-5
multi-machine collaborations (Boxer) 91-2
muscular movement, as aid to learning 123-4

National Council for Educational Technology (NCET) project 101, 103, 105, 197-8, 200
National Curriculum 178, 235

negative numbers, and calculator use 180-1, 185-6
non-arithmetical objects, programming 289-304
NOTES problem *see* problems
number routes 189-90
'numbers at the corners of a square' *see* problems

ontological differences 163-4
Open University 381, 383
optimisation problems *see* problems
orthogonality 358-60
overhead projector (OHPs) 101

palmtop computers, using Derive 307-23
see also personal technology
Paper Folding Test 219
parallel line task *see* problems
parameters, distinguishing from variables 268-70
Parental Involvement in the Core Curriculum (PICC) project 425
Paris University 410
PASCAL 257, 293-6
passion, mathematical 145-6
pedagogy
development of critical 135-50
role of calculator in 249-50
People Mover (Boxer) 84-6
personal computers (PCs)
connections with personal technology 100-1
outside mathematics departments 413-15
teaching calculus 409-11
personal organisers 97-8
personal technology, overview 95, 97-107

place value, and calculator use 179-80
pokability (in Boxer) 75-6
portable mathematics 444-7
primary pupils
access to computers 233-4
making maths videos 405-6, 423-33
using ARI-LAB 406, 449-67
using calculators 171, 173-93, 196-7, 237-9, 240-1, 242-4, 245-6, 246-7
using laptop computers 406, 435-48
using symbolic language 275-85
Prime Iterating Number Generators (PINGS) 326, 345-51
primitives in *Cabri-géomètre* 40-1, 55
printing, personal technology 100
problems
addition problems 183-5, 186-7, 188-9
algebraic problems 264-71
Bunny Hops 189-90
Chocolates problem 284-5
combine problems (ARI-LAB) 466
counting problems 177-80
equivalence problems 278
'Fill your screen' tasks 177, 181, 187-8
Frogs 424, 427-31
graph problems 202-5
HOELL problem (strings) 293-5
INVITE problem (Booleans) 296-7
'jigsaw' situation 36
market problems (ARI-LAB) 460-4, 466

Measurement of a Field task 280-3

Morton Brown problem 32-3

NOTES problem (strings) 290-1

'numbers at the corner of a square' 189

optimisation problems 333-42

parallel line task (*Cabri-géomètre*) 50, 55-65

reversal problems 183

wheels problems 439-43

programming

with Boxer 69-92

non-arithmetical objects 258, 289-304

progressive ideology 137, 149

Project AnA 275-85

Pythagoras' theorem, 157-8, 166-7

QR-factorization 359-60

quadratic functions

route to fractals 16-28

study of, with Derive 307-23

recursive systems 376-9

relations, algebraic, and spreadsheets 264-5, 282-3

representation *see* visual representation

research, need for 106

resolution process (ARI-LAB) 464-6

resource issues 5

access to technology 101-4, 231-5

reversal problems *see* problems

risk-taking 143

Rome University Mathematics Laboratory 409-11

Sarah Bonnell School 196

'scaffolding' 331-2

scaling (graphs) 208, 241, 319-20

scatter graphs (scattergrams)

new form of data analysis 172, 221, 222

produced by children 441-3, 445-6

scramblers 376-8

screens

as communication channel 85-6

display quality 100

electronic and mental 9, 95-6, 119-32

secondary students

access to technology 103, 232-3, 233-4

programming non-arithmetical objects 258, 289-304

using Boxer 12, 69-92

using *Cabri-géomètre* 12, 40-65

using *Derive* 307-23

using graphic calculators 171-2, 195-210, 247-8

using spreadsheets 257, 261-72

using symbolic language 275-85

see also A-level students

semantics and syntax 250, 318, 319, 464

situation-specific strategies 280-2

Sketchpad 158

'soft mastery' 140

software 7-8

looking *at*, *through*, and *back at* 96, 153-68

purpose and context 111-12

'tools' metaphor 7, 95, 112-17

see also individual software packages

spatial ability

enhanced by graphic calculators 172, 216-27

importance to mathematics 213-16, 226-7
spreadsheets 116, 158, 333
  and algebra 257, 258, 261-72, 275-85
  and applicable mathematics 365-84
  dimensionality 160-1, 162, 163-4, 369
statistical principles (Boxer) 72
strategies
  and calculator use 242-5
  empirical combinatorial 58
  iterative 244-5
  for number routes 189-90
  trial and improvement 205, 234-5, 264-5, 282
  visual and geometrical 55-63
strategy building environment (ARI-LAB) 453
strings 290-1
  pupils difficulties with 293-5
  successive approximation *see* trial
  and improvement strategies
supercalculators and linear algebra 326, 353-64
symbols
  algebraic 276-7, 289-404
  arithmetic 183-5, 451
  symbolic/algebraic and graphical/visual 130, 198, 202-3, 250, 322-3
syntax and semantics 250, 318, 319, 464
synthesis of equations 266

tasks
  in *Cabri-géomètre* 49-54
learners' construal of 125-6
  meaning of 126-7
  task-setting and power 148
  see also problems
teacher environment (ARI-LAB) 455
teachers
  attitude to calculators 174-5, 191-2
  challenges to 369-70
  and curriculum change 330-2
  collaboration 73, 89-91
  development of critical pedagogy 9-10, 96, 135-50
  didactic transposition 115, 121, 122
  didactical contract 63
  expectations of tasks 35-6, 44, 125
  importance of intervention 272, 442
  interaction with learners 354
  learning about laptops 435-6
  need for support 107, 447
  philosophy 8
  recommendations to, on using spreadsheets 383-4
  relinquishing control 224-6
  roles 199-200
Technical and Vocational Educational Initiative (TVEI) 101
technical features (personal technology) 100-1
techniques, effects of software on 113-14, 116, 154-6, 342
Technology in Mathematics Teaching (1993 conference) 1, 106-7
Teddy Bears Picnic *see* problems
'tools', as software metaphor 7, 95, 112-17

transformations (geometric) 49-51, 53-4
transposition didactique see didactic transposition
trial and improvement (trial-and-error) strategies 205, 234-5, 264-5, 282

unknowns
  algebraic and spreadsheet approaches 266-7, 282-3
  generalised arithmetic 318
users of mathematics
  engineers and mathematics 365-84
  mathematics outside mathematics department 411-15
  users, learners and mathematicians 7, 109-17, 154-5

variables
  distinguishing from parameters 268-70
  in spreadsheets 266-7, 278
  understanding concept 279, 317, 318-19
  visibility (Boxer) 74
vector spaces 355-8
vectors (Boxer) 86-8
videos *see* films and videos
visibility (in Boxer) 74-5
Visual Display of Quantitative Information, The (Tufte) 368-9
visual and geometrical strategies 55-63
visual mathematics 408
visual representation
  computer graphics 407-11, 413-14, 418-19
  dimensionality 368-9

visual representation environment (ARI-LAB) 453, 460-4, 455-8
  see also drawing; graphs; imagery; scatter graphs
visual-tactile mediation 249-50

Warwick University 435
wheels problems *see* problems
Worcester College of Higher Education 332
working *on* and working *through* 121, 163

Name Index

Abelson, H. 70
Ainley, J. 7, 309, 406, 435-48
Arieti, S. 130-1
Armstrong, P. 375
Artigue, M. 289, 300
Association of Teachers of Mathematics (ATM) 104, 336

Baenninger, M. 215
Baker, I. N. 24
Ball, D. 109
Banchoff, T. F. 408, 409
Barnsley, M.F. 31
Beardon, A. F. 33
Beeney, R. 126
Belenky, M. F. 137, 139, 140
Ben-Chaim, D. 215
Bergue, D. 52
Berry, J. W. 244
Berthelot, R. 63
Bissell, C. 8, 110, 326-7, 365-84
Bohm, J. 329
Booth, L. 262, 276, 279-80
Bottino, R. M. 406, 449-67
Boury, V. 53
Brandes, D. 137
Bridges, R. 375
Brissenden, T. 424
Brown, A.S. 92
Brooks, R. 24

Brousseau, G. 36, 52, 63
Brown, M. 32, 219, 309
Brown, Morton 6
Bruce, J. W. 21, 23
Bruner, J. S. 463
Buber, M. 9, 121
Bullett, S. R. 33
Burton, L. 1-10, 36, 149, 424
Buxton, Laurie 173

Callahan, M. J. 408
Capponi, B. 50, 261, 300
Capuzzo Dolcetta I. 409, 410, 411
Cartledge, C. M. 215
Chapman, D. A. 376, 380
Cheeseman, J. 248
Chevellard, Y. 115, 116, 121, 262, 263, 291-2
Chiappini, G. 263, 449-67
Christiansen, B. 125
Coates, D. 174, 180, 190, 191
Cobb, P. 451
Cockcroft, W. H. 235
Collins, A. 92
Collis, B. 215
Concus, P. 408
Cortes, A. 263
Cox, L. 202
Coxeter, H. S. M. 417
Craggs, C. 424

Crampin, M. 33
Culley, L. 201
Cuttle, M. L. 384

Davis, R. B. 36
Demana, F. 216, 232
Dennett, D. 128
Denvir, B. 6
Department of Education and Science (DES) 235
Dessart, D. J. 232, 236, 237-8
Dettori, G. 8, 257, 261-72
Dick, T. 232, 239, 248, 250
Dillon, C. R. 110, 368
diSessa, A. A. 6, 7, 12, 69-92
Douady, A. 27
Dowling, P. 110
Dreyfus, T. 451
Dubinsky, E. 217
Duffin, J. 196
Duguid, P., 92
Dunham, P. H. 215, 232, 239
Dunn, P. D. 231
Duren, P. 198

Eastwood, M. 232
Ekstrom, R. B. 219
Elkjaer, B. 196
Emerson, A. 425, 431
Emmer, M. 7, 405, 407-20
Ernest, P. 149n
Escher, M. C. 417

Falcone, M. 409, 410, 414-15, 419
Fatou, P. 24, 26
Feigenbaum, M. 23
Fennema, R. B. 214
Ferrari, P. L. 449-67
Fielker, D. 173, 232, 243, 249
Filloy, E. 276

Finzi Vita, S. 410
Fischbein, E. 44, 464
Fish, J. 129
Fitzgerald A. 231
Fletcher, T. 135
Flores, A. 216
Foxman, D. 233, 237, 240, 245, 246
Freire, P. 138

Garfunkel, S. A. 412-13
Garuti, R. 261-72
Gattegno, C. 128
Gaulin C. 233
Giblin, P. J. 21, 23
Ginnis, P. 137
Girling, M. 232
Giroux, H. 139, 148
Goldenberg, E. P. 115, 208
Goldin, G. 236
Goldstein, R. 329
Goodman, N. 130
Graham, A. 201, 232
Gray, E. 317
Green, D. 232
Griffin, P. 120, 153
Groves, S. 175, 190, 238, 242, 245, 248
Guillerault, M. 50

Hancock, R. 425
Harel, G. 217
Hart, K. 125, 219, 317
Harvey, B. 149
Harvey, J. G. 239
Hé, Y. 295
Healy, L. 111, 275
Hedren, R. 242, 245, 247
ter Hege, H. 247
Hembree, R. 232, 236, 237-8

Her Majesty's Inspectorate (HMI) 234
Herodotus 37
Hewitt, D. 157
Hillel, J. 44, 54
Hoffman, D. 408
Hooper, J. 232
Horowitz L. 216
Hoyles, C. 54, 112, 114, 115, 140, 196, 201, 277
Hubbard, J. H. 27
Hunter, M. 258-9, 307-23
Hurd, L.F. 31
Hurd, M. 320

Illich, I. 231
Irvine, S. H. 244

Jacques, I. 390
Jaworski, B. 1-10
Jindal, N. 201
Johnson, D. C. 186
Johnson, R. T. 201
Judd, C. 390
Juilfs, P. A. 216
Julia, G. 26
Julie, C. 202
Jürgens, H. 16, 23, 30

Kang, W. 121
Kaput, J. J. 250, 318
Kay, Alan 159
Keitel, C. 155
Kerslake, D. 219
Kieran, C. 44, 54, 289, 292, 303
Kilpatrick, J. 121
Kirshner, D. 250
Koçak, H. 408
Küchemann, D. 219, 276, 277, 317
Kutzler, B. 331

Laborde, C. 12, 35-65
Laborde, J. M. 458
Lacan, J. 130
Lagrange, J.-B. 258, 289-304
Laidlaw, D. 408
Lapointe, A. E. 233
LaTorre, D. 6, 8, 326, 353-64
Lave, J. 280
Lay, E. 77
Leinhardt, G. O. 198, 207
Lemut, E. 261-72
Lesh, R. 451
Lins, R. 280
Love, E. 7, 8, 95, 121, 125, 109-17

Mackinnon, N. 109, 112
Malara, N. 261
Mandelbrot, B. B. 24
Mariotti, A. 44
Marshall, P. 307-23
Mason, J. 6, 7, 9, 95-6, 112-13, 119-32, 153, 155,

159, 163, 167, 369
Matelski, J.P. 24
Mathematical Association (MA) 104

Maxwell, M. E. 239
McCaffrey, M. 201
McGee, M. G. 219
McLeod, D. 216
Meeks, W. 408
Monaghan, J. 307-23
Murray, J. 333

National Council for Educational Technology (NCET) 101, 197
National Foundation for Educational Research (NFER) 103
Netchitailova, L. 261-72

Nevile, L. 7, 96, 153-68, 369
Neville, B. 130
Newcombe, N. 215
Newman, D. 247
Noss, R. 112, 114, 115

Office for Standards in Education
     (OFSTED) 192, 233, 237
Oldknow, A. 6, 95, 97-107

Papert, S. 136-7, 432
Parzysz, B. 37
Pea, R. 36
Peitgen, H.-O. 16, 23, 30
Perks, P. 171, 173-93
Perry, W. G. 137, 139
Picciotto, H. 71
Pimm, D. 122
Ploger, D. 70, 71, 77
Plunkett, S. 232
Polya, G. 153
Pope, S. 232
Povey, H. 9, 96, 135-50
Pratt, D. 7, 406, 435-48

Quesada, A. R. 239

Rippon, P. J. 4, 11, 15-33
Roberts, D. M. 232, 236
Rogalski, J. 295
Rojano, T. 275, 283
Roper, T. 307-23
Rothery, A. 8, 325-6, 329-42
Ruthven, K. 6, 101, 171, 172, 197,
     198, 200, 205, 216, 231-50, 319

Salin, M. H. 63
Samurcay, R. 299
Saupe, D. 16, 23, 30
Schoenfeld, A. 44

School Curriculum and Assessment
     Authority (SCAA) 102
Schumacher, E. F. 231
Scrivener, S. 129
Sheath, G. 9, 405-6, 423-33
Sherman, J. 214
Shoaf-Grubbs, M. M. 172, 213-27
Shuard, H. 173, 177, 180, 182, 185,
     189, 238, 242, 244, 245, 248-9
Shumway, R. 232, 236, 247
Sigmon, K. 390
Silver, E. A. 216
Smart, T. 171-2, 195-210
Smith, R. 243
Steffe, L. P. 36
Stevick, E. 126
Stone, J. A. R. 372
Strasser, R. 458
Sutherland, R. 54, 111, 140, 257-8,
     262, 275-85, 299
Swiatek, G. 22

Tabor, J. 327, 387-404
Tahta, D. 126
Tall, D. 276, 317
Tartre, L. A. 214, 215
Taylor, L. J. C. 216
Terquem, O. 263
Thompson, A. 149n
Threadgill-Sowder, J. A. 216
Thurston, L. L. 214
Tohidi, N. E. 215
Tomlinson, P. 310
Tufte, E. R. 368-9
Turkle, S. 140

Underwood, G. 201
Usiskin, Z. 263

Vazquez, J. L. 215

Vonder Embser C. 250

Wain, G. T. 308, 310, 315
Waits, B. 216, 232
Walsh, A. 196-7, 200
Walther, G. 125
Watson, D. 233
Weiler, K. 148
Wheatley, C. L. 243, 244
Wheatley, G. H. 247
Wild, P. 9, 326, 345-51
Williams, D. 155
Willis, P. 149
Wood, T. 36

Young, G. S. 412-13

Elementary Linear Algebra with DERIVE:
an integrated text Hill R J, Keagy T A

This introduction to linear algebra fully integrates hand and Derive examples in the explanation of new topics. These develop progressively to expand the reader's appreciation of the theory and its applications. An accompanying disk provides all the Derive procedures used in the book. Linear algebra is an area of mathematics which blends an elegant theoretical structure with a rich supply of applications, but the study of linear algebra has historically been burdened with tedious calculations which have distracted the learner and discouraged the user.

In this text extensive use is made of the Derive symbolic computer software system to simplify calculations and allow the learner and user to focus on the beauty of the structure of linear algebra and its associated applications. The elementary properties of matrices, vector spaces, linear transformations, and determinants are presented in a style that promotes an understanding and appreciation for the order and logic of the theory without the unnecessary frustration of long, time consuming calculations.

Since this early material can be covered more efficiently, more emphasis can be spent on the study of eigenvalues, eigenvectors, matrix analysis, and applications in diverse areas such as least square approximations, Markof processes, difference equations, and linear systems of differential equations with constant coefficients. To support the learner, the text includes 180 example problems worked in detail, almost 600 Derive statements illustrating the use of the computer software, and more than 600 exercises with the solutions to most included in an appendix.

Chartwell-Bratt, ISBN 0-86238-403-6, 1995

Mathematical Activities with Computer Algebra:
a photocopiable resource book
Etchells T, Hunter M, Monaghan J, Pozzi S, Rothery A

This photocopiable resource book is the first of a new generation of support materials for computer algebra. Designed to be used with any computer algebra system, the authors go beyond mere button pressing and show how to harness the power of computer algebra systems for educational purposes. Concepts are illustrated, techniques and methods presented, and modelling and applications are explained. Appendices give overviews of DERIVE, Maple, Mathematica, Theorist (Math Plus) and the new TI-92 calculator.

Activity Worksheets, Help Sheets and Teaching Notes cover a wide range of mathematical topics at school and college level.

Topics covered include; functions and graphs, differentiation, integration, sequences and series, vectors and matrices, mechanics, trigonometry, numerical methods.

Activities include: Multiplying factors; Equation of a tangent; Taxing functions; The tile factory; Function and derivative - visualisation; The approximate derivative function; Sketching graphs; Pollution and population; Max cone; Optimising transport costs; Area under a curve; Enclosed areas; A function whose derivative is itself; Wine glass design; The limit of a sequence; Visualising Taylor approximations; Visualising matrix transformations; Blood groups; Circular motion; Swing safety; No turning back; Modelling the sine function; Solving equations with tangents.

Chartwell-Bratt, ISBN 0-86238-405-2, 1995

Discovering Geometry with a Computer
- using Cabri-Geometre Heinz Schumann and David Green

This book provides a wide ranging discussion of geometrical investigation aided by a computer. Although the book begins with an introduction to one particular software package - Cabri-

Geometre - to which it refers throughout, most of the activities suggested will be of interest to users of other software packages such as Geometer's Sketchpad and Geometry Inventor.

The topics covered in depth are: Learning geometry through interactive construction; Creating macros; Discovering theorems; Loci; Symmetry; Geometrical microworld design; Isoperimetric problems; Transformations; Twenty problems to investigate.

This 282 page book contains over 900 geometrical diagrams and hundreds of ideas for investigation are to be found throughout its pages.

Associated with the book is a disk (MS-DOS or Macintosh formats) containing over 400 Cabri-Geometre figures and macros designed to illustrate the contents of the book and to aid personal exploration and extension.

This is the first substantial work providing a thorough treatment of Cabri-Geometre. It will be of interest to teachers of mathematics in schools, to those involved with pre-service or in-service teacher education, and to mathematics lecturers in universities and colleges, and it will appeal to those who have fond memories of geometry from their school or college days.
Chartwell-Bratt, 282 pages, ISBN 0-86238-373-0, 1995

Taking a New Angle Chris Little and Rosamund Sutherland

This booklet presents some ideas for using Cabri-Geometre to explore elementary properties of angles at a point, between parallel lines, and in triangles and polygons. The activities may be photocopied.

CONTENTS: Introduction, Getting started with Cabri, Angles in Cabri, Angles around a point, Vertically opposite angles, Alternate angles, Angles in triangles, Angles in a polygon.

ACTIVITIES: Introduction to Cabri-Geometre, Constructing an angle, Angles around a point, Vertically opposite angles, Alternate angles, Corresponding angles, Finding alternate and corresponding angles, Opposite, alternate and corresponding angles, Angles in a triangle, Angles in polygons.
Chartwell-Bratt, 20 pages, ISBN 0-86238-377-3, 1995

Exploring Trigonometry Chris Little and Rosamund Sutherland

This booklet presents ideas and activities for using Cabri-Geometre to teach trigonometry. The activities may be photocopied.

CONTENTS: Introduction, Getting Started with Cabri (points,lines and circles, any equilateral triangle, angles on the circumference of a circle, any right-angled triangle) Trigonometry with Cabri (a paper construction for investigating SINE, constructing a triangle for exploring trigonometry,investigating SINE - the case of the 30 degree angle, is there a rule for any angle? introducing SINE of an angle, using the result of the investigation, calculating the SINE of an angle with a calculator, investigating the COSINE function, using COSINE and SINE) A concluding note.

ACTIVITIES: Introduction to Cabri-Geometre, Any equilateral triangle, angles on the circumference of a circle subtended by a diameter, Any right-angled triangle, Constructing a triangle for exploring trigonometry, Investigating the SINE function - the case of the 30 degree angle, Investigating the SINE function - is there a rule for any angle? Using SINE, Investigating the COSINE function - the case of the 30 degree angle, Investigating the COSINE function - is there a rule for any angle? Making up trigonometry problems,Using COSINE and SINE.
Chartwell-Bratt, 26 pages, ISBN 0-86238-376-5, 1995

Transforming Transformations

Chris Little and Rosamund Sutherland

This booklet presents ideas for using Cabri-Geometre to analyse and explore the plane isometric transformations - reflection, translation, and rotation. Constructing images of plane figures using pencil and paper methods (for example using rules and compasses, or by means of counting squares on squared paper or graph paper) is time-consuming and not particularly satisfying. Once the transformations are constructed in Cabri, however, both the transformation itself and the object can easily be varied, and the basic properties can be investigated. We think that there is a real benefit in asking pupils to construct their own transformation macros since this provokes them to analyse the transformations for themselves. As they construct the macro, they make sense of the properties of the transformation. Each transformation is treated in the same way. First pupils build a macro which constructs the image of a single point. They then use this to construct a macro for the image of a triangle. With this macro, the basic properties of the transformation are explored, including combinations of transformations.

CONTENTS: Introduction, Macro-constructions, Reflection (What is a reflection? Reflecting triangles, A puzzle) Rotation (What is a rotation? Rotating triangles), Translation, Two reflections.

ACTIVITIES: Macro-constructions, Using symmetrical point, Creating a macro to reflect a point, Reflecting triangles, Reflection, Creating a macro to rotate a point, Investigating rotation, Rotating triangles, Rotation, Creating a macro for translation, Vectors, Two reflections.

Chartwell-Bratt, 34 pages, ISBN 0-86238-378-1, 1995

Mathematics with Excel David Sjostrand
Makes good use of Excel's graphical features to illustrate and help visualisation of important mathematical concepts. Based on Excel v5 but usable with earlier versions and other spreadsheets. Also shows how to link Excel with DERIVE.

Chartwell-Bratt, 260 pages, ISBN 0-86238-361-7, 1994

Modelling with Spreadsheets Andrew Rothery
Usable with any spreadsheet. Explains the principles of modelling and spreadsheets. "Optimisation" is central theme. Clear and accessible.

Chartwell-Bratt, 63 pages, ISBN 0-86238-258-0, 1989

Teaching Mathematics with DERIVE   Josef Boehm (ed)
Proceedings of the 1992 Conference on the Didactics of Computer Algebra

Chartwell-Bratt, 298 pages, ISBN 0-86238-319-6, 1993

DERIVE in Education: Opportunities and Strategies
H Heugl, B Kutzler (eds)
Proceedings of the 1993 Conference on the Didactics of Computer Algebra

Chartwell-Bratt, 302 pages, ISBN 0-86238-351-X, 1994

DERIVE-based Investigations for Post-16 Core Mathematics, 2nd Ed A
J Watkins
Practical investigations using DERIVE. Extensively trialled.

Chartwell-Bratt, 102 pages, ISBN 0-86238-312-9, 1993

Fields of Physics by Finite Element Analysis G Backstrom
Using the software package PDEase (previousl SPDE - solver for partial differential equations)

Chartwell-Bratt, ISBN 0-86238-382-X, 1995

UNIVERSITY OF PLYMOUTH
LIBRARY SERVICES (EXMOUTH)
DOUGLAS AVENUE
EXMOUTH
DEVON EX8 2AT